UNITS A

AREAS B

SOLID BODIES C

ARITHMETIC D

FUNCTIONS OF A CIRCLE E

ANALYTICA GEOMETRY F

ATISTICS G

LCULUS H

CULUS I

ATIONS J

STATICS K

KINEMATICS L

DYNAMICS M

HYDRAULICS N

HEAT O

STRENGTH P

MACHINE PARTS Q

PRODUCTION ENGINEERING R

ELECTRICAL ENGINEERING S

RADIATION PHYSICS T

CHEMISTRY U

TABLES Z

ENGINEERING FORMULAS

by

Kurt Gieck
Reiner Gieck

Sixth Edition

McGraw-Hill, Inc.
New York St. Louis San Francisco Montreal Toronto

Library of Congress Cataloging-in-Publication Data

Gieck, Kurt + Reiner
Engineering formulas.

Translation of: Technische Formelsammlung.
Includes index.
1. Engineering – Tables. I. Title.
TA151.G4713 1986 620'.00212 85-23153
ISBN 0-07-023455-8

ISBN 0-07-023455-8

First published in the English Language under the title
A COLLECTION OF TECHNICAL FORMULAE

Sixth American edition
published by McGraw-Hill, Inc. in 1990
English translation by J. Walters

Printed in West Germany

Preface

The purpose of this collection of technical formulae is to provide a brief, clear and handy guide to the more important technical and mathematical formulae.

Since the book has been printed on one side od the page only, the facing pages are available for additional notes.

Each separate subject has been associated with a capital letter. The various formulae have been grouped under corresponding small letters and numbered serially. This method enablies the formulae used in any particular calculation to be indicated.

Preface

to the enlarged and revised 6th edition

A section on DIFFERENTIAL EQUATIONS has been included in the new section J; INTEGRAL CALCULUS is treated in an enlarged section I.

Vector Calculus	has been added to the ANALYTICAL GEOMETRY section.
Network and Installation	has been added to the ELECTRICAL ENGINEERING section.
Grinding and Extrusion	has been added to the PRODUCTION ENGINEERING section.

The sections HEAT and MACHINE PARTS together with the associated TABLES have been revised and brought up to date.

Kurt Gieck
Reiner Gieck

Reference to BS, DIN and VDE

BS · British Standards Institution
(Address: 2 Park St, LONDON W 1 A 2 BS

DIN · Deutsches Institut für Normung
(Address: D-1000 BERLIN 30, Postfach 1107)

VDI · Verein Deutscher Ingenieure
(Address: D-4000 DUESSELDORF 1, Postfach 1139).

Method of Presentation and Use of Units

Most of the equations clearly reveal the physical relationships which they describe and are valid regardless of the system of units employed, provided that they are consistent.

Some of the equations are empirical in origin and the units quoted must be used in the formula to obtain the correct result, these are mainly to be found in sections Q and R.

It is intended that the Stroud notation is used when evaluating the formulae i.e. both the quantity and the unit is substituted for a given symbol and the subsequent calculation involves manipulation of numbers and units together.

For example, taking equation I 23: $t = \frac{s}{v}$

if s (distance) $= 2{\cdot}8$ metres
v (speed) $= 8$ metres/second

then $t = \frac{2{\cdot}8 \text{ metres} \times \text{second}}{8 \text{ metres}}$

hence $t = 0{\cdot}35$ seconds (time)
cancelling the unit 'metres'

It is clear that t should have the units of time; if it does not, then it is obvious that an error has been made and the working should be checked. As a help, in many cases, the anticipated units are quoted using the abbreviation "EU", **E**xample-**U**nit.

When the numerical values and the units are included in the calculations, their equivalents or definitions are best written so that they are dimensionless and have the value of 1·0. In this form they are sometimes called "Unity Brackets" and their use can be illustrated in three ways:

with consistent units,

equation a 6

$$1\ \text{km} = 10^3\ \text{m} \qquad \text{becomes} \qquad 1 = \left[\frac{1\ \text{km}}{10^3\ \text{m}}\right]$$

equation a 62

$$12\ \text{in} = 1\ \text{ft} \qquad \text{becomes} \qquad 1 = \left[\frac{1\ \text{ft}}{12\ \text{in}}\right]$$

equation a 90

$$778{\cdot}6\ \text{ft lbf} = 1\ \text{Btu} \qquad \text{becomes} \qquad 1 = \left[\frac{778{\cdot}6\ \text{ft lbf}}{1\ \text{Btu}}\right]$$

for example, to convert 14·7 lbf/in^2 to lbf/ft^2

$$14{\cdot}7\ \frac{\text{lbf}}{\text{in}^2} = 14{\cdot}7\ \frac{\text{lbf}}{\text{in}^2}\left[\frac{12\ \text{in}}{1\ \text{ft}}\right]^2 = 14{\cdot}7 \times 144\ \frac{\text{lbf}}{\text{ft}^2} = 2117\ \frac{\text{lbf}}{\text{ft}^2}$$

in the conversion between different systems of units,

equation a 36

$$1\ \text{N} = 0{\cdot}102\ \text{kgf} \qquad \text{becomes} \qquad 1 = \left[\frac{0{\cdot}102\ \text{kgf}}{1\ \text{N}}\right]$$

equation a 65

$$1\ \text{m} = 3{\cdot}281\ \text{ft} \qquad \text{becomes} \qquad 1 = \left[\frac{1\ \text{m}}{3{\cdot}281\ \text{ft}}\right]$$

equation a 110

$$1\ \text{Btu/lb} = 0{\cdot}556\ \text{kcal/kg} \qquad \text{becomes} \qquad 1 = \left[\frac{0{\cdot}556\ \text{kcal lb}}{1\ \text{kg Btu}}\right]$$

For example, to convert 1000 kgf/cm^2 to S.I. units,

$$1000\ \frac{\text{kgf}}{\text{cm}^2} = 1000\ \frac{\text{kgf}}{\text{cm}^2}\left[\frac{9{\cdot}81\ \text{N}}{1\ \text{kgf}}\right]\left[\frac{10^4\ \text{cm}^2}{1\ \text{m}^2}\right]\left[\frac{1\ \text{MN}}{10^6\ \text{N}}\right]$$

$$= 98{\cdot}1\ \frac{\text{MN}}{\text{m}^2}$$

in the use of definitions:

1 lbf is the force required to accelerate a mass of 1 lb at the rate of 32·174 ft/s².

$$1 \text{ lbf} = 1 \text{ lb} \times 32{\cdot}174 \frac{\text{ft}}{\text{s}^2} \quad \text{becomes} \quad 1 = \left[\frac{32{\cdot}174 \text{ lb ft}}{1 \text{ s}^2 \text{ lbf}}\right]$$

Similarly, the Newton is defined by the equation

$$1 \text{ N} = 1 \text{ kg} \times \frac{1 \text{ m}}{\text{s}^2} \quad \text{which becomes} \quad 1 = \left[\frac{1 \text{ N s}^2}{1 \text{ kg m}}\right]$$

$$\text{and } 1 \text{ kgf} = 1 \text{ kg} \times 9{\cdot}81 \frac{\text{m}}{\text{s}^2} \quad \text{becomes} \quad 1 = \left[\frac{9{\cdot}81 \text{ kg m}}{1 \text{ kgf s}^2}\right]$$

For example, to find the force in S.I. units required to accelerate a mass of 3 lb at the rate of 2·5 ft/s², proceed as follows:

$$F = m\,a, \text{ equation m 1.}$$

$$F = 3 \text{ lb} \times 2{\cdot}5 \frac{\text{ft}}{\text{s}^2} \left[\frac{0{\cdot}4536 \text{ kg}}{1 \text{ lb}}\right]\left[\frac{1 \text{ m}}{3{\cdot}281 \text{ ft}}\right]\left[\frac{1 \text{ N s}^2}{1 \text{ kg m}}\right]$$

$$= \frac{3 \times 2{\cdot}5 \times 0{\cdot}4536}{3{\cdot}281} \text{ N} = 1{\cdot}036 \text{ N}$$

which is a unit of force.

Base Quantities and Base Units of the International System of Measurement

base quantity		base unit	
name	symbol *(italic letters)*	name	symbol (vertical letters)
length	l	metre	m
mass	m	kilogram	kg
time	t	second	s
electric current	I	ampere	A
absolute temperature	T	kelvin	K
amount of substance	n	mole	mol
light intensity	I_V	candela	cd

Old units are put in ( ) brackets

List of symbols

Space and time

- α, β, γ angles
- Ω solid angle
- b, B breadth
- d, D diameter (diagonal)
- h, H height
- l, L length
- p pitch
- r, R radius
- s distance covered, perimeter
- t thickness
- u, U circumference
- A area, cross section
- A_m generated surface
- A_o surface area
- V volume
- t time
- v velocity, linear
- ω velocity, angular
- a acceleration, linear
- α acceleration, angular
- g acceleration, gravitational

Periodical and related phenomens

- T period
- f frequency
- n rotational speed
- ω angular frequency
- λ wavelength
- c velocity of light

Mechanics

- m mass
- ϱ density
- F force, direct force
- f, σ direct stress
- q, τ shear stress
- p normal pressure
- ε extension, strain
- E modulus of elasticity (Young's modulus)
- G modulus of rigidity (shear modulus)
- M bending moment
- T torsional moment, torque
- Z modulus of section
- Q shear force, shear load
- V vertical reaction
- W weight or load, work
- w uniformly distributed load
- I moment of inertia, second moment of area
- I_p polar moment of inertia
- J torsion constant
- Z modulus of section
- μ coefficient of sliding friction
- μ_o coefficient of static friction
- μ_q coefficient of friction of a radial bearing
- μ_l coefficient of friction of a longitudinal bearing
- f coefficient of rolling friction
- η dynamic viscosity
- ν kinematic viscosity
- P power
- η efficiency

Heat

T absolute temperature
t temperature
α linear coefficient of expansion
γ cubic coefficient of expansion
Φ heat current or flow
φ density of heat flow
q quantity of heat per unit mass
Q quantity of heat
c_p specific heat at constant pressure
c_v specific heat at constant volume
γ ratio of c_p to c_v
R gas constant
λ thermal conductivity
α heat transfer coefficient
k coefficient of heat transmission
C radiation constant
v specific volume

Electricity and magnetism

I current
J current density
V, U voltage
U_q source voltage
R resistance
G conductance
Q quantity of electricity (charge)
C capacitance
D dielectric displacement
E electric field strength
Φ magnetic flux
B magnetic induction
L inductance
H magn. field strength
Θ circulation (magnetic potential)
V magnetic voltage
R_m magnetic resistance
Λ magnetic conductance
δ length of air gap
α temperature coefficient of resistance
γ conductivity
ϱ resistivity
ε permittivity, dielectric constant
ε_0 absolute permittivity
ε_r relative permittivity
N number of turns
μ permeability
μ_0 absolute permeability
μ_r relative permeability
p number of pairs of poles
z number of conductors
Q quality, figure of merit
δ loss angle
Z impedance
X reactance
P_s apparent power
P_q reactive power
C_M moment constant

Light and related electromagnetic radiations

I_e radiant intensity
I_v luminous intensity
Φ_e radiant power, radiant flux
Φ_v luminous flux
Q_e radiant energy
Q_v quantity of light
E_e irrediance
E_v illuminance
H_e radiant exposure
H_v light exposure
L_e radiance
L_v luminance
c velocity of light
n refractive index
f focal length
D refractive power

UNITS

Decimal multiples and fractions of units

da	=	deca	=	10^{1}	d	=	deci	=	10^{-1}
h	=	hecto	=	10^{2}	c	=	centi	=	10^{-2}
k	=	kilo	=	10^{3}	m	=	milli	=	10^{-3}
M	=	mega	=	10^{6}	μ	=	micro	=	10^{-6}
G	=	giga	=	10^{9}	n	=	nano	=	10^{-9}
T	=	tera	=	10^{12}	p	=	pico	=	10^{-12}
P	=	peta	=	10^{15}	f	=	femto	=	10^{-15}
E	=	exa	=	10^{18}	a	=	atto	=	10^{-18}

Units of length

			m	μm	mm	cm	dm	km
a 1	1 m	=	1	10^{6}	10^{3}	10^{2}	10	10^{-3}
a 2	1 μm	=	10^{-6}	1	10^{-3}	10^{-4}	10^{-5}	10^{-9}
a 3	1 mm	=	10^{-3}	10^{3}	1	10^{-1}	10^{-2}	10^{-6}
a 4	1 cm	=	10^{-2}	10^{4}	10	1	10^{-1}	10^{-5}
a 5	1 dm	=	10^{-1}	10^{5}	10^{2}	10	1	10^{-4}
a 6	1 km	=	10^{3}	10^{9}	10^{6}	10^{5}	10^{4}	1

Units of length (continued)

			mm	μm	nm	(Å) 1)	pm	(mÅ) 2)
a 7	1 mm	=	1	10^{3}	10^{6}	10^{7}	10^{9}	10^{10}
a 8	1 μm	=	10^{-3}	1	10^{3}	10^{4}	10^{6}	10^{7}
a 9	1 nm	=	10^{-6}	10^{-3}	1	10	10^{3}	10^{4}
a 10	(1 Å)	=	10^{-7}	10^{-4}	10^{-1}	1	10^{2}	10^{3}
a 11	1 pm	=	10^{-9}	10^{-6}	10^{-3}	10^{-2}	1	10
a 12	(1 mÅ)	=	10^{-10}	10^{-7}	10^{-4}	10^{-3}	10^{-1}	1

Units of area

			m^2	$μm^2$	mm^2	cm^2	dm^2	km^2
a 13	1 m^2	=	1	10^{12}	10^{6}	10^{4}	10^{2}	10^{-6}
a 14	1 $μm^2$	=	10^{-12}	1	10^{-6}	10^{-8}	10^{-10}	10^{-18}
a 15	1 mm^2	=	10^{-6}	10^{6}	1	10^{-2}	10^{-4}	10^{-12}
a 16	1 cm^2	=	10^{-4}	10^{8}	10^{2}	1	10^{-2}	10^{-10}
a 17	1 dm^2	=	10^{-2}	10^{10}	10^{4}	10^{2}	1	10^{-8}
a 18	1 km^2	=	10^{6}	10^{18}	10^{12}	10^{10}	10^{8}	1

1) Å = Ångström | 2) 1 mÅ = 1 XE = 1 X-unit

UNITS

Units of volume

	m^3	mm^3	cm^3	dm^3 [1]	km^3	
1 m^3 =	1	10^9	10^6	10^3	10^{-9}	a 19
1 mm^3 =	10^{-9}	1	10^{-3}	10^{-6}	10^{-18}	a 20
1 cm^3 =	10^{-6}	10^3	1	10^{-3}	10^{-15}	a 21
1 dm^3 =	10^{-3}	10^6	10^3	1	10^{-12}	a 22
1 km^3 =	10^9	10^{18}	10^{15}	10^{12}	1	a 23

Units of mass

	kg	mg	g	dt	t = Mg	
1 kg =	1	10^6	10^3	10^{-2}	10^{-3}	a 24
1 mg =	10^{-6}	1	10^{-3}	10^{-8}	10^{-9}	a 25
1 g =	10^{-3}	10^3	1	10^{-5}	10^{-6}	a 26
1 dt =	10^2	10^8	10^5	1	10^{-1}	a 27
1 t = 1 Mg =	10^3	10^9	10^6	10	1	a 28

Units of time

	s	ns	μs	ms	min	
1 s =	1	10^9	10^6	10^3	$16{\cdot}66\times10^{-3}$	a 29
1 ns =	10^{-9}	1	10^{-3}	10^{-6}	$16{\cdot}66\times10^{-12}$	a 30
1 μs =	10^{-6}	10^3	1	10^{-3}	$16{\cdot}66\times10^{-9}$	a 31
1 ms =	10^{-3}	10^6	10^3	1	$16{\cdot}66\times10^{-6}$	a 32
1 min =	60	60×10^9	60×10^6	60×10^3	1	a 33
1 h =	3600	$3{\cdot}6\times10^{12}$	$3{\cdot}6\times10^9$	$3{\cdot}6\times10^6$	60	a 34
1 d =	$86{\cdot}4\times10^3$	$86{\cdot}4\times10^{12}$	$86{\cdot}4\times10^9$	$86{\cdot}4\times10^6$	1440	a 35

Units of force (gravitational force also)

	N [2]	kN	MN	(kgf)	(dyn)	
1 N =	1	10^{-3}	10^{-6}	0·102	10^5	a 36
1 kN =	10^3	1	10^{-3}	$0{\cdot}102\times10^3$	10^8	a 37
1 MN =	10^6	10^3	1	$0{\cdot}102\times10^6$	10^{11}	a 38

[1] 1 dm^3 = 1 l = 1 liter

[2] 1 N = 1 kg m/s^2 = 1 Newton

Units of pressure

			Pa	N/mm²	bar	(kgf/cm²)	(torr)
a 39	1 Pa = N/m²	=	1	10^{-6}	10^{-5}	$1{\cdot}02\times10^{-5}$	0·0075
a 40	1 N/mm²	=	10^6	1	10	10·2	$7{\cdot}5\times10^3$
a 41	1 bar	=	10^5	0·1	1	1·02	750
a 42	(1 kgf/cm² = 1 at)	=	98 100	$9{\cdot}81\times10^{-2}$	0·981	1	736
a 43	(1 torr) [1]	=	133	$0{\cdot}133\times10^{-3}$	$1{\cdot}33\times10^{-3}$	$1{\cdot}36\times10^{-3}$	1

Units of work

			J	kW h	(kgf m)	(kcal)	(hp h)
a 44	1 J [2]	=	1	$0{\cdot}278\times10^{-6}$	0·102	$0{\cdot}239\times10^{-3}$	$0{\cdot}378\times10^{-6}$
a 45	1 kW h	=	$3{\cdot}60\times10^6$	1	367×10^3	860	1·36
a 46	(1 kgf m)	=	9·81	$2{\cdot}72\times10^{-6}$	1	$2{\cdot}345\times10^{-3}$	$3{\cdot}70\times10^{-6}$
a 47	(1 kcal)	=	4186·8	$1{\cdot}16\times10^{-3}$	426·9	1	$1{\cdot}58\times10^{-3}$
a 48	(1 hp h)	=	$2{\cdot}65\times10^6$	0·736	$0{\cdot}27\times10^6$	632	1

Units of power

			W	kW	(kgf m/s)	(kcal/h)	(hp)
a 49	1 W [3]	=	1	10^{-3}	0·102	0·860	$1{\cdot}36\times10^{-3}$
a 50	1 kW	=	1000	1	102	860	1·36
a 51	(1 kgf m/s)	=	9·81	$9{\cdot}81\times10^{-3}$	1	8·43	$13{\cdot}3\times10^{-3}$
a 52	(1 kcal h)	=	1·16	$1{\cdot}16\times10^{-3}$	0·119	1	$1{\cdot}58\times10^{-3}$
a 53	(1 hp)	=	736	0·736	75	632	1

Unit of mass for jewels

a 54 1 carat = 200 mg = $0{\cdot}2\times10^{-3}$kg = 1/5000 kg

Unit of fineness for precious metals

a 55 24 carat ≙ 1000·00 ‰ | 18 carat ≙ 750·00 ‰

a 56 14 carat ≙ 583·33 ‰ | 8 carat ≙ 333·33 ‰

Units of temperature

a 57 $T = \left(\frac{t}{°\mathrm{C}} + 273{\cdot}15\right)\mathrm{K} = \frac{5}{9}\cdot\frac{T_R}{\mathrm{Rank}}\,\mathrm{K}$

a 58 $T_R = \left(\frac{t_F}{°\mathrm{F}} + 459{\cdot}67\right)\mathrm{Rank} = \frac{9}{5}\cdot\frac{T}{\mathrm{K}}\,\mathrm{Rank}$

a 59 $t = \frac{5}{9}\left(\frac{t_F}{°\mathrm{F}} - 32\right)°\mathrm{C} = \left(\frac{T}{\mathrm{K}} - 273{\cdot}15\right)°\mathrm{C}$

a 60 $t_F = \left(\frac{9}{5}\cdot\frac{t}{°\mathrm{C}} + 32\right)°\mathrm{F} = \left(\frac{T_R}{\mathrm{Rank}} - 459{\cdot}67\right)°\mathrm{F}$

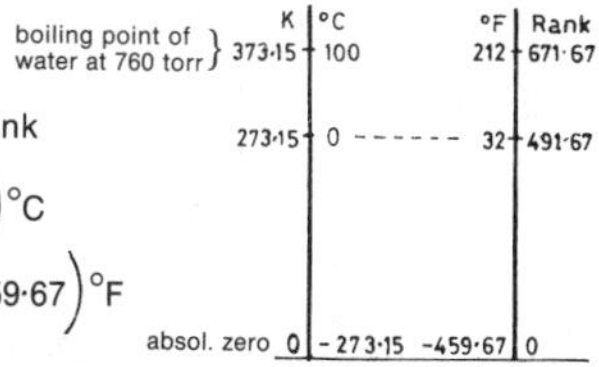

T, T_R, t and t_F are the temperatures in the scales for Kelvin, Rankine, Celsius, Fahrenheit.

[1] 1 torr = 1/760 atm = 1·333 22 mbar = 1 mm Hg at t = 0 °C

[2] 1 J = N m = 1 W s | [3] 1 W = 1 J/s = 1 N m/s

Conversion, Anglo-American to metric units

Units of length

	in	ft	yd	mm	m	km	
1 in =	1	0·08333	0·02778	25·4	0·0254	–	a 61
1 ft =	12	1	0·3333	304·8	0·3048	–	a 62
1 yd =	36	3	1	914·4	0·9144	–	a 63
1 mm =	0·03937	3281×10^{-6}	1094×10^{-6}	1	0·001	10^{-6}	a 64
1 m =	39·37	3·281	1·094	1000	1	0·001	a 65
1 km =	39370	3281	1094	10^6	1000	1	a 66

Units of area

	sq in	sq ft	sq yd	cm^2	dm^2	m^2	
1 sq in =	1	$6{\cdot}944 \times 10^{-3}$	$0{\cdot}772 \times 10^{-3}$	6·452	0·06452	$64{\cdot}5 \times 10^{-5}$	a 67
1 sq ft =	144	1	0·1111	929	9·29	0·0929	a 68
1 sq yd =	1296	9	1	8361	83·61	0·8361	a 69
1 cm^2 =	0·155	$1{\cdot}076 \times 10^{-3}$	$1{\cdot}197 \times 10^{-4}$	1	0·01	0·0001	a 70
1 dm^2 =	15·5	0·1076	0·01196	100	1	0·001	a 71
1 m^2 =	1550	10·76	1·196	10000	100	1	a 72

Units of volume

	cu in	cu ft	cu yd	cm^3	dm^3	m^3	
1 cu in =	1	$5{\cdot}786 \times 10^{-4}$	$2{\cdot}144 \times 10^{-5}$	16·39	0·01639	$1{\cdot}64 \times 10^{-5}$	a 73
1 cu ft =	1728	1	0·037	28316	28·32	0·0283	a 74
1 cu yd =	46656	27	1	764555	764·55	0·7646	a 75
1 cm^3 =	0·06102	3532×10^{-8}	$1{\cdot}31 \times 10^{-6}$	1	0·001	10^{-6}	a 76
1 dm^3 =	61·02	0·03532	0·00131	1000	1	0·001	a 77
1 m^3 =	61023	35·32	1·307	10^6	1000	1	a 78

Units of mass

	dram	oz	lb	g	kg	Mg	
1 dram =	1	0·0625	0·003906	1·772	0·00177	$1{\cdot}77 \times 10^{-6}$	a 79
1 oz =	16	1	0·0625	28·35	0·02832	$28{\cdot}3 \times 10^{-6}$	a 80
1 lb =	256	16	1	453·6	0·4531	$4{\cdot}53 \times 10^{-4}$	a 81
1 g =	0·5643	0·03527	0·002205	1	0·001	10^{-6}	a 82
1 kg =	564·3	35·27	2·205	1000	1	0·001	a 83
1 Mg =	$564{\cdot}3 \times 10^3$	35270	2205	10^6	1000	1	a 84

continued A 5

continued from A 4

Units of work

		ft lb	kgf m	J = W s	kW h	kcal	Btu
a 85	1 ft lb =	1	0·1383	1·356	$376·8 \times 10^{-9}$	324×10^{-6}	$1·286 \times 10^{-3}$
a 86	1 kgf m =	7·233	1	9·807	$2·725 \times 10^{-6}$	$2·344 \times 10^{-3}$	$9·301 \times 10^{-3}$
a 87	1 J = 1 W s =	0·7376	0·102	1	$277·8 \times 10^{-9}$	239×10^{-6}	$948·4 \times 10^{-6}$
a 88	1 kW h =	$2·655 \times 10^{6}$	$367·1 \times 10^{3}$	$3·6 \times 10^{6}$	1	860	3413
a 89	1 kcal =	$3·087 \times 10^{3}$	426·9	4187	$1·163 \times 10^{-3}$	1	3·968
a 90	1 Btu =	778·6	107·6	1055	293×10^{-6}	0·252	1

Units of power

		hp	kgf m/s	J/s = W	kW	kcal/s	Btu/s
a 91	1 hp =	1	76·04	745·7	0·7457	0·1782	0·7073
a 92	1 kgf m/s =	$13·15 \times 10^{-3}$	1	9·807	$9·807 \times 10^{-3}$	$2·344 \times 10^{-3}$	$9·296 \times 10^{-3}$
a 93	1 J/s = 1 W =	$1·341 \times 10^{-3}$	0·102	1	10^{-3}	239×10^{-6}	$948·4 \times 10^{-6}$
a 94	1 kW =	1·341	102	1000	1	0·239	0·9484
a 95	1 kcal/s =	5·614	426·9	4187	4·187	1	3·968
a 96	1 Btu/s =	1·415	107·6	1055	1·055	0·252	1

Other units

a 97	1 mil = 10^{-3} in	=	0·0254 mm
a 98	1 sq mil = 10^{-6} sq in	=	645·2 μm^2
a 99	1 yard = 3 ft	=	0·914 m
a100	1 English mile = 1760 yds	=	1609 m
a101	1 Nautical mile	=	1852 m
a102	1 Geographical mile	=	7420 m
a103	1 long ton = 2240 lb	=	1·016 Mg
a104	1 short ton (US) = 2000 lb	=	0·9072 Mg
a105	1 long ton = 2240 lbf	=	9·96 MN
a106	1 short ton (US) = 2000 lbf	=	9·00 MN
a107	1 Imp. gallon (Imperial gallon)	=	4·546 dm^3
a108	1 US gallon	=	3·785 dm^3
a109	1 BTU/ft^3 = 9·547 kcal/m^3	=	39·964 kJ/m^3
a110	1 BTU/lb = 0·556 kcal/kg	=	2·327 kJ/kg
a111	1 lbf/ft^2 = 4·882 kgf/m^2	=	47·8924 N/m^2
a112	1 lbf/in^2 (p.s.i.) = 0·0703 kgf/cm^2	=	0·6896 N/cm^2
a113	1 chain = 22 yds	=	20·11 m
a114	1 Hundredweight (GB) (cwt) = 112 lbf	=	498 kN
a115	1 Quarter (GB) = 28 lbf	=	124·5 kN
a116	1 Stone (GB) = 14 lbf	=	62·3 kN

AREAS

square

b 1 $A = a^2$

b 2 $a = \sqrt{A}$

b 3 $d = a\sqrt{2}$

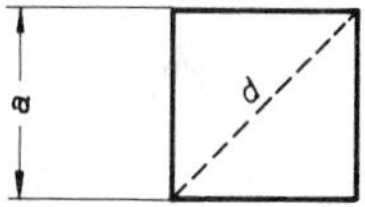

rectangle

b 4 $A = a\,b$

b 5 $d = \sqrt{a^2 + b^2}$

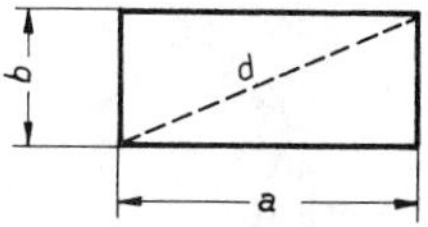

parallelogram

b 6 $A = a\,h = a\,b\,\sin\alpha$

b 7 $d_1 = \sqrt{(a + h\cot\alpha)^2 + h^2}$

b 8 $d_2 = \sqrt{(a - h\cot\alpha)^2 + h^2}$

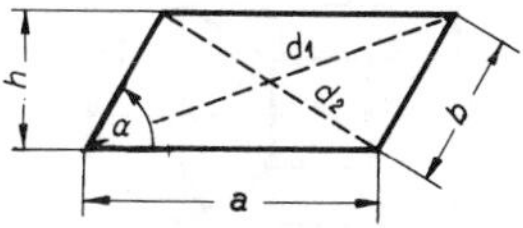

trapezium

b 9 $A = \dfrac{a + b}{2}\,h = m\,h$

b 10 $m = \dfrac{a + b}{2}$

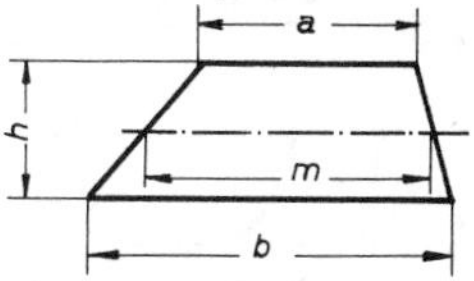

triangle

b 11 $A = \dfrac{a \cdot h}{2} = \varrho\,s$

b 12 $= \sqrt{s(s-a)(s-b)(s-c)}$

b 13 $s = \dfrac{a + b + c}{2}$

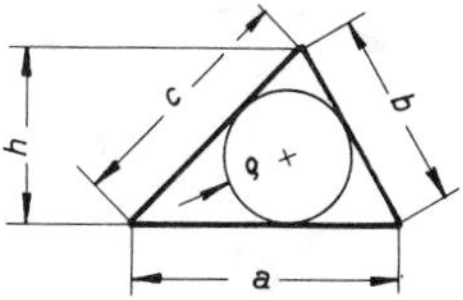

AREAS

equilateral triangle

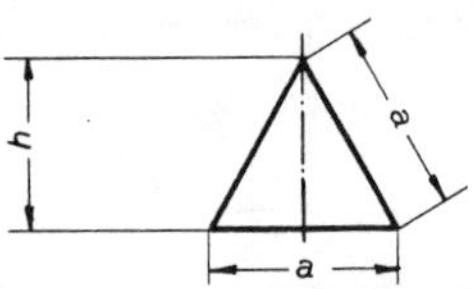

$A = \frac{a^2}{4}\sqrt{3}$ b 14

$h = \frac{a}{2}\sqrt{3}$ b 15

pentagon

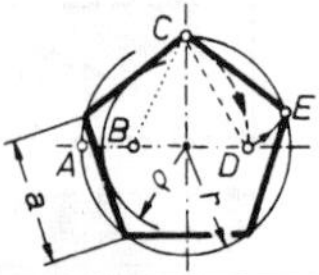

$A = \frac{5}{8} r^2 \sqrt{10 + 2\sqrt{5}}$ b 16

$a = \frac{1}{2} r \sqrt{10 - 2\sqrt{5}}$ b 17

$\varrho = \frac{1}{4} r \sqrt{6 + 2\sqrt{5}}$ b 18

construction:
$\overline{AB} = 0{\cdot}5\,r$, $\overline{BC} = \overline{BD}$, $\overline{CD} = \overline{CE}$

hexagon

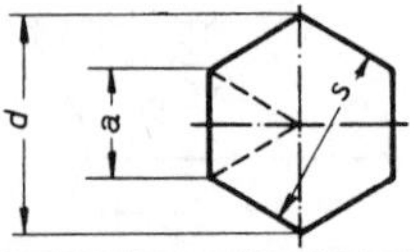

$A = \frac{3}{2} a^2 \sqrt{3}$ b 19

$d = 2\,a$ b 20

$= \frac{2}{\sqrt{3}} s \approx 1{\cdot}155\,s$ b 21

$s = \frac{\sqrt{3}}{2} d \approx 0{\cdot}866\,d$ b 22

octagon

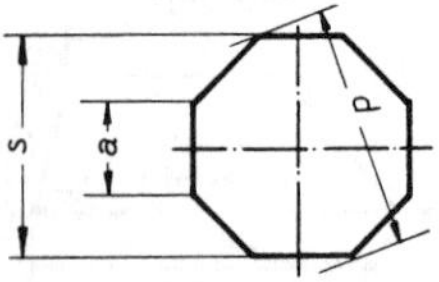

$A = 2\,a\,s \approx 0{\cdot}83\,s^2$ b 23

$= 2\,s\sqrt{d^2 - s^2}$ b 24

$a = s \tan 22{\cdot}5^0 \approx 0{\cdot}415\,s$ b 25

$s = d \cos 22{\cdot}5^0 \approx 0{\cdot}924\,d$ b 26

$d = \frac{s}{\cos 22{\cdot}5^0} \approx 1{\cdot}083\,s$ b 27

polygon

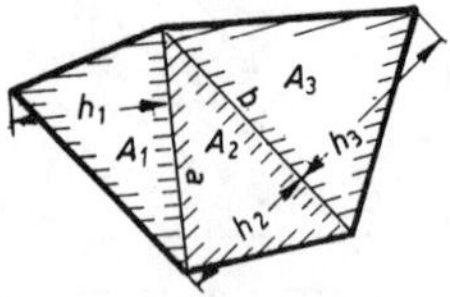

$A = A_1 + A_2 + A_3$ b 28

$= \frac{a\,h_1 + b\,h_2 + b\,h_3}{2}$ b 29

circle

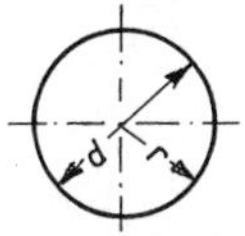

b 30 $A = \frac{\pi}{4} d^2 = \pi r^2$

b 31 $\approx 0{\cdot}785\, d^2$

b 32 $U = 2 \pi r = \pi d$

annulus

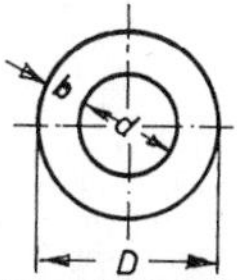

b 33 $A = \frac{\pi}{4} (D^2 - d^2)$

b 34 $= \pi\ (d + b) b$

b 35 $b = \frac{D - d}{2}$

sector of a circle

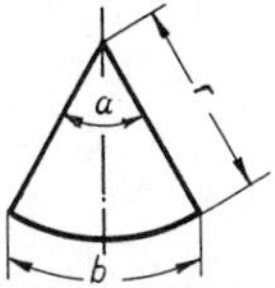

b 36 $A = \frac{\pi}{360^0} r^2\ \alpha = \frac{\hat{\alpha}}{2} r^2$

b 37 $= \frac{b\, r}{2}$

b 38 $b = \frac{\pi}{180^0}\ r\ \alpha$

b 39 $\hat{\alpha} = \frac{\pi}{180^0}\ \alpha$ ($\hat{\alpha} = \alpha$ in circular measure)

segment of a circle

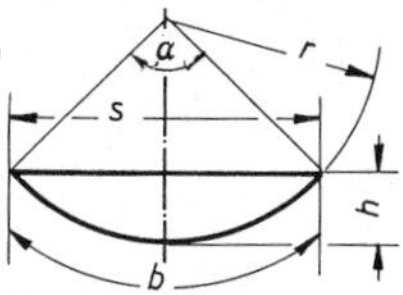

b 40 $s = 2\, r\ \sin \frac{\alpha}{2}$

b 41 $A = \frac{h}{6\, s} (3\, h^2 + 4\, s^2) = \frac{r^2}{2} (\hat{\alpha} - \sin \alpha)$

b 42 $r = \frac{h}{2} + \frac{s^2}{8\, h}$

b 43 $h = r(1 - \cos \frac{\alpha}{2}) = \frac{s}{2} \tan \frac{\alpha}{4}$

b 44 $\hat{\alpha}$ see formula b 39

ellipse

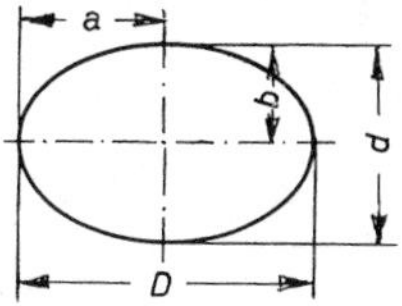

b 45 $A = \frac{\pi}{4} D\ d = \pi\, a\, b$

b 46 $U \approx \pi\ \frac{D + d}{2}$

b 47 $= \pi (a+b) \left[1 + \left(\frac{1}{2}\right)^2 \lambda^2 + \left(\frac{1}{2} \times \frac{1}{4}\right)^2 \lambda^4 + \right.$

$+ \left(\frac{1}{2} \times \frac{1}{4} \times \frac{3}{6}\right)^2 \lambda^6 + \left(\frac{1}{2} \times \frac{1}{4} \times \frac{3}{6} \times \frac{5}{8}\right)^2 \lambda^8 +$

$\left. + \left(\frac{1}{2} \times \frac{1}{4} \times \frac{3}{6} \times \frac{5}{8} \times \frac{7}{10}\right)^2 \lambda^{10} + \quad \ldots \right]$, where $\lambda = \frac{a - b}{a + b}$

SOLID BODIES

cube

c 1 $V = a^3$

c 2 $A_0 = 6\,a^2$

c 3 $d = \sqrt{3}\,a$

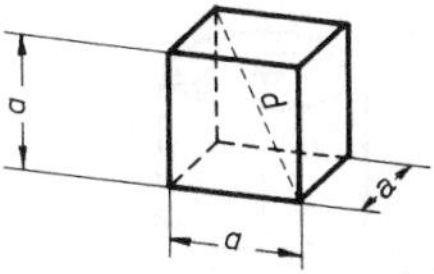

cuboid

c 4 $V = a\,b\,c$

c 5 $A_0 = 2(a\,b + a\,c + b\,c)$

c 6 $d = \sqrt{a^2 + b^2 + c^2}$

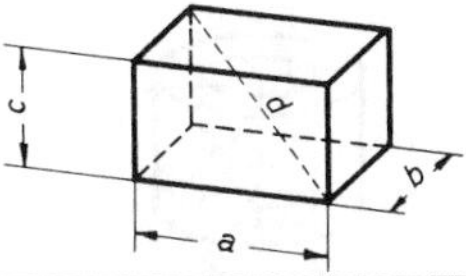

parallelepiped

c 7 $V = A_1\,h$

(Cavalieri principle)

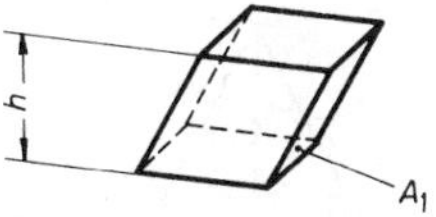

pyramid

c 8 $V = \dfrac{A_1\,h}{3}$

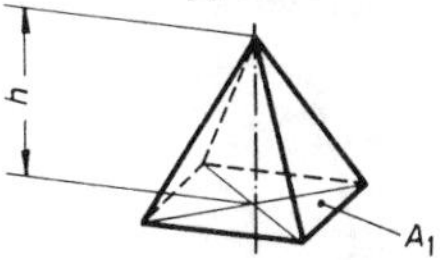

frustum of pyramid

c 9 $V = \dfrac{h}{3}\left(A_1 + A_2 + \sqrt{A_1\,A_2}\right)$

c 10 $\approx h\,\dfrac{A_1 + A_2}{2}$ (for $A_1 \approx A_2$)

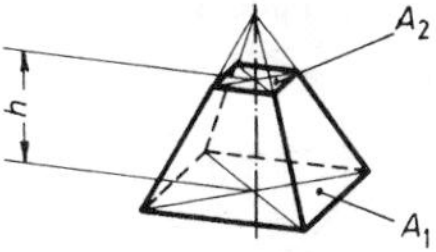

C 2 SOLID BODIES

cylinder

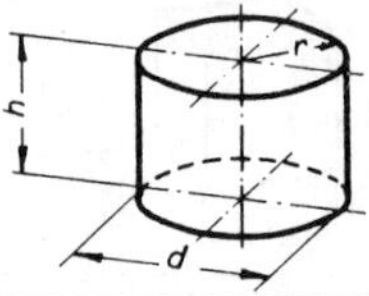

$$V = \frac{\pi}{4} d^2 h$$ c 11

$$A_m = 2 \pi r h$$ c 12

$$A_o = 2 \pi r (r + h)$$ c 13

hollow cylinder

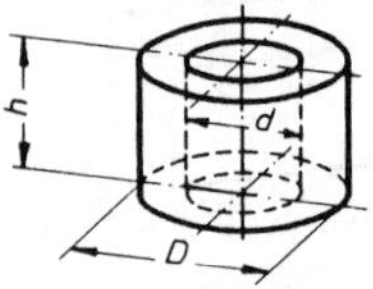

$$V = \frac{\pi}{4} h (D^2 - d^2)$$ c 14

cone

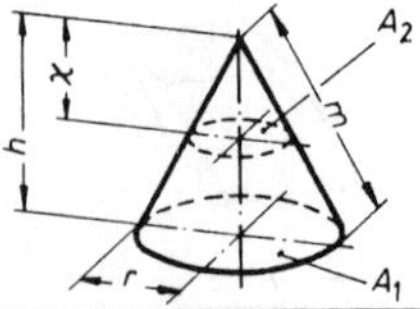

$$V = \frac{\pi}{3} r^2 h$$ c 15

$$A_m = \pi r m$$ c 16

$$A_o = \pi r (r + m)$$ c 17

$$m = \sqrt{h^2 + r^2}$$ c 18

$$A_2 : A_1 = x^2 : h^2$$ c 19

frustum of cone

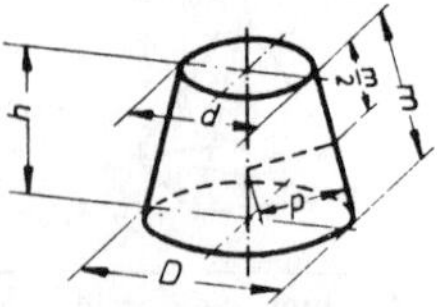

$$V = \frac{\pi}{12} h (D^2 + Dd + d^2)$$ c 20

$$A_m = \frac{\pi}{2} m (D + d) = 2 \pi p m$$ c 21

$$m = \sqrt{\left(\frac{D - d}{2}\right)^2 + h^2}$$ c 22

sphere

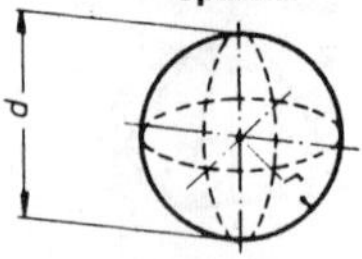

$$V = \frac{4}{3} \pi r^3 = \frac{1}{6} \pi d^3$$ c 23

$$\approx 4 \cdot 189 \, r^3$$ c 24

$$A_o = 4 \pi r^2 = \pi d^2$$ c 25

zone of a sphere

c 26 $V = \frac{\pi}{6} h (3a^2 + 3b^2 + h^2)$

c 27 $A_m = 2 \pi r h$

c 28 $A_o = \pi(2 r h + a^2 + b^2)$

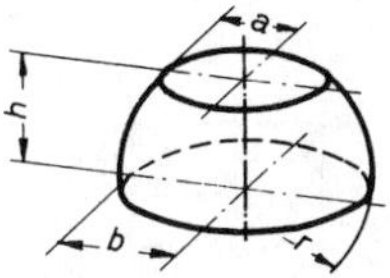

segment of a sphere

c 29 $V = \frac{\pi}{6} h (\frac{3}{4} s^2 + h^2)$

$= \pi h^2 (r - \frac{h}{3})$

c 30 $A_m = 2 \pi r h$

c 31 $= \frac{\pi}{4} (s^2 + 4 h^2)$

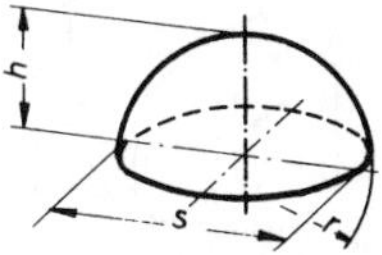

sector of a sphere

c 32 $V = \frac{2}{3} \pi r^2 h$

c 33 $A_o = \frac{\pi}{2} r (4 h + s)$

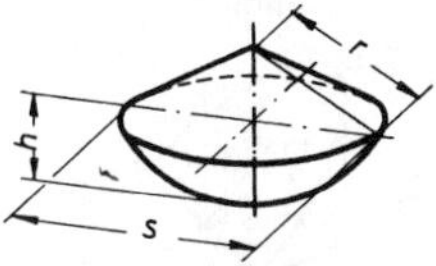

sphere with cylindrical boring

c 34 $V = \frac{\pi}{6} h^3$

c 35 $A_o = 2 \pi h (R + r)$

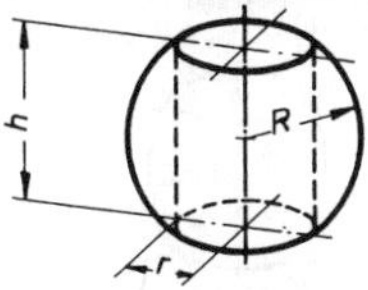

sphere with conical boring

c 36 $V = \frac{2}{3} \pi r^2 h$

c 37 $A_o = 2 \pi r \left(h + \sqrt{r^2 - \frac{h^2}{4}}\right)$

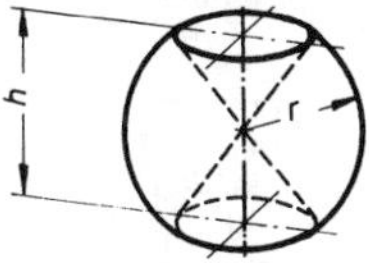

SOLID BODIES

torus

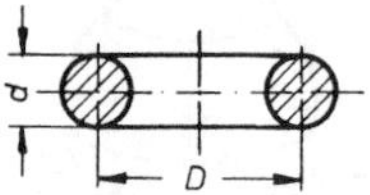

$$V = \frac{\pi^2}{4} D d^2$$ c 38

$$A_o = \pi^2 D d$$ c 39

sliced cylinder

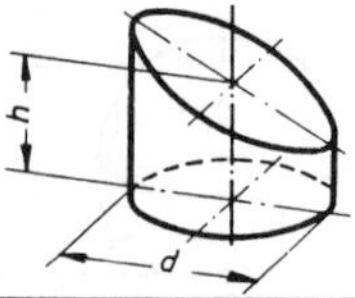

$$V = \frac{\pi}{4} d^2 h$$ c 40

ungula

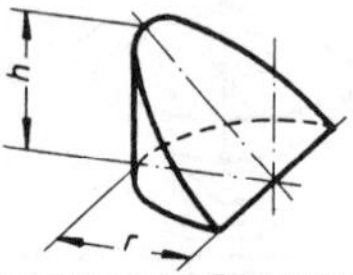

$$V = \frac{2}{3} r^2 h$$ c 41

$$A_m = 2 r h$$ c 42

$$A_o = A_m + \frac{\pi}{2} r^2 + \frac{\pi}{2} r \sqrt{r^2 + h^2}$$ c 43

barrel

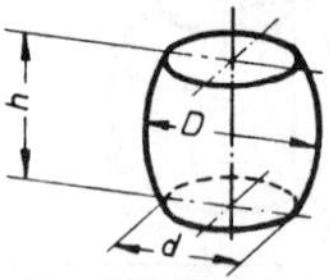

$$V \approx \frac{\pi}{12} h (2 D^2 + d^2)$$ c 44

prismoid

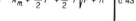

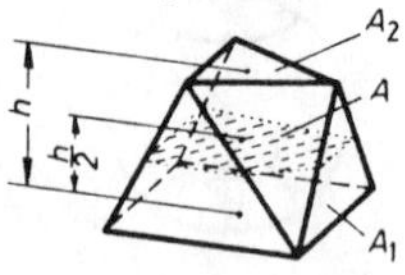

$$V = \frac{h}{6} (A_1 + A_2 + 4 A)$$ c 45

This formula may be used for calculations involving solids shown in fig. C1 ... C3 and thus spheres and parts of spheres.

Rules for powers and roots

	general	numerical examples
d 1	$p\,a^n \pm q\,a^n = (p \pm q)a^n$	$3a^4 + 4a^4 = 7a^4$
d 2	$a^m\,a^n = a^{m+n}$	$a^8 \times a^4 = a^{12}$
d 3	$a^m/a^n = a^{m-n}$	$a^8/a^2 = a^{8-2} = a^6$
d 4	$(a^m)^n = (a^n)^m = a^{mn}$	$(a^3)^2 = (a^2)^3 = a^{2\times 3} = a^6$
d 5	$a^{-n} = 1/a^n$	$a^{-4} = 1/a^4$
d 6	$\dfrac{a^n}{b^n} = \left(\dfrac{a}{b}\right)^n$	$\dfrac{a^3}{b^3} = \left(\dfrac{a}{b}\right)^3$
d 7	$p\sqrt[n]{a} \pm q\sqrt[n]{a} = (p \pm q)\sqrt[n]{a}$	$4\sqrt[3]{x} + 7\sqrt[3]{x} = 11\sqrt[3]{x}$
d 8	$\sqrt[n]{a\,b} = \sqrt[n]{a} \times \sqrt[n]{b}$	$\sqrt[4]{16 \times 81} = \sqrt[4]{16} \times \sqrt[4]{81}$
d 9	$\dfrac{\sqrt[n]{a}}{\sqrt[n]{b}} = \sqrt[n]{\dfrac{a}{b}} = \left(\dfrac{a}{b}\right)^{\frac{1}{n}}$	$\dfrac{\sqrt{8}}{\sqrt{2}} = \sqrt{4} = 2$
d10	$\sqrt[nx]{a^{mx}} = \sqrt[n]{a^m}$	$\sqrt[6]{a^8} = \sqrt[3]{a^4}$
d11	$\sqrt[n]{a^m} = \left(\sqrt[n]{a}\right)^m = a^{\frac{m}{n}}$ +)	$\sqrt[4]{a^3} = \left(\sqrt[4]{a}\right)^3 = a^{\frac{3}{4}}$
d12	$\sqrt{-a} = \mathrm{i}\sqrt{a}$	$\sqrt{-9} = \mathrm{i}\sqrt{9} = \mathrm{i}\,3$

+) Not applicable to special calculations } e.g. $\sqrt{(-2)^2} = \sqrt{4} = +2$; $(\sqrt{-2})^2 = -2$

Exponents of powers and roots have to be non-dimensional quantities!

Quadratic equation (equation of the second degree)

d13 Normal form $x^2 + px + q = 0$

d14 Solutions $x_1 \; ; \; x_2 = -\dfrac{p}{2} \pm \sqrt{\dfrac{p^2}{4} - q}$

d15 Vieta's rule $p = -(x_1 + x_2) \; ; \quad q = x_1\,x_2$

Iterative calculation of an n-th root

d16 When $x = \sqrt[n]{A}$, then $x = \dfrac{1}{n}\left[(n-1)x_0 + \dfrac{A}{x_0^{\,n-1}}\right]$,

where x_0 is the initially estimated value of x. Repeatedly inserting the obtained x as a new value of x_0 gradually increases the accuracy of x.

Expansion of general algebraic expressions

d 17 $(a \pm b)^2 = a^2 \pm 2ab + b^2$

d 18 $(a \pm b)^3 = a^3 \pm 3a^2 b + 3ab^2 \pm b^3$

d 19 $(a + b)^n = a^n + \frac{n}{1} a^{n-1} b + \frac{n(n-1)}{1 \cdot 2} a^{n-2} b^2 + $
$$+ \frac{n(n-1)(n-2)}{1 \times 2 \times 3} a^{n-3} b^3 + \ldots b^n$$

d 20 $(a + b + c)^2 = a^2 + 2ab + 2ac + b^2 + 2bc + c^2$

d 21 $(a - b + c)^2 = a^2 - 2ab + 2ac + b^2 - 2bc + c^2$

d 22 $a^2 - b^2 = (a + b)(a - b)$

d 23 $a^3 + b^3 = (a + b)(a^2 - ab + b^2)$

d 24 $a^3 - b^3 = (a - b)(a^2 + ab + b^2)$

d 25 $a^n - b^n = (a - b)(a^{n-1} + a^{n-2} b + a^{n-3} b^2 + \ldots$
$$\ldots + ab^{n-2} + b^{n-1})$$

Binomial theorem

d 26 $(a + b)^n = \binom{n}{0} a^n + \binom{n}{1} a^{n-1} b + \binom{n}{2} a^{n-2} b^2 + \binom{n}{3} a^{n-3} b^3 + \ldots$ *)

d 27 $\binom{n}{k} = \frac{n(n-1)(n-2) \ldots (n-k+1)}{1 \times 2 \times 3 \ldots k}$ *)

*) n must be a whole number

d 28 $(a + b)^4 = 1a^4 + \frac{4}{1} a^{4-1} b + \frac{4 \times 3}{1 \times 2} a^{4-2} b^2 + \frac{4 \times 3 \times 2}{1 \times 2 \times 3} a^{4-3} b^3 + b^4$
$$= a^4 + 4a^3 b + 6a^2 b^2 + 4a \times b^3 + b^4$$

Diagrammatic solution

d 29 **Coefficient – Pascal triangle**

$(a + b)^0$							1						
$(a + b)^1$						1		1					
$(a + b)^2$					1		2		1				
$(a + b)^3$				1		3		3		1			
$(a + b)^4$			1		4		6 ↓	4			1		
$(a + b)^5$		1		5 ↓	10			10		5		1	
$(a + b)^6$	1		6		15		20		15		6		1

Continue with each line starting and finishing with 1. The second and penultimate numbers should be the exponents, the others the sum of those to the right and left immediately above them.

d 30 **Exponents:** The sum of the exponents a and b in each separate term is equal to the binomial exponent n. As the power of a decreases the power of b increases.

d 31 **Signs:** $(a+b)$ is always positive
$(a-b)$ is initially positive and changes from term to term.

d 32 **Examples:**

$$(a + b)^5 = a^5 + 5a^4 b + 10a^3 b^2 + 10a^2 b^3 + 5ab^4 + b^5$$
$$(a - b)^5 = +a^5 - 5a^4 b + 10a^3 b^2 - 10a^2 b^3 + 5ab^4 - b^5$$

Proper fraction rational function

$$y(x) = \frac{P(x)}{Q(x)} = \frac{a_0 + a_1 x + a_2 x^2 + \ldots + a_m x^m}{b_0 + b_1 x + b_2 x^2 + \ldots + b_n x^n} \qquad n>m$$

n and m whole numbers

Coefficients a_ν, b_μ can be real or complex. If n_i are the zeros of the denominator $Q(x)$ the factorized form from $y(x)$ is:

d 33 $$y(x) = \frac{P(x)}{Q(x)} = \frac{P(x)}{a(x-n_1)^{k1}\,(x-n_2)^{k2}\ldots(x-n_q)^{kq}}$$

where real or complex zeros of $Q(x)$ can occur $k_1, k_2 \ldots k_q$ times; a is a constant factor.

Partial fraction expansion

For easy manipulation of $y(x)$, e.g. for integration, the expansion $y(x)$ into partial fractions is often appropriate

d 34 $$y(x) = \frac{P(x)}{Q(x)} = \frac{A_{11}}{x-n_1} + \frac{A_{12}}{(x-n_1)^2} + \ldots + \frac{A_{1k1}}{(x-n_1)^{k1}} + \frac{A_{21}}{x-n_2} + \frac{A_{22}}{(x-n_2)^2} + \ldots + \frac{A_{2k2}}{(x-n_2)^{k2}} + \ldots + \frac{A_{q1}}{x-n_q} + \frac{A_{q2}}{(x-n_q)^2} + \ldots + \frac{A_{qkq}}{(x-n_q)^{kq}}$$

Complex zeros occur in pairs (conjugate complex numbers) when $Q(x)$ has real coefficients. For expansion these pairs are combined to real partial fractions. If in d 33 the zeros $n_2 = \bar{n}_1$ (n_2 is conjugate complex to n_1) and if due to the pairwise occurrence $k_1 = k_2 = k$, the partial fractions of d 34 with the constants $A_{11} \ldots A_{2k2}$ can be combined to the following partial fractions:

d 35 $$\frac{B_{11}x + C_{11}}{x^2+ax+b} + \frac{B_{12}x + C_{12}}{(x^2+ax+b)^2} + \ldots + \frac{B_{1k}x + C_{1k}}{(x^2+ax+b)^k}$$

To obtain the constants A_{11} to A_{qkq} resp. B_{11}, C_{11} to B_{1k}, C_{1k} coefficients of equal power in x at the left side of the equation are compared with those at the right side after having converted to the common denominator $Q(x)$.

Example:

$$y(x) = \frac{2x-1}{(x+1-2i)(x+1+2i)(x+1)^2} = \frac{2x-1}{Q(x)} = \frac{B_{11}x+C_{11}}{x^2+2x+5} + \frac{A_{q1}}{x+1} + \frac{A_{q2}}{(x+1)^2}$$

$$\frac{2x-1}{Q(x)} = \frac{B_{11}x(x+1)^2 + C_{11}(x+1)^2 + A_{q1}(x+1)(x^2+2x+5) + A_{q2}(x^2+2x+5)}{Q(x)}$$

$$2x-1 = (A_{q1}+B_{11})x^3 + (3A_{q1}+A_{q2}+2B_{11}+C_{11})x^2 + (7A_{q1}+2A_{q2}+B_{11}+2C_{11})x + 5A_{q1}+5A_{q2}+C_{11}$$

Comparison of coefficients between left and right side:

$$B_{11} = -1/2; \quad C_{11} = 1/4; \quad A_{q1} = 1/2; \quad A_{q2} = -3/4.$$

If there are single zeros n_i, the constants A_{11}, A_{21} … A_{q1} of equation d 34 can be obtained by:

d 36 $$A_{11} = P(n_1)/Q'(n_1); \quad A_{21} = P(n_2)/Q'(n_2); \ldots A_q = P(n_q)/Q'(n_q)$$

General

	system	log to the base of	terminology
37	$\log_a$	a	log to base a
38	$\log_{10} = \lg$	10	common log
39	$\log_e = \ln$	e	natural log
40	$\log_2 = \mathrm{lb}$	2	log to base 2

The symbols in $\log_a x = b$ are called:
- a base
- x antilogarithm
- b logarithm (log)

Rules for logarithmic calculations

41 $$\log_a (x\,y) = \log_a x + \log_a y$$

42 $$\log_a \frac{x}{y} = \log_a x - \log_a y$$

43 $$\log_a x^n = n \times \log_a x$$

44 $$\log_a \sqrt[n]{x} = \frac{1}{n} \log_a x$$

Exponential equation

45 $$a^x = b = e^{x \ln a}$$

46 hence $$x = \frac{\log b}{\log a} \quad \Big| \quad a = \sqrt[x]{b}$$

Conversion of logarithms

47 $$\log x = \lg e \times \ln x = 0{\cdot}434\,294 \times \ln x$$

48 $$\ln x = \frac{\lg x}{\lg e} = 2{\cdot}302\,585 \times \lg x$$

49 $$\mathrm{ld}\ x = 1{\cdot}442\,695 \times \ln x = 3{\cdot}321\,928 \times \lg x$$

Base of the natural logs e = 2·71828183 . . .

Key to common logarithm of a number

50 $\lg 0{\cdot}01 = -2.$ or $8.\ \ldots -10$

51 $\lg 0{\cdot}1 = -1.$ or $9.\ \ldots -10$

52 $\lg 1 = 0.$

53 $\lg 10 = 1.$

54 $\lg 100 = 2.$

etc.

Note: The antilogarithm always has to be a non-dimensional quantity.

Permutations

An ordered selection or arrangement of r out of n things is called a "permutation" of the n things taken r at a time. The number of these permutations is denoted by:

55 $$P_r^n = n(n-1)(n-2)\dots(n-r+1), \qquad n \geqq r$$

If $r = n$, this becomes

$$P_n^n = P_n = n(n-1)(n-2)\dots 1 = n!\ ^{*)}$$

Example: The $n = 3$ things a, b, c can be permutated with each other (i.e. 3 at a time) in the following 6 ways:

$abc \quad bac \quad cab$

$acb \quad bca \quad cba$ — Here $r = n = 3$.

56 $$P_3 = 3! = 1 \times 2 \times 3 = 6.$$

Special case: The number of permutations of n things taken all together incorporating n_1 of one sort, n_2 of another sort and n_r of a rth sort is:

57 $$P_{n,r} = \frac{n!}{n_1! \times n_2! \times \dots n_r!}$$

Example: The $n = 3$ things a, a, b can be permutated the following 3 ways:

$aab \quad aba \quad baa$ — Here $n = 3$, $n_1 = 2$, $n_2 = 1$.

$$P_{3,2} = \frac{3!}{2! \times 1!} = \frac{1 \times 2 \times 3}{1 \times 2 \times 1} = 3.$$

Combinations

A selection of r out n things without regard to order is called "combination" of n things taken r at a time. The number of these combinations is denoted by

58 $$C_r^n = \frac{n!}{r!(n-r)!} = \binom{n}{r}\ ^{**)}$$

Example: The $n = 3$ things a, b, c taken together give only the one combination abc. Here $n = 3$, $r = 3$.

Hence $$C_3^3 = \binom{3}{3} = \frac{3 \times 2 \times 1}{1 \times 2 \times 3} = 1.$$

The table on page D 6 compares combinations and permutations (with and without the things repeating).

---

*) $n!$ is pronounced „n factorial"

**) Symbol usual for binomial coefficient (see d 27)

Combinations and Permutations

(Explanations see D 5)

	d 62	d 61	d 60	d 59
	Number of combinations without repeating regardless of the positions of things	Number of combinations with repeating regardless of the positions of things	Number of permutations without repeating taking into account the positions of things	Number of permutations with repeating taking into account the positions of things
Formula	$C_r^n = \frac{n!}{r!(n-r)!} = \binom{n}{r}$ *) $= \frac{n(n-1)\cdots(n-r+1)}{r!}$	${}^wC_r^n = \binom{n+r-1}{r}$ *) $= \frac{n(n+1)\cdots(n+r-1)}{r!}$	$P_r^n = \frac{n!}{(n-r)!} = \binom{n}{r} r!$ *) $= n(n-1)\cdots(n-r+1)$	${}^wP_r^n = n^r$
Explanation of symbols	C: Number of possible combinations n: Number of things given r: Number of things selected from n given things		P: Number of possible permutations	
Examples: Given	$n = 3$ things a, b, c	$r = 2$ things selected from the 3 given things		
Examples: Possibilities	· ab ac · · bc · · ·	aa ab ac · bb bc · · cc	· ab ac ba · bc ca cb ·	aa ab ac ba bb bc ca cb cc
Examples: Calculation of the number of possibilities	$C_2^3 = \frac{3!}{2!(3-2)!}$ $= \binom{3}{2} = \frac{3 \times 2}{1 \times 2} = 3$	${}^wC_2^3 = \binom{3+2-1}{2}$ $= \binom{4}{2}$ $= \frac{4 \times 3}{1 \times 2} = 6$	$P_2^3 = \binom{3}{2} 2!$ $= \frac{3!}{(3-2)!} = \frac{3!}{1!}$ $= \frac{1 \times 2 \times 3}{1} = 6$	${}^wP_2^3 = 3^2$ $= 9$
Examples: Note	*ab* and *ba* for example are the same combination		*ab* and *ba* for example are different permutations	

w : with repeating

*) calculation according to d 27

Second order determinants

d 63 $$a_{11}\,x + a_{12}\,y = r_1 \qquad a_{21}\,x + a_{22}\,y = r_2 \qquad D = \begin{vmatrix} a_{11} & a_{12} \\ a_{21} & a_{22} \end{vmatrix} = a_{11}\,a_{22} - a_{21}\,a_{12}$$

insert r column in place of

column x	column y
d 64 $D_1 = \begin{vmatrix} r_1 & a_{12} \\ r_2 & a_{22} \end{vmatrix} = r_1\,a_{22} - r_2\,a_{12}$	$D_2 = \begin{vmatrix} a_{11} & r_1 \\ a_{21} & r_2 \end{vmatrix} = r_2\,a_{11} - r_1\,a_{21}$
$x = \dfrac{D_1}{D}$	$y = \dfrac{D_2}{D}$

Third order determinants (Sarrus rule)

d 65 $$\begin{aligned} a_{11}\,x + a_{12}\,y + a_{13}\,z &= r_1 \\ a_{21}\,x + a_{22}\,y + a_{23}\,z &= r_2 \\ a_{31}\,x + a_{32}\,y + a_{33}\,z &= r_3 \end{aligned}$$

d 66 $$D = \begin{vmatrix} a_{11} & a_{12} & a_{13} \\ a_{21} & a_{22} & a_{23} \\ a_{31} & a_{32} & a_{33} \end{vmatrix}\begin{matrix} a_{11} & a_{12} \\ a_{21} & a_{22} \\ a_{31} & a_{32} \end{matrix} \begin{aligned} &= a_{11}\,a_{22}\,a_{33} + a_{12}\,a_{23}\,a_{31} \\ &\quad + a_{13}\,a_{21}\,a_{32} - a_{13}\,a_{22}\,a_{31} \\ &\quad - a_{11}\,a_{23}\,a_{32} - a_{12}\,a_{21}\,a_{33} \end{aligned}$$

insert r column for x column:

d 67 $$D_1 = \begin{vmatrix} r_1 & a_{12} & a_{13} \\ r_2 & a_{22} & a_{23} \\ r_3 & a_{32} & a_{33} \end{vmatrix}\begin{matrix} r_1 & a_{12} \\ r_2 & a_{22} \\ r_3 & a_{32} \end{matrix} \begin{aligned} &= r_1\,a_{22}\,a_{33} + a_{12}\,a_{23}\,r_3 \\ &\quad + a_{13}\,r_2\,a_{32} - a_{13}\,a_{22}\,r_3 \\ &\quad - r_1\,a_{23}\,a_{32} - a_{12}\,r_2\,a_{33} \end{aligned}$$

determine D_2 and D_3 similarly by replacing the y- and z-column by the r-column:

d 68 $$x = \frac{D_1}{D} \qquad y = \frac{D_2}{D} \qquad z = \frac{D_3}{D}$$

continued on D 8

Determinants of more than the 2nd order:

(The Sarrus Rule, see D 7, may be used for determinants of higher order than the 3rd).

By adding or subtracting suitable multiples of two rows or columns, endeavour to obtain zero values. Expand the determinant starting from the row or column containing most zeros. Alternate the signs of terms, starting with a_{11} as +.

Example:

d 69

$$\begin{matrix} \overset{+}{a}_{11} & \overset{-}{a}_{12} & \overset{+}{a}_{13} & \overset{-}{0} \\ a_{21} & a_{22} & a_{23} & \overset{+}{a}_{24} \\ a_{31} & a_{32} & a_{33} & \overset{-}{a}_{34} \\ a_{41} & a_{42} & a_{43} & \overset{+}{0} \end{matrix}$$

Expand on 4th column:

d 70

$$a_{24}\begin{vmatrix} \overset{+}{a}_{11} & \overset{-}{a}_{12} & \overset{+}{a}_{13} \\ a_{31} & a_{32} & a_{33} \\ a_{41} & a_{42} & a_{43} \end{vmatrix} \qquad -a_{34}\begin{vmatrix} \overset{+}{a}_{11} & \overset{-}{a}_{12} & \overset{+}{a}_{13} \\ a_{21} & a_{22} & a_{23} \\ a_{41} & a_{42} & a_{43} \end{vmatrix}$$

Further expand as:

d 71

$$D = a_{24}\left(a_{11}\begin{vmatrix} a_{32} & a_{33} \\ a_{42} & a_{43} \end{vmatrix} - a_{12}\begin{vmatrix} a_{31} & a_{33} \\ a_{41} & a_{43} \end{vmatrix} + a_{13}\begin{vmatrix} a_{31} & a_{32} \\ a_{41} & a_{42} \end{vmatrix}\right) - a_{34}\left(\cdot\ \cdot\ \cdot\right)$$

To form the determinants D_1, D_2, . . . (see D 7) substitute the r column for the first, second, . . . column of D, and evaluate in the same way as for D.

For determinant of the nth order, find $u_{1 \ldots n}$ from the formulae:

d 72

$$u_1 = \frac{D_1}{D}, \quad u_2 = \frac{D_2}{D}, \quad \ldots \; u_n = \frac{D_n}{D}$$

Note: For determinants of the nth order continue until determinants of the 3rd order have been obtained.

Arithmetic series

The sequence 1, 4, 7, 10 etc. is called an arithmetic series. (The difference between two consecutive terms is constant).

d 73 Formulae: $a_n = a_1 + (n - 1)\,d$

$$s_n = \frac{n}{2}(a_1 + a_n) = a_1\,n + \frac{n(n-1)\,d}{2}$$

Arithmetic mean: Each term of an arithmetic series is the arithmetic mean a_m of its adjacent terms a_{m-1} and a_{m+1}.

d 74 Thus, the *m*th term is $a_m = \frac{a_{m-1} + a_{m+1}}{2}$ for $1 < m < n$

(e.g. in the above series $a_3 = \frac{4 + 10}{2} = 7$)

Geometric series

The sequence 1, 2, 4, 8 etc is called a geometric series. (The quotient of two consecutive terms is constant).

d 75 Formulae: $a_n = a_1\,q^{n-1}$

$$s_n = a_1 \frac{q^n - 1}{q - 1} = \frac{q \times a_n - a_1}{q - 1}$$

Geometric mean: Each term of a geometric series is the geometric mean a_m of its adjacent terms a_{m-1} and a_{m+1}.

d 76 Thus, the *m*th term is $a_m = \sqrt{a_{m-1} \times a_{m+1}}$ for $1 < m < n$

(e.g. in the above series) $a_3 = \sqrt{2 \times 8} = 4$)

For infinite geometric series ($n \to \infty$; $|q| < 1$) the following statements apply

d 77 $a_n = \lim_{n \to \infty} a_n = 0$; $s_n = \lim_{n \to \infty} s_n = a_1 \frac{1}{1 - q}$

Decimal-geometric series

Application for calculation of standardized number-series

Quotient of two consecutive terms is called „progressive ratio φ".

d 78 $\varphi = \sqrt[b]{10}.$ $b \geq 1$, integer.

b determines the number of terms or number of standardized numbers of a series within one decade. The values of the terms which should be rounded up, are calculated according to d 77:

d 79 $a_n = a_1(\sqrt[b]{10})^{n-1} = a_n(10^{1/b})^{n-1}$ $n = 1 \ldots b$

Starting with $a_1 = 1$ or $a_1 = 10$ or $a_1 = 100$ or ...

Examples:

b	designation	note
6, 12, 24, ...	E 6, E 12, E 24, ...	intern. E-series, see Z 22
5, 10, 20, ...	R 5, R 10, R 20, ...	DIN-series, see R 1

a_1 : initial term
a_n : final term
d : difference between two consecutive terms
n : number of terms
s_n : sum to n terms
q : quotient of two consecutive terms

Binominal series

d 80 $$f(x) = (1 \pm x)^{\alpha} = 1 \pm \binom{\alpha}{1}x + \binom{\alpha}{2}x^2 \pm \binom{\alpha}{3}x^3 + \dots$$

α may be either positive or negative, a whole number or a fraction.

Expansion of the binomial coefficient:

$$\binom{\alpha}{n} = \frac{\alpha(\alpha-1)(\alpha-2)(\alpha-3)\dots(\alpha-n+1)}{1\times 2\times 3\dots\times n}$$

Examples:

		for
d 81	$\frac{1}{1 \pm x} = (1 \pm x)^{-1} = 1 \mp x + x^2 \mp x^3 + \dots$	$\lvert x\rvert<1$
d 82	$\sqrt{1 \pm x} = (1 \pm x)^{\frac{1}{2}} = 1 \pm \frac{1}{2}x - \frac{1}{8}x^2 \pm \frac{1}{16}x^3 - \dots$	$\lvert x\rvert<1$
d 83	$\frac{1}{\sqrt{1 \pm x}} = (1 \pm x)^{-\frac{1}{2}} = 1 \mp \frac{1}{2}x + \frac{3}{8}x^2 \mp \frac{5}{16}x^3 + \dots$	$\lvert x\rvert<1$

Taylor series

d 84 $$f(x) = f(a) + \frac{f'(a)}{1!}(x-a) + \frac{f''(a)}{2!}(x-a)^2 + \dots$$

putting $a = 0$ gives the MacLaurin series:

d 85 $$f(x) = f(0) + \frac{f'(0)}{1!}x + \frac{f''(0)}{2!}x^2 + \dots$$

Examples:

		for
d 86	$e^x = 1 + \frac{x}{1!} + \frac{x^2}{2!} + \frac{x^3}{3!} + \dots$	all x
d 87	$a^x = 1 + \frac{x \ln a}{1!} + \frac{(x \ln a)^2}{2!} + \frac{(x \ln a)^3}{3!} + \dots$	all x
d 88	$\ln x = 2\left[\frac{x-1}{x+1} + \frac{1}{3}\left(\frac{x-1}{x+1}\right)^3 + \frac{1}{5}\left(\frac{x-1}{x+1}\right)^5 + \dots\right]$	$x > 0$
d 89	$\ln(1+x) = x - \frac{x^2}{2} + \frac{x^3}{3} - \frac{x^4}{4} + \frac{x^5}{5} - \dots$	$-1 < x$, $x \leqq +1$
d 90	$\ln 2 = 1 - \frac{1}{2} + \frac{1}{3} - \frac{1}{4} + \frac{1}{5} - \dots$	

continued on D 11

Taylor series
(continued)

Examples

		for
d 91	$\sin x = x - \frac{x^3}{3!} + \frac{x^5}{5!} - \frac{x^7}{7!} + \ldots$	all x
d 92	$\cos x = 1 - \frac{x^2}{2!} + \frac{x^4}{4!} - \frac{x^6}{6!} + \ldots$	all x
d 93	$\tan x = x + \frac{1}{3}x^3 + \frac{2}{15}x^5 + \frac{17}{315}x^7 + \ldots$	$\lvert x\rvert < \frac{\pi}{2}$
d 94	$\cot x = \frac{1}{x} - \frac{1}{3}x - \frac{1}{45}x^3 - \frac{2}{945}x^5 - \ldots$	$0 < \lvert x\rvert$ $\lvert x\rvert < \pi$
d 95	$\arcsin x = x + \frac{1}{2}\frac{x^3}{3} + \frac{1\times3}{2\times4}\frac{x^5}{5} + \frac{1\times3\times5}{2\times4\times6}\frac{x^7}{7} + \ldots$	$\lvert x\rvert \leqq 1$
d 96	$\arccos x = \frac{\pi}{2} - \arcsin x$	$\lvert x\rvert \leqq 1$
d 97	$\arctan x = x - \frac{x^3}{3} + \frac{x^5}{5} - \frac{x^7}{7} + \frac{x^9}{9} - \ldots$	$\lvert x\rvert \leqq 1$
d 98	$\operatorname{arccot} x = \frac{\pi}{2} - \arctan x$	$\lvert x\rvert \leqq 1$
d 99	$\sinh x = x + \frac{x^3}{3!} + \frac{x^5}{5!} + \frac{x^7}{7!} + \frac{x^9}{9!} + \ldots$	all x
d100	$\cosh x = 1 + \frac{x^2}{2!} + \frac{x^4}{4!} + \frac{x^6}{6!} + \frac{x^8}{8!} + \ldots$	all x
d101	$\tanh x = x - \frac{1}{3}x^3 + \frac{2}{15}x^5 - \frac{17}{315}x^7 + \ldots$	$\lvert x\rvert < \frac{\pi}{2}$
d102	$\coth x = \frac{1}{x} + \frac{1}{3}x - \frac{1}{45}x^3 + \frac{2}{945}x^5 - \ldots$	$0 < \lvert x\rvert$ $\lvert x\rvert < \pi$
d103	$\operatorname{arsinh} x = x - \frac{1}{2}\frac{x^3}{3} + \frac{1\times3}{2\times4}\frac{x^5}{5} - \frac{1\times3\times5}{2\times4\times6}\frac{x^7}{7} + \ldots$	$\lvert x\rvert < 1$
d104	$\operatorname{arcosh} x = \ln 2x - \frac{1}{2}\frac{1}{2x^2} - \frac{1\times3}{2\times4}\frac{1}{4x^4} - \frac{1\times3\times5}{2\times4\times6}\frac{1}{6x^6} + \ldots$	$\lvert x\rvert > 1$
d105	$\operatorname{artanh} x = x + \frac{x^3}{3} + \frac{x^5}{5} + \frac{x^7}{7} + \frac{x^9}{9} + \ldots$	$\lvert x\rvert < 1$
d106	$\operatorname{arcoth} x = \frac{1}{x} + \frac{1}{3x^3} + \frac{1}{5x^5} + \frac{1}{7x^7} + \ldots$	$\lvert x\rvert > 1$

Fourier series

General: Each periodic function $f(x)$, which can be subdivided into a finite number of continuous intervals within its period $-\pi \le x \le \pi$, may be expanded in this interval into convergent series of the following form ($x = \omega t$):

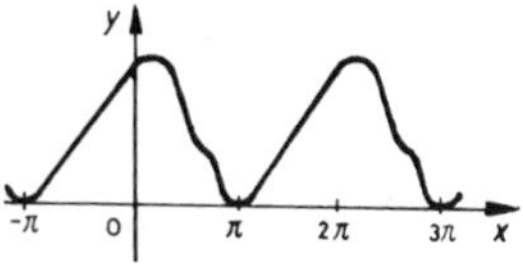

d107 $$f(x) = \frac{a_0}{2} + \sum_{n=1}^{\infty}\left[a_n \cos(nx) + b_n \sin(nx)\right]$$

The various coefficients can be calculated by:

d108 $$a_k = \frac{1}{\pi}\int_{-\pi}^{\pi} f(x)\cos(kx)\,dx \qquad b_k = \frac{1}{\pi}\int_{-\pi}^{\pi} f(x)\sin(kx)\,dx$$

with the index $k = 0, 1, 2 \ldots$ 2, ...

Simplified calculation of coefficient for symmetr. waveforms:

Even function: $f(x) = f(-x)$

d109 $$a_k = \frac{2}{\pi}\int_{0}^{\pi} f(x)\cos(kx)\,dx$$

with the index $k = 0, 1, 2 \ldots$

d110 $b_k = 0$

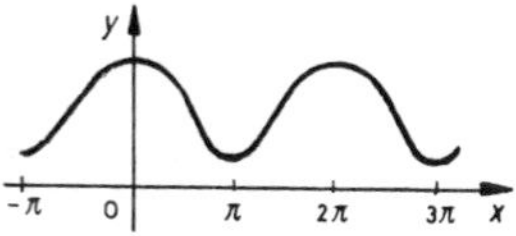

Odd function: $f(x) = -f(-x)$

d111 $a_k = 0$

d112 $$b_k = \frac{2}{\pi}\int_{0}^{\pi} f(x)\sin(kx)\,dx$$

with the index $k = 0, 1, 2, \ldots$

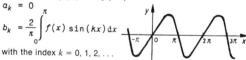

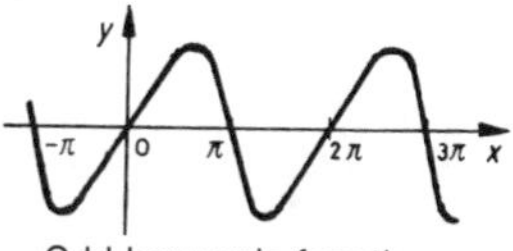

	Even-harmonic functions	Odd-harmonic functions
d113	$f(x) = f(-x)$ and	$f(x) = -f(\ x)$ and
d114	$f(\frac{\pi}{2} + x) = -f(\frac{\pi}{2} - x)$ give:	$f(\frac{\pi}{2} + x) = -f(\frac{\pi}{2} - x)$ give:
d115	$a_k = \frac{4}{\pi}\int_{0}^{\pi/2} f(x)\cos(kx)\,dx$ for $k = 1, 3, 5, \ldots$	$b_k = \frac{4}{\pi}\int_{0}^{\pi/2} f(x)\sin(kx)\,dx$ for $k = 1, 3, 5, \ldots$
d116	$a_k = 0$ for $k = 0, 2, 4, \ldots$	$a_k = 0$ for $k = 0, 1, 2, \ldots$
d117	$b_k = 0$ for $k = 1, 2, 3, \ldots$	$b_k = 0$ for $k = 2, 4, 6, \ldots$

Table of Fourier expansions

d118 $y = a \quad \text{for} \quad 0 < x < \pi$

d119 $y = -a \quad \text{for} \quad \pi < x < 2\pi$

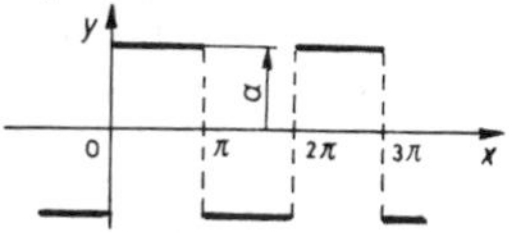

d120 $$y = \frac{4a}{\pi}\left[\sin x + \frac{\sin(3x)}{3} + \frac{\sin(5x)}{5} + \ldots\right]$$

d121 $y = a \quad \text{for} \quad \alpha < x < \pi - \alpha$

d122 $y = -a \quad \text{for} \quad \pi + \alpha < x < 2\pi - \alpha$

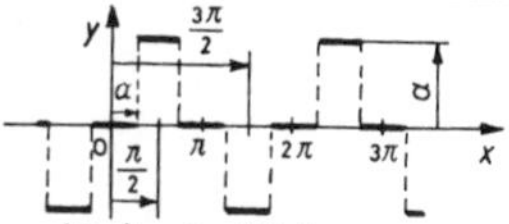

d123 $$y = \frac{4a}{\pi}\left[\cos\alpha \sin x + \frac{1}{3}\cos(3\alpha)\sin(3x) + \frac{1}{5}\cos(5\alpha)\sin(5x) + \ldots\right]$$

d124 $y = a \quad \text{for} \quad \alpha < x < 2\pi - \alpha$

d125 $y = f(2\pi + x)$

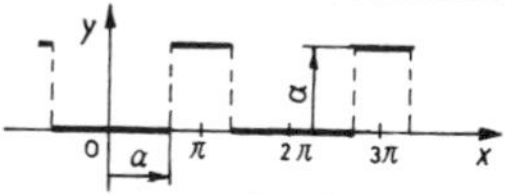

d126 $$y = \frac{2a}{\pi}\left[\frac{\pi-\alpha}{2} - \frac{\sin(\pi-\alpha)}{1}\cos x + \frac{\sin 2(\pi-\alpha)}{2}\cos(2x) - \frac{\sin 3(\pi-\alpha)}{3}\cos(3x) + \ldots\right]$$

d127 $y = ax/b \quad \text{for} \quad 0 \leqq x \leqq b$

d128 $y = a \quad \text{for} \quad b \leqq x \leqq \pi - b$

d129 $y = a(\pi - x)/b \quad \text{for} \quad \pi - b \leqq x \leqq \pi$

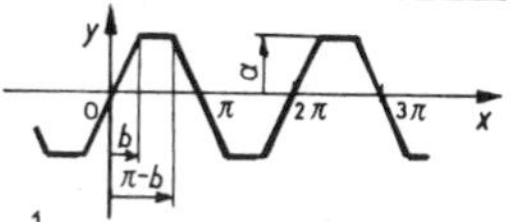

d130 $$y = \frac{4}{\pi}\,\frac{a}{b}\left[\frac{1}{1^2}\sin b \sin x + \frac{1}{3^2}\sin(3b)\sin(3x) + \frac{1}{5^2}\sin(5b)\sin(5x) + \ldots\right]$$

d131 $y = \dfrac{ax}{2\pi} \quad \text{for} \quad 0 < x < 2\pi$

d132 $y = f(2\pi + x)$

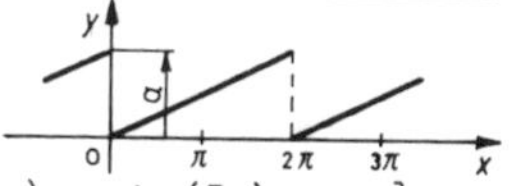

d133 $$y = \frac{a}{2} - \frac{a}{\pi}\left[\frac{\sin x}{1} + \frac{\sin(2x)}{2} + \frac{\sin(3x)}{3} + \ldots\right]$$

continued on D 14

Continuation of D 13

d134 $y = 2ax/\pi \quad \text{for} \quad 0 \leqq x \leqq \pi/2$

d135 $y = 2a(\pi - x)/\pi \quad \text{for} \quad \pi/2 \leqq x \leqq \pi$

d136 $y = -f(\pi + x)$

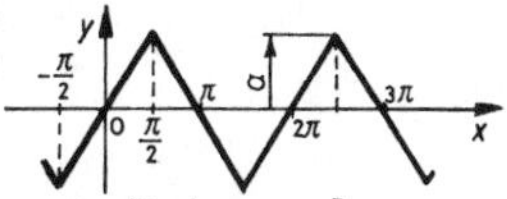

d137 $$y = \frac{8}{\pi^2}\, a\left[\sin x - \frac{\sin(3x)}{3^2} + \frac{\sin(5x)}{5^2} - \ldots\right]$$

d138 $y = ax/\pi \quad \text{for} \quad 0 \leqq x \leqq \pi$

d139 $y = a(2\pi - x)/\pi \quad \text{for} \quad \pi \leqq x \leqq 2\pi$

d140 $y = f(2\pi + x)$

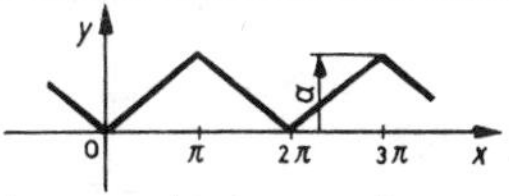

d141 $$y = \frac{a}{2} - \frac{4a}{\pi^2}\left[\frac{\cos x}{1^2} + \frac{\cos(3x)}{3^2} + \frac{\cos(5x)}{5^2} + \ldots\right]$$

d142 $y = a \sin x \quad \text{for} \quad 0 \leqq x \leqq \pi$

d143 $y = -a \sin x \quad \text{for} \quad \pi \leqq x \leqq 2\pi$

d144 $y = f(\pi + x)$

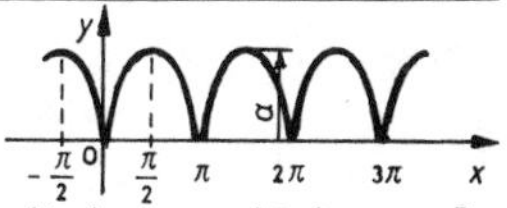

d145 $$y = \frac{2a}{\pi} - \frac{4a}{\pi}\left[\frac{\cos(2x)}{1 \times 3} + \frac{\cos(4x)}{3 \times 5} + \frac{\cos(6x)}{5 \times 7} + \ldots\right]$$

d146 $y = 0 \quad \text{for} \quad 0 \leqq x \leqq \pi/2$

d147 $y = a \sin\left(x - \frac{\pi}{2}\right) \quad \text{for} \quad \frac{\pi}{2} \leqq x \leqq \frac{3\pi}{2}$

d148 $y = f(2\pi + x)$

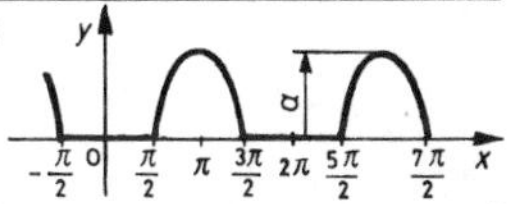

d149 $$y = \frac{2a}{\pi}\left[\frac{1}{2} - \frac{\pi}{4}\cos x + \frac{\cos(2x)}{2^2 - 1} - \frac{\cos(4x)}{4^2 - 1} + \frac{\cos(6x)}{6^2 - 1} - \ldots\right]$$

d150 $y = x^2 \quad \text{for} \quad -\pi \leqq x \leqq \pi$

d151 $y = f(-x) = f(2\pi + x)$

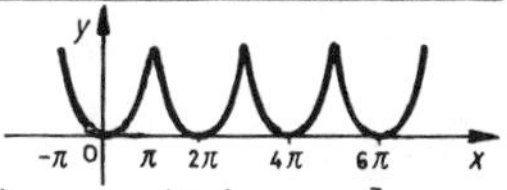

d152 $$y = \frac{\pi^2}{3} - 4\left[\frac{\cos x}{1^2} - \frac{\cos(2x)}{2^2} + \frac{\cos(3x)}{3^2} - \ldots\right]$$

d153 $y = ax/\pi \quad \text{for} \quad 0 \leqq x \leqq \pi$

d154 $y = f(2\pi + x)$

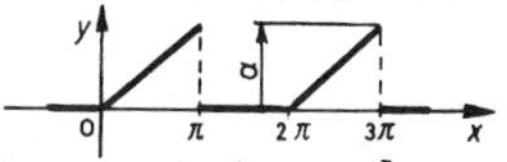

d155 $$y = \frac{a}{4} - \frac{2a}{\pi^2}\left[\frac{\cos x}{1^2} + \frac{\cos(3x)}{3^2} + \frac{\cos(5x)}{5^2} + \ldots\right] + \frac{a}{\pi}\left[\frac{\sin x}{1} - \frac{\sin(2x)}{2} + \frac{\sin(3x)}{3} - \ldots\right]$$

General

The Fourier-Transformation $\boldsymbol{F}\{s(t)\}$ based on the Fourier integral converts the time function $s(t)$ in a continuous spectrum (spectral density) $S(\omega)$ in a way that frequency ω corresponds to the spectral density. $s(t)$ must have the following characteristics:

a) be piecewise smooth in a defined number of finite intervals

d 156 b) have defined values at the jumps $s(t+0)$ and $s(t-0)$, so that the value is equal to the average

d 157 $$s(t) = 1/2\,[s(t-0) + s(t+0)]$$

d 158 c) $\int_{-\infty}^{+\infty} |s(t)|\, dt$ must be absolutely convergent.

The inverse Fourier transformation $\boldsymbol{F}^{-1}\{S(\omega)\}$ gives the time function $s(t)$.

Definitions

d 159 $$\boldsymbol{F}\{s(t)\} = S(\omega) = \int_{-\infty}^{+\infty} s(t)\, e^{-i\omega t}\, dt; \qquad i = \sqrt{-1}$$

d 160 $$\boldsymbol{F}^{-1}\{S(\omega)\} = s(t) = \frac{1}{2\pi}\int_{-\infty}^{+\infty} S(\omega)\, e^{i\omega t}\, d\omega; \qquad i = \sqrt{-1}$$

d 161 Spectral energy $$\int_{-\infty}^{+\infty} |s(t)|^2\, dt = \frac{1}{2\pi}\int_{-\infty}^{+\infty} |S(\omega)|^2\, d\omega$$

Calculation rules

d 162 Time translation $$\boldsymbol{F}\{s(t-\tau)\} = S(\omega)\, e^{-i\omega\tau}; \quad i = \sqrt{-1}$$

d 163 Convolution $$s_1(t) * s_2(t) = \int_{-\infty}^{+\infty} s_1(\tau)\, s_2(t-\tau)\, d\tau$$

d 164 $$= \int_{-\infty}^{+\infty} s_2(\tau)\, s_1(t-\tau)\, d\tau$$

d 165 $$\boldsymbol{F}\{s_1(t) * s_2(t)\} = S_1(\omega)\, S_2(\omega)$$

d 166 $$\boldsymbol{F}\{s(t)\} = S(\omega)$$

d 167 $$\boldsymbol{F}\{s(at)\} = \frac{1}{|a|} S\left(\frac{\omega}{a}\right) \qquad a \text{ real} > 0$$

d 168 $$\boldsymbol{F}\{s_1(t) + s_2(t)\} = S_1(\omega) + S_2(\omega)$$

continued on D 16

continued from D 15

Using equation d 159 calculated spectral densities are given for some important time functions. Correspondence between time function and spectral density:

d 169 $$s(t) = \frac{1}{2\pi}\int_{-\infty}^{\infty} S(\omega)\, e^{i\omega t}\, d\omega \; ; \qquad S(\omega) = \int_{-\infty}^{\infty} s(t)\, e^{-i\omega t}\, dt$$

Time function $s(t)$	Spectral density $S(\omega)$
d 170 **R**ectangle function $A\,R_T(t)$ 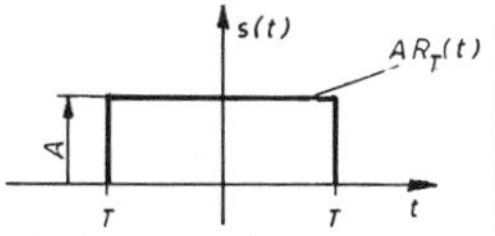	$2AT\,\sin(\omega T)/(\omega T)$ 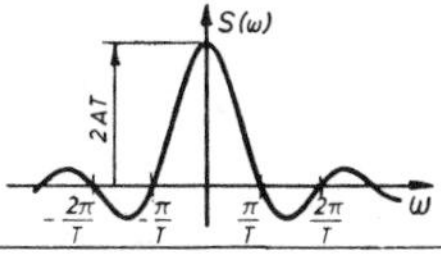
d 171 Dirac-Pulse $A\,\delta(t)$ d 172 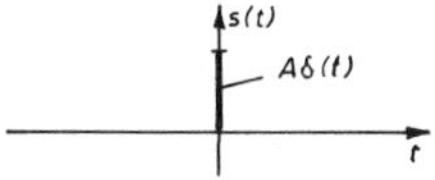	$S(\omega) = A$ (spectral density is constant over ω)
d 173 Rectangle funct. with change of polarity $A\,R_{T/2}(t-T/2) - A\,R_{T/2}(t+T/2)$ d 174 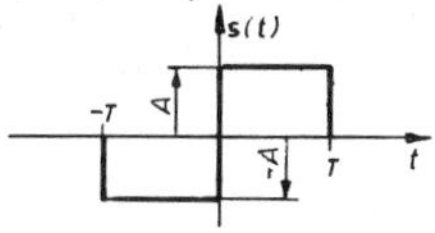	$S(\omega) = -j\,2AT\,\dfrac{\sin^2\frac{\omega T}{2}}{\frac{\omega T}{2}}$
d 175 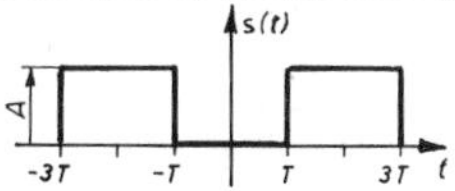	$S(\omega) = 4AT\,\cos(2\omega T)\dfrac{\sin(\omega T)}{\omega T}$
176/177 $s(t) = \dfrac{A}{\pi}\omega_0 \dfrac{\sin(\omega_0 t)}{\omega_0 t}$ with $\omega_0 = \dfrac{2\pi}{T}$ 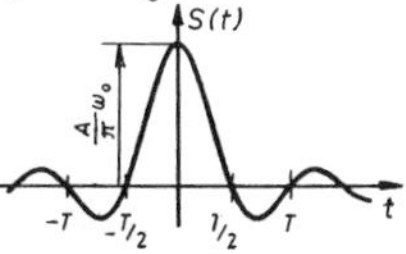	$S(\omega) = A\,R_{\omega_0}(\omega)$ (**R**ectangle function) 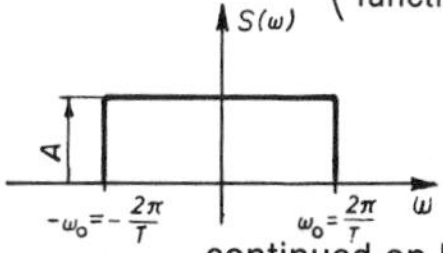

continued on D 17

continued from D 16

	Time function $s(t)$	Spectral density $S(\omega)$
d 178	Triangle function $A\ D_T(t)$	
d 179	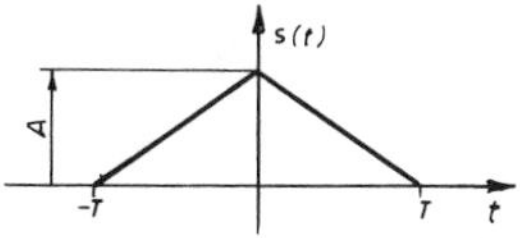	$S(\omega) = \left(\frac{\sin(T\omega/2)}{T\omega/2}\right)^2 A\,T$ 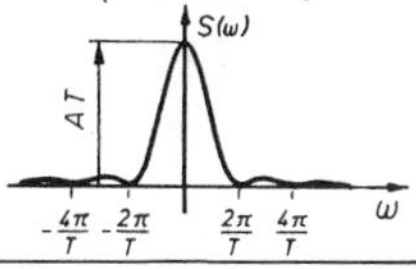
	Modulated rectangle	
d 180	$A\,R_T(t)\ \cos(\omega_o t)$ with $\omega_o = \frac{2\pi}{T_o} = \frac{2\pi}{\alpha\,T}$	
d 181	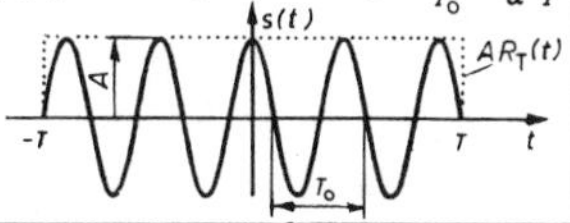	$S(\omega) = A\,\frac{\sin T(\omega+\omega_o)}{\omega+\omega_o} + A\,\frac{\sin T(\omega-\omega_o)}{\omega-\omega_o}$
d 182	Gaussian-pulse $A\ e^{-a^2t^2}$	
d 183	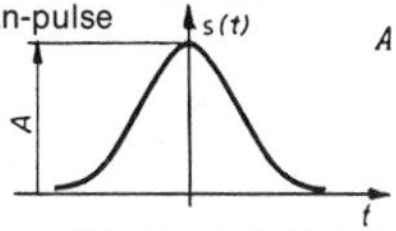	$S(\omega) = \frac{A}{a}\sqrt{\pi}\ e^{\frac{-\omega^2}{4a^2}}$
d 184	cos-pulse $A\ \cos(\omega_o t)$ with $\omega_o = \frac{2\pi}{T}$	
d 185	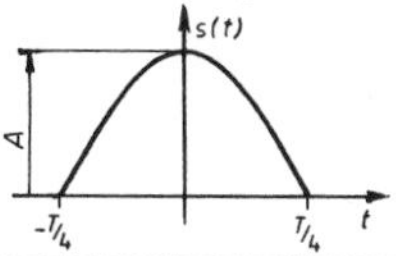	$S(\omega) = \frac{A\ T}{\pi} \times \frac{\cos\left(\frac{T}{4}\omega\right)}{1-\left(\frac{T}{2\pi}\omega\right)^2}$
d 186	$\cos^2$-pulse $A^2\ \cos^2(\omega_o t)$ with $\omega_o = \frac{2\pi}{T}$	
d 187	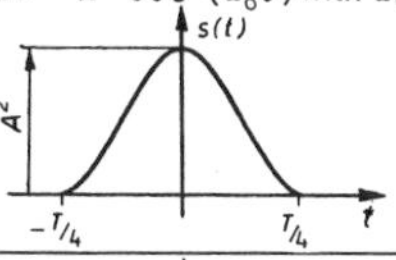	$S(\omega) = \frac{A\ T}{4} \times \frac{\sin\left(\omega\frac{T}{4}\right)}{\left(\omega\frac{T}{4}\right)} \times \frac{1}{1-\frac{T^2\ \omega^2}{16\pi^2}}$
d 188	Exponential-pulse	
d 189	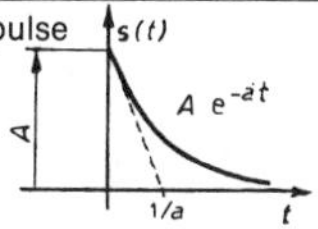	$S(\omega) = \frac{A}{j\,\omega + a}$

General: The Laplace-Transformation $\boldsymbol{L}\{f(t)\}$ based on the Integral-function

d 190 $$F(p) = \int_0^\infty f(t)\ e^{-pt}\ dt$$

converts the time function $f(t)$, which has to be zero for $t<0$ and which must be given completely for $t \geq 0$, into a picture function. The part e^{-pt} in d 190 is used as an attention factor

d 191 to get convergency of the integral for as many time functions as possible; here is $p = \sigma + i\omega$ with $\sigma \geq 0$ a complex operation variable. In this picture-domain differential equations can be solved and unique, non periodical processes (e. g. oscillating) can be handled; the desired time behaviour is reached finally by inverse transformation in the t-domain (see D 20).

Definitions

d 192/193 $$\boldsymbol{L}\{f(t)\} = F(p) = \int_0^\infty f(t)\ e^{-pt} dt \qquad \boldsymbol{L}^{-1}\{F(p)\} = f(t) = \frac{1}{2\pi i}\int_{\sigma_0 - i\infty}^{\sigma_0 + i\infty} F(p)\, e^{pt}\, dp$$

abbreviated description: $f(t) \circ\!\!-\!\!\!-\!\!\bullet F(p)$ | abbreviated description: $F(p) \bullet\!\!-\!\!\!-\!\!\circ f(t)$

Calculation rules (operation rules)

d 194	Linearity	$\boldsymbol{L}\{f_1(t) + f_2(t)\}$	$= F_1(p) + F_2(p)$
d 195		$\boldsymbol{L}\{c\ f_1(t)\}$	$= c\ F_1(p)$
d 196	Translation	$\boldsymbol{L}\{f(t-a)\}$	$= e^{-ap}\ F(p)$
d 197	Convolution	$f_1(t) * f_2(t)$	$= \int_0^t f_1(t-\tau)\ f_2(\tau)\ d\tau$
d 198			$= \int_0^t f_1(\tau)\ f_2(t-\tau)\ d\tau$
d 199		$f_1(t) * f_2(t) \circ\!\!-\!\!\!-\!\!\bullet$	$F_1(p)\ F_2(p)$
d 200	Variable transform.	$\boldsymbol{L}\left\{\frac{1}{a}\ f\left(\frac{t}{a}\right)\right\}$	$= F(a\ p)$
d 201	Differen-tiation	$\boldsymbol{L}\{f'(t)\}$	$= p\ F(p) - f(0^+)$
d 202		$\boldsymbol{L}\{f''(t)\}$	$= p^2 F(p) - p\ f(0^+) - f'(0^+)$
d 203		$\boldsymbol{L}\{f^n(t)\}$	$= p^n\ F(p) - \sum_{k=0}^{n-1} f^{(k)}(0^+)\, p^{n-k-1}$
d 204	Integration	$\boldsymbol{L}\left\{\int_0^t f(t)\ dt\right\}$	$= \frac{1}{p} F(p)$

Application of the *L*-Transformation to differential equations

Scheme

d 205
d 206

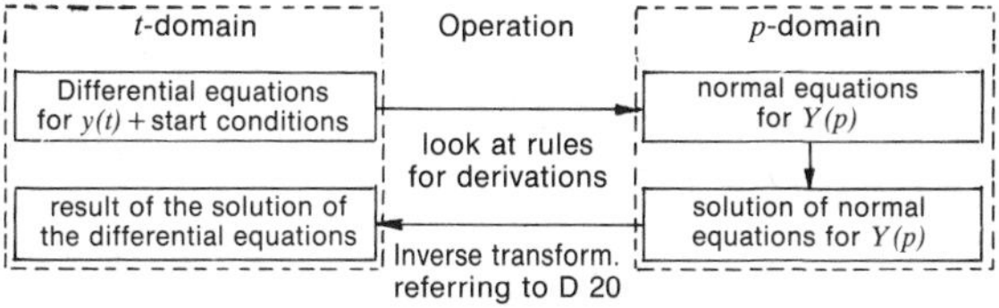

d 207 Difficulty of the solution of the differential equations is transferred to the inverse transformation. This can be simplified by expansion from $Y(p)$ into partial fractions (see D 3) or into such partial functions, for which in D 20 conversions are given back in the time domain.

Example: $2y' + y \;\circ\!\!-\!\!\bullet\; f(t)$; $\quad f(t)$ is startfunction

$y(0^+) = 2 \triangleq$ start condition

with d 201, d 205: $2p\,Y(p) - 2y(0^+) + Y(p) = F(p)$

with d 206: $y(t) \circ\!\!-\!\!\!-\!\!\!-\!\!\bullet\; Y(p) = \dfrac{F(p) + 2y(0^+)}{1 + 2p} = \dfrac{1/p + 2y(0^+)}{1 + 2p}$

According to $f(t) \circ\!\!-\!\!\bullet F(p)$ there are different solutions for $y(t)$. [Here $f(t)$ is assumed to be a step function. In this case referring to d 213 $F(p) = 1/p$).]

Application of D 3: $\quad Y(p) = \dfrac{1}{p(1+2p)} + \dfrac{2y(0^+)}{1+2p} = \dfrac{1}{p} - \dfrac{2}{1+2p} + \dfrac{2y(0^+)}{1+2p}$

after D 20: $\quad y(t) = 1 - 2\,\dfrac{1}{2}\,e^{-t/2} + 2 \times 2\,\dfrac{1}{2}\,e^{-t/2} = 1 + e^{-t/2}$

Application of the convolution rule to the *L*-Transformation on linear networks

The originate function $f_1(t)$ is changed to a response $y(t)$ after having passed a network. The network is defined by its transfer function $F_2(p)$. $F_2(p)$ has the inverse transformate $f_2(t)$.

d 208

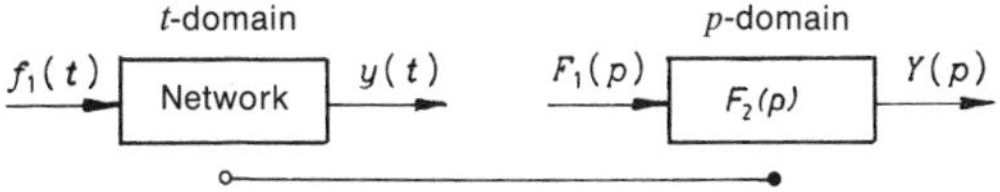

d 209 $y(t) = f_1(t) * f_2(t) \;\circ\!\!-\!\!\!-\!\!\!-\!\!\bullet\; Y(p) = F_1(p) \times F_2(p)$

For a given network the response $y(t)$ depends on $f_1(t)$. $y(t)$ can be found by d 205. After having found $Y(p)$ calculation is continued on line d 206. Complete inverse transformation to the t-domain is possible when $F_2(p)$ is given as proper fraction rational function in p and when **L**-Transformate $F_1(p)$ is given in D 20.

Table of correlation

d 210 $$F(p) = \int_0^{\infty} f(t)\, e^{-pt}\, dt; \qquad f(t) = \frac{1}{2\pi i}\int_{\sigma_0 - i\infty}^{\sigma_0 + i\infty} F(p)\, e^{pt}\, dp$$

with $p = i\omega = i\,2\pi f$; $i = \sqrt{-1}$

No.	p-domain Laplace transf. $F(p)$	t-domain original-function $f(t)$
d 211, d 212	1	$\delta(t)$ ≙ Dirac
d 213, d 214	$1/p$	1 for $t > 0$; 0 for $t < 0$ (Step-function)
d 215, d 216	$1/p^2$	t
d 217, d 218	$1/p^n$	$\frac{t^{n-1}}{(n-1)!}$
d 219, d 220	$1/(p-a)$	$\exp(at)$
d 221, d 222	$1/(p-a)^2$	$t\ \exp(at)$
d 223, d 224	$\frac{a}{p(p-a)}$	$\exp(at) - 1$
d 225, d 226	$\frac{1}{1+T\ p}$	$\frac{1}{T}\exp(-t/T)$
d 227, d 228	$\frac{a}{p^2-a^2}$	$\sinh(at)$
d 229, d 230	$\frac{p}{p^2-a^2}$	$\cosh(at)$
d 231, d 232	$\frac{k}{p^2+k^2}$	$\sin(kt)$
d 233, d 234	$\frac{p}{p^2+k^2}$	$\cos(kt)$
d 235, d 236	$\frac{1}{(p^2+k^2)^2}$	$\frac{1}{2k^3}\sin(kt) - \frac{1}{2k^2}\,t\cos(kt)$
37/238	$\frac{p}{(p^2+k^2)^2}$	$\frac{1}{2k}\sin(kt)$

No.	p-domain Laplace transf. $F(p)$	t-domain original function $f(t)$
d 211 – d 214	$\frac{p^2}{(p^2+k^2)^2}$	$\frac{1}{2k}\sin(kt) + \frac{t}{2}\cos(kt)$
d 215 – d 217	$\frac{p^3}{(p^2+k^2)^2}$	$\cos(kt) - \frac{k}{2}t\sin(kt)$
d 218 – d 220	$\frac{1}{(p-a)(p-b)}$	for $b \neq a$: $\frac{e^{bt}-e^{at}}{b-a}$
d 221, d 222	$\frac{1}{(p+a)^2+k^2}$	$\frac{1}{k}e^{-at}\sin(kt)$
d 223, d 224	$\frac{1}{\sqrt{p}}$	$\frac{1}{\sqrt{\pi\ t}}$
d 225, d 226	$\frac{1}{p\sqrt{p}}$	$2\sqrt{\frac{t}{\pi}}$
d 227, d 228	$\sqrt{p}$	$-1/(2\sqrt{\pi}\ t^{3/2})$
d 229	$p\sqrt{p}$	$3/(4\sqrt{\pi}\ t^{5/2})$
d 230	$\ln\frac{p+b}{p+a}$	$\frac{1}{t}\left(e^{-at} - e^{-bt}\right)$
d 231, d 232	$\arctan(a/p)$	$1/t\ \sin(at)$
d 233, d 234	for $a > 0$: $e^{-a\sqrt{p}}$	$\frac{a}{2t\sqrt{\pi t}}\,e^{\frac{-a^2}{4t}}$
d 235, d 236	for $a \geqslant 0$: $\frac{1}{p}e^{-a\sqrt{p}}$	$\operatorname{erfc}\frac{a}{2\sqrt{t}}$
37/238	$\frac{1}{\sqrt{p^2+k^2}}$	$J_0(kt)$ Bessel function

Complex numbers

General

$z = r e^{\widehat{\varphi}} = a + \mathrm{i}\, b$
a = real part of z
b = imaginary part of z
r = absolute value of z
= or modulus of z
φ = argument of z
a and b are real

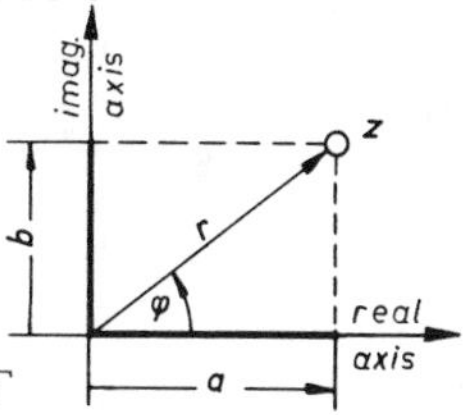

d 239 $$\mathrm{i} = \sqrt{-1}$$

d 240 $\mathrm{i}^1 = +\mathrm{i}$ | $\mathrm{i}^{-1} = -\mathrm{i}$

d 241 $\mathrm{i}^2 = -1$ | $\mathrm{i}^{-2} = -1$

d 242 $\mathrm{i}^3 = -\mathrm{i}$ | $\mathrm{i}^{-3} = +\mathrm{i}$

d 243 $\mathrm{i}^4 = +1$ | $\mathrm{i}^{-4} = +1$

d 244 $\mathrm{i}^5 = +\mathrm{i}$ | $\mathrm{i}^{-5} = -\mathrm{i}$

etc.

Note: In electrical engineering the letter j is used for i to avoid confusion.

In the Cartesian coordinate system:

d 245 $$z = a + \mathrm{i}b$$

d 246 $$z_1 + z_2 = (a_1 + a_2) + \mathrm{i}(b_1 + b_2)$$

d 247 $$z_1 - z_2 = (a_1 - a_2) + \mathrm{i}(b_1 - b_2)$$

d 248 $$z_1 \times z_2 = (a_1 a_2 - b_1 b_2) + \mathrm{i}(a_1 b_2 + a_2 b_1)$$

d 249 $$\frac{z_1}{z_2} = \frac{a_1 a_2 + b_1 b_2}{a_2^2 + b_2^2} + \mathrm{i}\frac{-a_1 b_2 + a_2 b_1}{a_2^2 + b_2^2}$$

d 250 $$a^2 + b^2 = (a + \mathrm{i}b)(a - \mathrm{i}b)$$

d 251 $$\sqrt{a \pm \mathrm{i}b} = \sqrt{\frac{a + \sqrt{a^2 + b^2}}{2}} \pm \mathrm{i}\sqrt{\frac{-a + \sqrt{a^2 + b^2}}{2}}$$

Where $a_1 = a_2$ and $b_1 = b_2$, then $z_1 = z_2$

continued on D 22

Complex numbers
(continued)

In the polar coordinate system:

d 252 $$z = r(\cos\varphi + i\sin\varphi) = a + ib$$

d 253 $$r = +\sqrt{a^2 + b^2}$$

d 254 $$\varphi = \arctan\frac{b}{a}$$

d 255 $$\sin\varphi = \frac{b}{r} \quad \Big| \quad \cos\varphi = \frac{a}{r} \quad \Big| \quad \tan\varphi = \frac{b}{a}$$

d 256 $$z_1 \times z_2 = r_1 \times r_2\left[\cos(\varphi_1 + \varphi_2) + i\sin(\varphi_1 + \varphi_2)\right]$$

d 257 $$\frac{z_1}{z_2} = \frac{r_1}{r_2}\left[\cos(\varphi_1 - \varphi_2) + i\sin(\varphi_1 - \varphi_2)\right] \quad (z_2 \neq 0)$$

d 258 $$z^n = r^n\left[\cos(n\varphi) + i\sin(n\varphi)\right] \quad (n > 0 \text{ integer})$$

d 259 $$\sqrt[n]{z} = \left|\sqrt[n]{r}\right|\left(\cos\frac{\varphi + 2\pi k}{n} + i\sin\frac{\varphi + 2\pi k}{n}\right)$$

d 260 $$\sqrt[n]{1} = \cos\frac{2\pi k}{n} + i\sin\frac{2\pi k}{n} \quad (n\text{th roots of unity})$$

In formula d 259 and d 260 $k = 0, 1, 2, \ldots, n-1$

d 261 $$e^{i\varphi} = \cos\varphi + i\sin\varphi$$

d 262 $$e^{-i\varphi} = \cos\varphi - i\sin\varphi = \frac{1}{\cos\varphi + i\sin\varphi}$$

d 263 $$\left|e^{-i\varphi}\right| = \sqrt{\cos^2\varphi + \sin^2\varphi} = 1$$

d 264 $$\cos\varphi = \frac{e^{i\varphi} + e^{-i\varphi}}{2} \quad \Big| \quad \sin\varphi = \frac{e^{i\varphi} - e^{-i\varphi}}{2i}$$

d 265 $$\ln z = \ln|r| + i(\varphi + 2\pi k) \quad (k = 0, \pm1, \pm2, \ldots)$$

If $r_1 = r_2$ and $\varphi_1 = \varphi_2 + 2\pi k$, then $z_1 = z_2$

Note: φ should be measured along the arc,
k is any arbitrary whole number

Compound interest calculation

d 266 $$k_n = k_0 \times q^n$$

d 267 $$n = \frac{\lg \frac{k_n}{k_0}}{\lg q}$$

$$q = \sqrt[n]{\frac{k_n}{k_0}}$$

Annuity interest calculation

d 268 $$k_n = k_0 \times q^n - r \times q \frac{q^n - 1}{q - 1}$$

d 269 $$r = \frac{(k_0 \times q^n - k_n)(q-1)}{(q^n - 1)q}$$

d 270 $$n = \frac{\lg \frac{r \times q - k_n(q-1)}{r \times q - k_0(q-1)}}{\lg q}$$

Where $k_n = 0$, we get the "redemption formulae"

Deposit calculation

(savings bank formula)

d 271 $$k_n = k_0 \times q^n + r \times q \frac{q^n - 1}{q - 1}$$

d 272 $$r = \frac{(k_n - k_0 \times q^n)(q-1)}{(q^n - 1)q}$$

d 273 $$n = \frac{\lg \frac{k_n(q-1) + r \times q}{k_0(q-1) + r \times q}}{\lg q}$$

Letters

k_0 : initial capital
k_n : capital after n years
r : annual pensions (withdrawals)

n : number of years
q : $1 + p$
p : rate of interest (e.g. 0·06 at 6%)

d 274 $$x = \frac{b\ c}{a}$$

d 275 $$a : b = c : x$$

x : 4th proportional

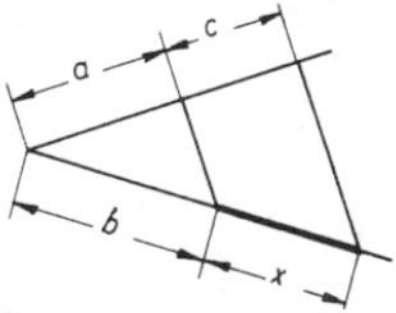

d 276 $$x = \frac{b^2}{a}$$

d 277 $$a : b = b : x$$

x : 3rd proportional

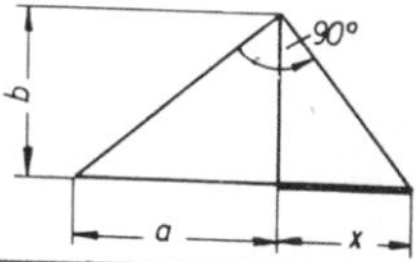

d 278 $$x = \sqrt{a\ b}$$

d 279 $$a : x = x : b$$

x : mean proportional

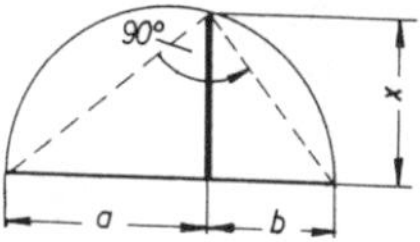

d 280 $$x^2 = a^2 + b^2$$

d 281 or $$x = \sqrt{a^2 + b^2}$$

x : hypothenuse of a right-angled triangle

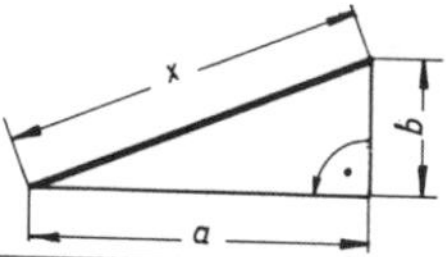

d 282 $$x = \frac{a}{2}\sqrt{3}$$

x : height of an equilateral triangle

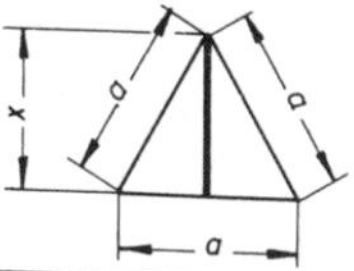

d 283 $$x = \frac{a}{2}(\sqrt{5} - 1)$$

d 284 $$\approx a\ 0{\cdot}618$$

d 285 $$a : x = x : (a-x)$$

x : larger section of a repeatedly subdivided line (golden section)

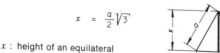

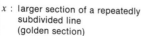

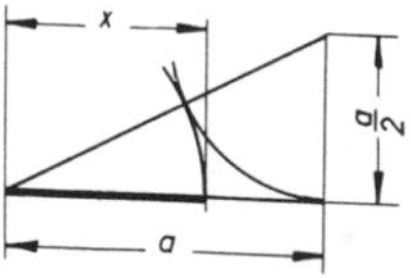

Circular and angular measure of a plane angle

Circular measure

Circular measure is the ratio of the distance d measured along the arc to the radius r.

It is given the unit "radian" which has no dimensions.

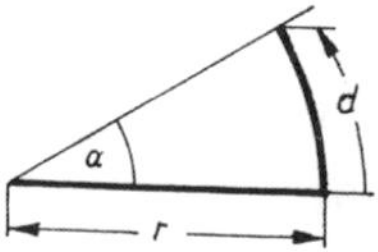

e 1 $$\alpha = \frac{d}{r} \text{ (rad)}$$

Unit: rad

Angular measure

Angular measure is obtained by dividing the angle subtended at the centre of a circle into 360 equal divisions known as "degrees".

Unit: °

e 2 A degree is divided into 60 minutes (unit: '),
e 3 a minute is divided into 60 seconds (unit: ").

Relation between circular and angular measure

By considering a circle, it may be seen that

e 4 $$360^\circ = 2\pi \text{ radians}$$

e 5 or $$1 \text{ rad} = 57 \cdot 2958^\circ$$

e 6

degrees	0°	30°	45°	60°	75°	90°	180°	270°	360°
radians	0	$\frac{\pi}{6}$	$\frac{\pi}{4}$	$\frac{\pi}{3}$	$\frac{5}{12}\pi$	$\frac{\pi}{2}$	π	$\frac{3}{2}\pi$	2π
	0	0·52	0·79	1·05	1·31	1·57	3·14	4·71	6·28

FUNCTIONS OF A CIRCLE

General terms

Right angled triangle

e 7 $\sin \alpha = \dfrac{\text{opposite}}{\text{hypotenuse}} = \dfrac{a}{c}$

e 8 $\cos \alpha = \dfrac{\text{adjacent}}{\text{hypotenuse}} = \dfrac{b}{c}$

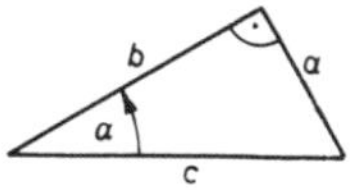

e 9 $\tan \alpha = \dfrac{\text{opposite}}{\text{adjacent}} = \dfrac{a}{b}$ $\cot \alpha = \dfrac{\text{adjacent}}{\text{opposite}} = \dfrac{b}{a}$

Functions of the more important angles

e 10

angle α	0°	30°	45°	60°	75°	90°	180°	270°	360°
sin α	0	0,500	0,707	0,866	0,966	1	0	−1	0
cos α	1	0,866	0,707	0,500	0,259	0	−1	0	1
tan α	0	0,577	1,000	1,732	3,732	∞	0	∞	0
cot α	∞	1,732	1,000	0,577	0,268	0	∞	0	∞

Relations between sine and cosine functions

Basic equations

e 13 Sine function $y = A \sin (k\alpha - \varphi)$

e 14 Cosine function $y = A \cos (k\alpha - \varphi)$

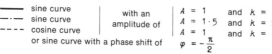

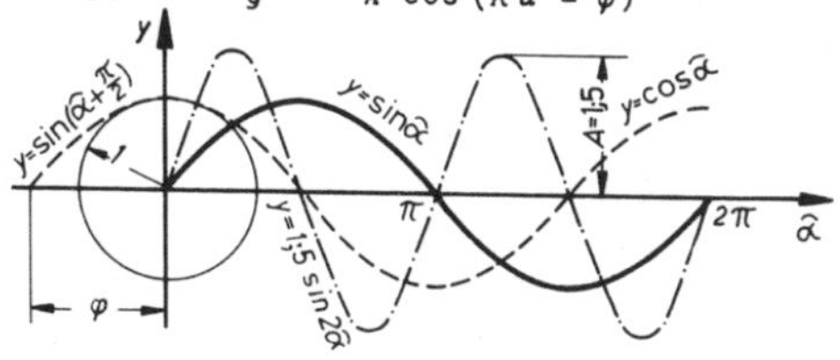

——— sine curve	with an amplitude of	$A = 1$	and	$k = 1$
—·— sine curve		$A = 1 \cdot 5$	and	$k = 2$
- - - - cosine curve		$A = 1$	and	$k = 1$
or sine curve with a phase shift of		$\varphi = -\dfrac{\pi}{2}$		

FUNCTIONS OF A CIRCLE

Quadrants

E 3

e 15	$\sin(90^\circ - \alpha) = +\cos\alpha$		$\sin(90^\circ + \alpha) = +\cos\alpha$	
e 16	$\cos(\;"\;) = +\sin\alpha$		$\cos(\;"\;) = -\sin\alpha$	
e 17	$\tan(\;"\;) = +\cot\alpha$		$\tan(\;"\;) = -\cot\alpha$	
e 18	$\cot(\;"\;) = +\tan\alpha$		$\cot(\;"\;) = -\tan\alpha$	
e 19	$\sin(180^\circ - \alpha) = +\sin\alpha$		$\sin(180^\circ + \alpha) = -\sin\alpha$	
e 20	$\cos(\;"\;) = -\cos\alpha$		$\cos(\;"\;) = -\cos\alpha$	
e 21	$\tan(\;"\;) = -\tan\alpha$		$\tan(\;"\;) = +\tan\alpha$	
e 22	$\cot(\;"\;) = -\cot\alpha$		$\cot(\;"\;) = +\cot\alpha$	
e 23	$\sin(270^\circ - \alpha) = -\cos\alpha$		$\sin(270^\circ + \alpha) = -\cos\alpha$	
e 24	$\cos(\;"\;) = -\sin\alpha$		$\cos(\;"\;) = +\sin\alpha$	
e 25	$\tan(\;"\;) = +\cot\alpha$		$\tan(\;"\;) = -\cot\alpha$	
e 26	$\cot(\;"\;) = +\tan\alpha$		$\cot(\;"\;) = -\tan\alpha$	
e 27	$\sin(360^\circ - \alpha) = -\sin\alpha$		$\sin(360^\circ + \alpha) = +\sin\alpha$	
e 28	$\cos(\;"\;) = +\cos\alpha$		$\cos(\;"\;) = +\cos\alpha$	
e 29	$\tan(\;"\;) = -\tan\alpha$		$\tan(\;"\;) = +\tan\alpha$	
e 30	$\cot(\;"\;) = -\cot\alpha$		$\cot(\;"\;) = +\cot\alpha$	
e 31	$\sin(-\alpha) = -\sin\alpha$		$\sin(\alpha \pm n \times 360^\circ) = +\sin\alpha$	
e 32	$\cos(\;"\;) = +\cos\alpha$		$\cos(\;"\;) = +\cos\alpha$	
e 33	$\tan(\;"\;) = -\tan\alpha$		$\tan(\alpha \pm n \times 180^\circ) = +\tan\alpha$	
e 34	$\cot(\;"\;) = -\cot\alpha$		$\cot(\;"\;) = +\cot\alpha$	

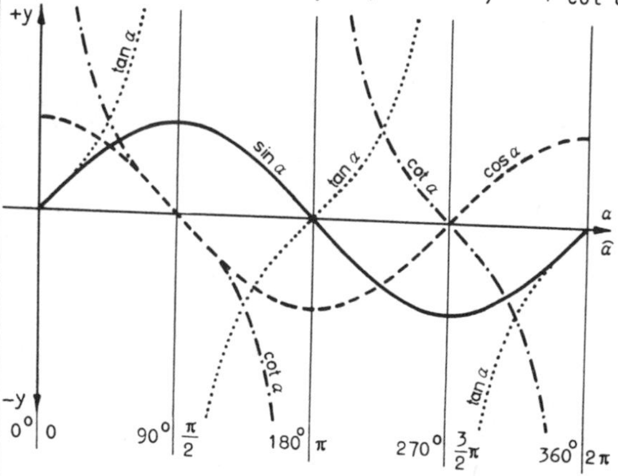

FUNCTIONS OF A CIRCLE

Trigonometric conversions

Basic identities

e 35 $\sin^2\alpha + \cos^2\alpha = 1$ | $\tan\alpha \ \cot\alpha = 1$

e 36 $1 + \tan^2\alpha = \dfrac{1}{\cos^2\alpha}$ | $1 + \cot^2\alpha = \dfrac{1}{\sin^2\alpha}$

Sum and difference of angles

e 37 $\sin(\alpha \pm \beta) = \sin\alpha \ \cos\beta \pm \cos\alpha \ \sin\beta$

e 38 $\cos(\alpha \pm \beta) = \cos\alpha \ \cos\beta \mp \sin\alpha \ \sin\beta$

e 39 $\tan(\alpha \pm \beta) = \dfrac{\tan\alpha \pm \tan\beta}{1 \mp \tan\alpha \ \tan\beta}$; $\cot(\alpha \pm \beta) = \dfrac{\cot\alpha \ \cot\beta \mp 1}{\pm \cot\alpha + \cot\beta}$

Sum and difference of functions of angles

e 40 $\sin\alpha + \sin\beta = 2\sin\dfrac{\alpha+\beta}{2} \ \cos\dfrac{\alpha-\beta}{2}$

e 41 $\sin\alpha - \sin\beta = 2\cos\dfrac{\alpha+\beta}{2} \ \sin\dfrac{\alpha-\beta}{2}$

e 42 $\cos\alpha + \cos\beta = 2\cos\dfrac{\alpha+\beta}{2} \ \cos\dfrac{\alpha-\beta}{2}$

e 43 $\cos\alpha - \cos\beta = -2\sin\dfrac{\alpha+\beta}{2} \ \sin\dfrac{\alpha-\beta}{2}$

e 44 $\tan\alpha \pm \tan\beta = \dfrac{\sin(\alpha \pm \beta)}{\cos\alpha \ \cos\beta}$

e 45 $\cot\alpha \pm \cot\beta = \dfrac{\sin(\beta \pm \alpha)}{\sin\alpha \ \sin\beta}$

e 46 $\sin\alpha \ \cos\beta = \frac{1}{2}\sin(\alpha+\beta) + \frac{1}{2}\sin(\alpha-\beta)$

e 47 $\cos\alpha \ \cos\beta = \frac{1}{2}\cos(\alpha+\beta) + \frac{1}{2}\cos(\alpha-\beta)$

e 48 $\sin\alpha \ \sin\beta = \frac{1}{2}\cos(\alpha-\beta) - \frac{1}{2}\cos(\alpha+\beta)$

e 49 $\tan\alpha \ \tan\beta = \dfrac{\tan\alpha + \tan\beta}{\cot\alpha + \cot\beta} = -\dfrac{\tan\alpha - \tan\beta}{\cot\alpha - \cot\beta}$

e 50 $\cot\alpha \ \cot\beta = \dfrac{\cot\alpha + \cot\beta}{\tan\alpha + \tan\beta} = -\dfrac{\cot\alpha - \cot\beta}{\tan\alpha - \tan\beta}$

e 51 $\cot\alpha \ \tan\beta = \dfrac{\cot\alpha + \tan\beta}{\tan\alpha + \cot\beta} = -\dfrac{\cot\alpha - \tan\beta}{\tan\alpha - \cot\beta}$

Sum of 2 harmonic oscillations of the same frequency

e 52 $a\sin(\omega t + \varphi_1) + b\cos(\omega t + \varphi_2) = \sqrt{c^2 + d^2}\ \sin(\omega t + \varphi)$

with $c = a\sin\varphi_1 + b\cos\varphi_2$; $d = a\cos\varphi_1 - b\sin\varphi_2$

$\varphi = \arctan\dfrac{c}{d}$ and $\varphi = \arcsin\dfrac{c}{\sqrt{c^2 + d^2}}$ } both conditions must be satisfied

Ratios between simple, double and half angles

	$\sin\alpha =$	$\cos\alpha =$	$\tan\alpha =$	$\cot\alpha =$
e 53	$\cos(90^\circ-\alpha)$	$\sin(90^\circ-\alpha)$	$\cot(90^\circ-\alpha)$	$\tan(90^\circ-\alpha)$
e 54	$\sqrt{1-\cos^2\alpha}$	$\sqrt{1-\sin^2\alpha}$	$\dfrac{1}{\cot\alpha}$	$\dfrac{1}{\tan\alpha}$
e 55	$2\sin\dfrac{\alpha}{2}\cos\dfrac{\alpha}{2}$	$\cos^2\dfrac{\alpha}{2}-\sin^2\dfrac{\alpha}{2}$	$\dfrac{\sin\alpha}{\cos\alpha}$	$\dfrac{\cos\alpha}{\sin\alpha}$
e 56	$\dfrac{\tan\alpha}{\sqrt{1+\tan^2\alpha}}$	$\dfrac{\cot\alpha}{\sqrt{1+\cot^2\alpha}}$	$\dfrac{\sin\alpha}{\sqrt{1-\sin^2\alpha}}$	$\dfrac{\cos\alpha}{\sqrt{1-\cos^2\alpha}}$
e 57	$\sqrt{\cos^2\alpha-\cos 2\alpha}$	$1-2\sin^2\dfrac{\alpha}{2}$	$\sqrt{\dfrac{1}{\cos^2\alpha}-1}$	$\sqrt{\dfrac{1}{\sin^2\alpha}-1}$
e 58	$\sqrt{\dfrac{1-\cos 2\alpha}{2}}$	$\sqrt{\dfrac{1+\cos 2\alpha}{2}}$	$\dfrac{\sqrt{1-\cos^2\alpha}}{\cos\alpha}$	$\dfrac{\sqrt{1-\sin^2\alpha}}{\sin\alpha}$
e 59	$\dfrac{1}{\sqrt{1+\cot^2\alpha}}$	$\dfrac{1}{\sqrt{1+\tan^2\alpha}}$		
e 60	$\dfrac{2\tan\frac{\alpha}{2}}{1+\tan^2\frac{\alpha}{2}}$	$\dfrac{1-\tan^2\frac{\alpha}{2}}{1+\tan^2\frac{\alpha}{2}}$	$\dfrac{2\tan\frac{\alpha}{2}}{1-\tan^2\frac{\alpha}{2}}$	$\dfrac{\cot^2\frac{\alpha}{2}-1}{2\cot\frac{\alpha}{2}}$

	$\sin 2\alpha =$	$\cos 2\alpha =$	$\tan 2\alpha =$	$\cot 2\alpha =$
e 61	$2\sin\alpha\cos\alpha$	$\cos^2\alpha-\sin^2\alpha$	$\dfrac{2\tan\alpha}{1-\tan^2\alpha}$	$\dfrac{\cot^2\alpha-1}{2\cot\alpha}$
e 62		$2\cos^2\alpha-1$	$\dfrac{2}{\cot\alpha-\tan\alpha}$	$\frac{1}{2}\cot\alpha-\frac{1}{2}\tan\alpha$
e 63		$1-2\sin^2\alpha$		

	$\sin\dfrac{\alpha}{2} =$	$\cos\dfrac{\alpha}{2} =$	$\tan\dfrac{\alpha}{2} =$	$\cot\dfrac{\alpha}{2} =$
e 64			$\dfrac{\sin\alpha}{1+\cos\alpha}$	$\dfrac{\sin\alpha}{1-\cos\alpha}$
e 65	$\sqrt{\dfrac{1-\cos\alpha}{2}}$	$\sqrt{\dfrac{1+\cos\alpha}{2}}$	$\dfrac{1-\cos\alpha}{\sin\alpha}$	$\dfrac{1+\cos\alpha}{\sin\alpha}$
e 66			$\sqrt{\dfrac{1-\cos\alpha}{1+\cos\alpha}}$	$\sqrt{\dfrac{1+\cos\alpha}{1-\cos\alpha}}$

Oblique angle triangle

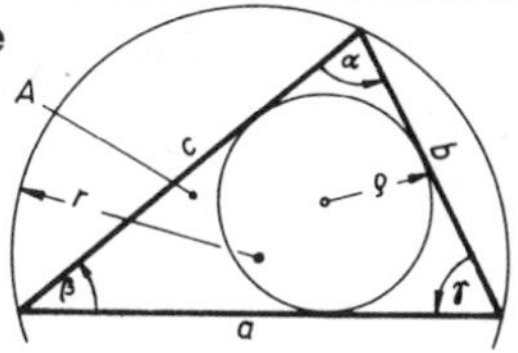

Sine Rule

e 67 $$\sin\alpha : \sin\beta : \sin\gamma = a : b : c$$

e 68 $$a = \frac{b}{\sin\beta}\sin\alpha = \frac{c}{\sin\gamma}\sin\alpha$$

e 69 $$b = \frac{a}{\sin\alpha}\sin\beta = \frac{c}{\sin\gamma}\sin\beta$$

e 70 $$c = \frac{a}{\sin\alpha}\sin\gamma = \frac{b}{\sin\beta}\sin\gamma$$

Cosine Rule

e 71 $$a^2 = b^2 + c^2 - 2\,b\,c\cos\alpha$$

e 72 $$b^2 = c^2 + a^2 - 2\,a\,c\cos\beta$$

e 73 $$c^2 = a^2 + b^2 - 2\,a\,b\cos\gamma$$

(for obtuse angles the cosine is negative)

Tangent Rule

e 74 $$\frac{a+b}{a-b} = \frac{\tan\frac{\alpha+\beta}{2}}{\tan\frac{\alpha-\beta}{2}} \quad\Big|\quad \frac{a+c}{a-c} = \frac{\tan\frac{\alpha+\gamma}{2}}{\tan\frac{\alpha-\gamma}{2}} \quad\Big|\quad \frac{b+c}{b-c} = \frac{\tan\frac{\beta+\gamma}{2}}{\tan\frac{\beta-\gamma}{2}}$$

Half-angle Rule

e 75 $$\tan\frac{\alpha}{2} = \frac{\varrho}{s-a} \quad\Big|\quad \tan\frac{\beta}{2} = \frac{\varrho}{s-b} \quad\Big|\quad \tan\frac{\gamma}{2} = \frac{\varrho}{s-c}$$

Area, radius of incircle and circumcircle

e 76 $$A = \frac{1}{2}\,b\,c\sin\alpha = \frac{1}{2}\,a\,c\sin\beta = \frac{1}{2}\,a\,b\sin\gamma$$

e 77 $$A = \sqrt{s(s-a)(s-b)(s-c)} = \varrho\,s$$

e 78 $$\varrho = \sqrt{\frac{(s-a)(s-b)(s-c)}{s}}$$

e 79 $$r = \frac{1}{2}\,\frac{a}{\sin\alpha} = \frac{1}{2}\,\frac{b}{\sin\beta} = \frac{1}{2}\,\frac{c}{\sin\gamma}$$

e 80 $$s = \frac{a+b+c}{2}$$

Inverse circular functions

Definitions

		function $y =$		
	arcsin x	arccos x	arctan x	arccot x
e 81 identical with	$x = \sin y$	$x = \cos y$	$x = \tan y$	$x = \cot y$
e 82 defined within	$-1 \leqq x \leqq +1$	$-1 \leqq x \leqq +1$	$-\infty < x < +\infty$	$-\infty < x < +\infty$
e 83 principal value	$-\frac{\pi}{2} \leqq y \leqq +\frac{\pi}{2}$	$\pi \geqq y \geqq 0$	$-\frac{\pi}{2} < y < +\frac{\pi}{2}$	$\pi > y > 0$

Basic properties

e 84 $\arccos x = \frac{\pi}{2} - \arcsin x$ | $\operatorname{arccot} x = \frac{\pi}{2} - \arctan x$

Ratios between inverse circular functions [1]

	Arcsin x =	Arccos x =	Arctan x =	Arccot x =
e 85	* $\operatorname{Arccos}\sqrt{1-x^2}$	* $\operatorname{Arcsin}\sqrt{1-x^2}$	$\operatorname{Arcsin}\frac{x}{\sqrt{1+x^2}}$	* $\operatorname{Arcsin}\frac{1}{\sqrt{1+x^2}}$
e 86	$\operatorname{Arctan}\frac{x}{\sqrt{1-x^2}}$	* $\operatorname{Arctan}\frac{\sqrt{1-x^2}}{x}$	* $\operatorname{Arccos}\frac{1}{\sqrt{1+x^2}}$	$\operatorname{Arccos}\frac{x}{\sqrt{1+x^2}}$
e 87	* $\operatorname{Arccot}\frac{\sqrt{1-x^2}}{x}$	$\operatorname{Arccot}\frac{x}{\sqrt{1-x^2}}$	* $\operatorname{Arccot}\frac{1}{x}$	* $\operatorname{Arctan}\frac{1}{x}$

For the defined x values cf. e 82:

e 88 $\arcsin(-x) = -\arcsin x$ | $\arccos(-x) = \pi - \arccos x$

e 89 $\arctan(-x) = -\arctan x$ | $\operatorname{arccot}(-x) = \pi - \operatorname{arccot} x$

Addition theorems

e 90 $\arcsin a \pm \arcsin b = \arcsin\left(a\sqrt{1-b^2} \pm b\sqrt{1-a^2}\right)$

e 91 $\arccos a \pm \arccos b = \arccos\left(a\,b \mp \sqrt{1-a^2}\,\sqrt{1-b^2}\right)$

e 92 $\arctan a \pm \arctan b = \arctan\frac{a \pm b}{1 \mp a\,b}$

e 93 $\operatorname{arccot} a \pm \operatorname{arccot} b = \operatorname{arccot}\frac{a\,b \mp 1}{b \pm a}$

[1] The formulas marked with * apply for $x \geq 0$

Straight line

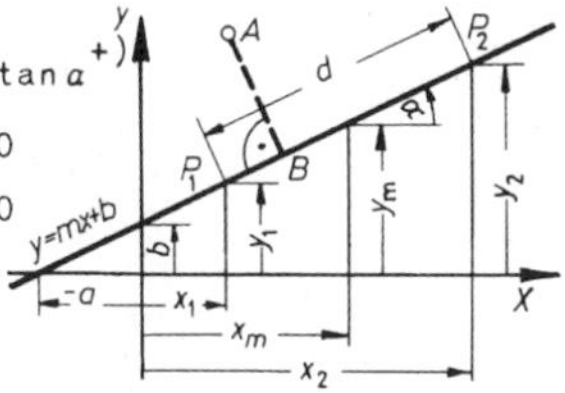

f 1 **Equation** $y = mx + b$

f 2 **Gradient** $m = \dfrac{y_2 - y_1}{x_2 - x_1} = \tan\alpha$ +)

Interc. form for $a \neq 0;\ b \neq 0$

f 3 $\dfrac{x}{a} + \dfrac{y}{b} - 1 = 0$

Gradient m_l of perpendicular $\overline{AB}$

f 4 $m_l = \dfrac{-1}{m}$

Line joining two points $P_1(x_1, y_1)$ and $P_2(x_2, y_2)$

f 5 $\dfrac{y - y_1}{x - x_1} = \dfrac{y_2 - y_1}{x_2 - x_1}$

Line through one point $P_1(x_1, y_1)$ and gradient m

f 6 $y - y_1 = m(x - x_1)$

f 7 **Distance between two points** $d = \sqrt{(x_2 - x_1)^2 + (y_2 - y_1)^2}$ +)

Mid point of a line joining two points

f 8 $x_m = \dfrac{x_1 + x_2}{2}$ | $y_m = \dfrac{y_1 + y_2}{2}$

Point of intersection of two straight lines (see diagram triangle)

f 9 $x_3 = \dfrac{b_2 - b_1}{m_1 - m_2}$ | $y_3 = m_1 x_3 + b_1 = m_2 x_3 + b_2$

Angle of intersection φ **of two straight lines**

f10 $\tan\varphi = \dfrac{m_2 - m_1}{1 + m_2 \cdot m_1}$ +) (see diagram-triangle)

Triangle

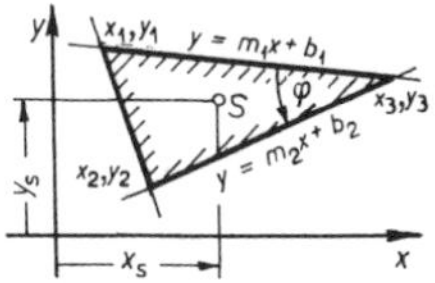

Centroid S

f11 $x_S = \dfrac{x_1 + x_2 + x_3}{3}$

f12 $y_S = \dfrac{y_1 + y_2 + y_3}{3}$

Area

f13 $A = \dfrac{(x_1 y_2 - x_2 y_1) + (x_2 y_3 - x_3 y_2) + (x_3 y_1 - x_1 y_3)}{2}$

+) Where x and y have same dimension and are represented in equal scales (see also h 1).

ANALYTICAL GEOMETRY

Circle, Parabola

F 2

Circle

Circle equation

centre		
	at the origin	elsewhere
f 14	$x^2 + y^2 = r^2$	$(x-x_0)^2 + (y-y_0)^2 = r^2$

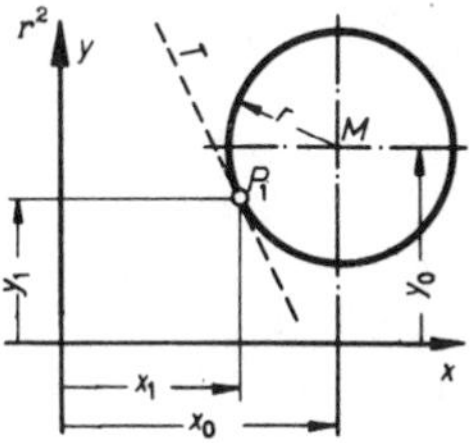

Basic equation

f 15 $$x^2 + y^2 + ax + by + c = 0$$

Radius of circle

f 16 $$r = \sqrt{x_0^2 + y_0^2 - c}$$

Coordinates of the centre M

f 17 $$x_0 = -\frac{a}{2} \qquad y_0 = -\frac{b}{2}$$

Tangent T at point $P_1\ (x_1, y_1)$

f 18 $$y = \frac{r^2 - (x-x_0)(x_1-x_0)}{y_1 - y_0} + y_0$$

Parabola

Parabola equation (by converting to this equation the vertex and parameter p may be ascertained)

	vertex		parabola	
	at the origin	elsewhere	open at	F: focus L: directrix
f 19	$x^2 = 2py$	$(x-x_0)^2 = 2p(y-y_0)$	top	S: tangent at
f 20	$x^2 = -2py$	$(x-x_0)^2 = -2p(y-y_0)$	bottom	the vertex

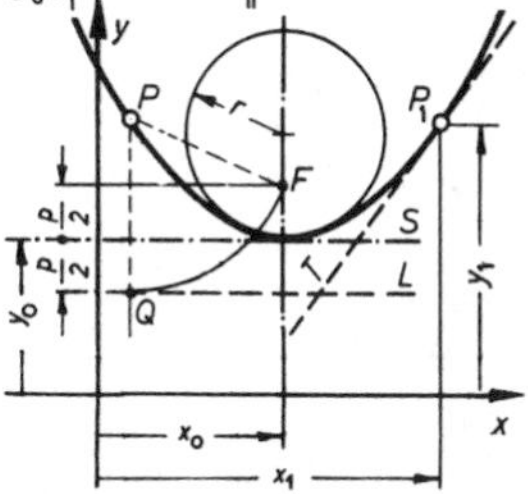

Basic equation

f 21 $$y = ax^2 + bx + c$$

f 22 **Vertex radius** $\qquad r = p$

f 23 **Basic property** $\qquad \overline{PF} = \overline{PQ}$

Tangent T at point $P_1\ (x_1, y_1)$

f 24 $$y = \frac{2(y_1-y_0)(x-x_1)}{x_1 - x_0} + y_1$$

Hyperbola

Hyperbolic equation

	point of intersection of asymptotes at the origin	point of intersection of asymptotes elsewhere
f 25	$\frac{x^2}{a^2} - \frac{y^2}{b^2} - 1 = 0$	$\frac{(x - x_0)^2}{a^2} - \frac{(y - y_0)^2}{b^2} - 1 = 0$

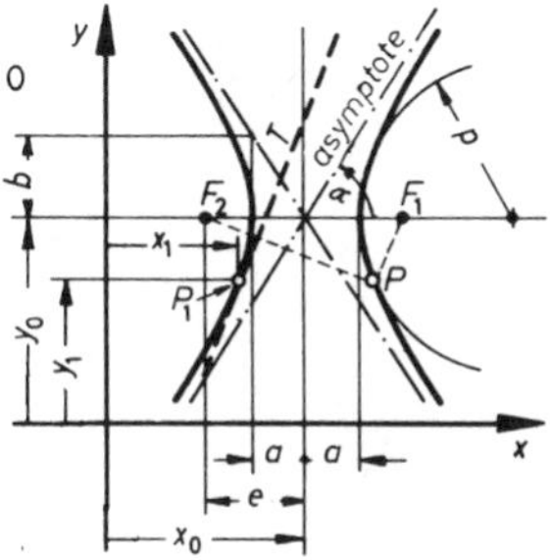

Basic equation

f 26 $Ax^2 + By^2 + Cx + Dy + E = 0$

Basic property

f 27 $\overline{F_2 P} - \overline{F_1 P} = 2a$

Eccentricity

f 28 $e = \sqrt{a^2 + b^2}$ +)

Gradient of asymptotes

f 29 $\tan\alpha = m = \pm \frac{b}{a}$ +)

Vertex radius $p = \frac{b^2}{a}$

f 30 **Tangent** T at $P_1\,(x_1, y_1)$ $y = \frac{b^2}{a^2} \, \frac{(x_1 - x_0)(x - x_1)}{y_1 - y_0} + y_1$

Rectangular hyperbola

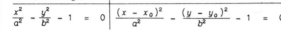

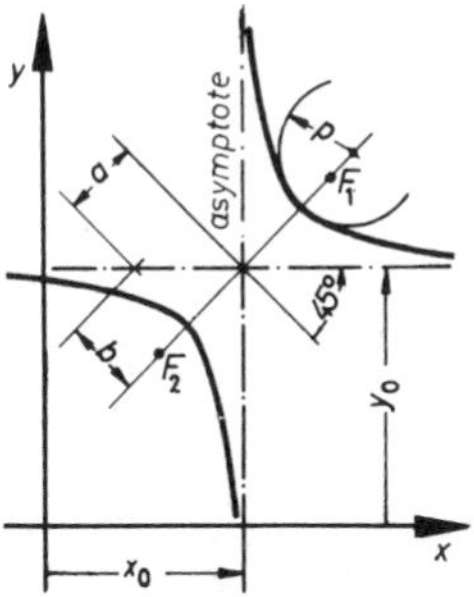

Explanation in a rectangular hyperbola $a = b$ thus

Gradient of asymptotes

f 31 $\tan\alpha \overset{+)}{=} m = \pm 1$ $(\alpha = 45^\circ)$

Equation (for asymptotes parallel to x and y axes):

	point of intersection of asymptotes at the origin	point of intersection of asymptotes elsewhere
f 32	$x\,y = c^2$	$(x - x_0)(y - y_0) = c^2$

Vertex radius

f 33 $p = a$ (parameter)

+) Conditions according to note on page F 1

Ellipse

Ellipse equation

point of intersection of axes at the origin	elsewhere
$\frac{x^2}{a^2} + \frac{y^2}{b^2} - 1 = 0$	$\frac{(x - x_0)^2}{a^2} + \frac{(y - y_0)^2}{b^2} - 1 = 0$

f 34

Vertex radii

f 35 $r_N = \frac{b^2}{a}$ | $r_H = \frac{a^2}{b}$

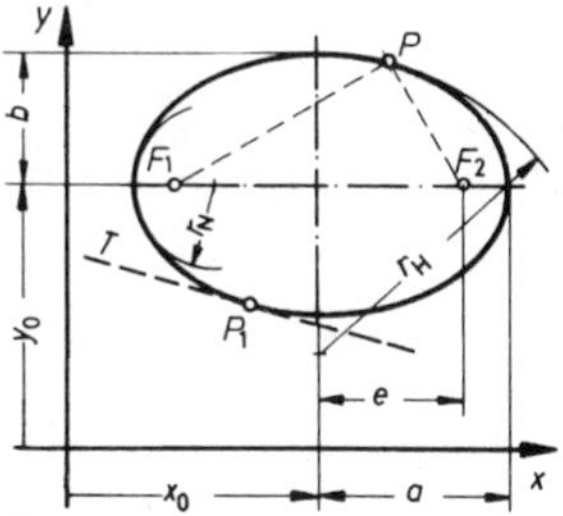

Eccentricity

f 36 $e = \sqrt{a^2 - b^2}$

Basic property

f 37 $\overline{F_1 P} + \overline{F_2 P} = 2a$

Tangent T at P_1 $(x_1; y_1)$

f 38 $y = -\frac{b^2}{a^2} \frac{(x_1 - x_0)(x - x_1)}{y_1 - y_0} + y_1$

Note: F_1 *and* F_2 are focal points

Exponential curve

Basic equation

f 39 $y = a^x$

Here a is a positive constant $\neq 1$, and x is a number.

Note:
All exponential curves pass through the point $x = 0$; $y = 1$.

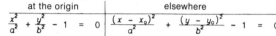

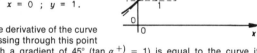

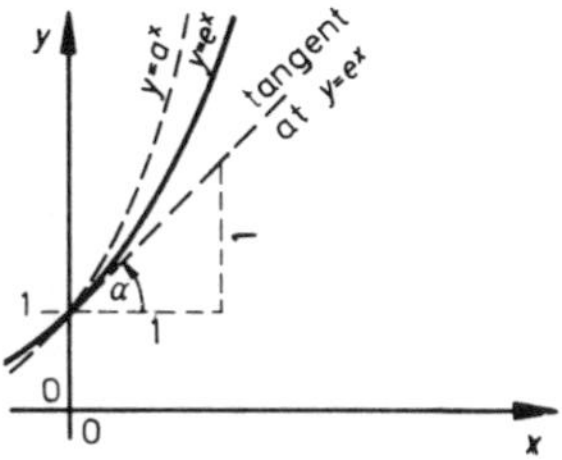

The derivative of the curve passing through this point with a gradient of 45° (tan α +) = 1) is equal to the curve itself. The constant a now becomes e (Euler number) and is the base of the natural log.

$$e = 2 \cdot 718281828459$$

+) Conditions according to note on page F 1

Hyperbolic functions

Definition

f 40 $\sinh x = \dfrac{e^{x} - e^{-x}}{2}$ +)

f 41 $\cosh x = \dfrac{e^{x} + e^{-x}}{2}$

f 42 $\tanh x = \dfrac{e^{x} - e^{-x}}{e^{x} + e^{-x}} = \dfrac{e^{2x} - 1}{e^{2x} + 1}$

f 43 $\coth x = \dfrac{e^{x} + e^{-x}}{e^{x} - e^{-x}} = \dfrac{e^{2x} + 1}{e^{2x} - 1}$

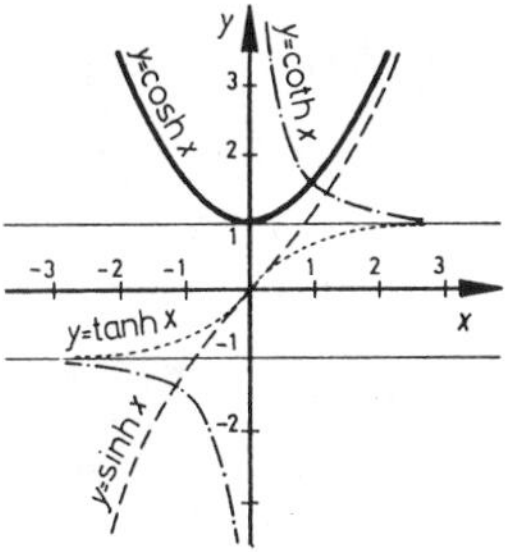

Basic properties

f 44 $\cosh^2 x - \sinh^2 x = 1$

f 45 $\tanh x \times \coth x = 1$

f 46 $\tanh x = \dfrac{\sinh x}{\cosh x}$ | $1 - \tanh^2 x = \dfrac{1}{\cosh^2 x}$ | $1 - \coth^2 x = \dfrac{-1}{\sinh^2 x}$

Ratios between hyperbolic functions

	$\sinh x =$	$\cosh x =$	$\tanh x =$	$\coth x =$
f 47	$\pm\sqrt{\cosh^2 x - 1}$ *	$\sqrt{\sinh^2 x + 1}$	$\dfrac{\sinh x}{\sqrt{\sinh^2 x + 1}}$	$\dfrac{\sqrt{\sinh^2 x + 1}}{\sinh x}$
f 48	$\dfrac{\tanh x}{\sqrt{1 - \tanh^2 x}}$	$\dfrac{1}{\sqrt{1 - \tanh^2 x}}$	$\pm\dfrac{\sqrt{\cosh^2 x - 1}}{\cosh x}$ *	$\pm\dfrac{\cosh x}{\sqrt{\cosh^2 x - 1}}$ *
f 49	$\pm\dfrac{1}{\sqrt{\coth^2 x - 1}}$ *	$\dfrac{\lvert\coth x\rvert}{\sqrt{\coth^2 x - 1}}$	$\dfrac{1}{\coth x}$	$\dfrac{1}{\tanh x}$

f 50 $\sinh(-x) = -\sinh x$ | $\cosh(-x) = +\cosh x$

f 51 $\tanh(-x) = -\tanh x$ | $\coth(-x) = -\coth x$

Addition theorems

f 52 $\sinh(a \pm b) = \sinh a \times \cosh b \pm \cosh a \times \sinh b$

f 53 $\cosh(a \pm b) = \cosh a \times \cosh b \pm \sinh a \times \sinh b$

f 54 $\tanh(a \pm b) = \dfrac{\tanh a \pm \tanh b}{1 \pm \tanh a \times \tanh b}$

f 55 $\coth(a \pm b) = \dfrac{\coth a \times \coth b \pm 1}{\coth a \pm \coth b}$

+) Exponent x always has to be non-dimensional quantity

* Sign + for $x > 0$; – for $x < 0$

Inverse hyperbolic functions

Definition

	function $y =$			
	arsinh x	arcosh x	artanh x	arcoth x
f 56 identical with	$x = \sinh y$	$x = \cosh y$	$x = \tanh y$	$x = \coth y$
f 57 logarithmic equivalents	$= \ln(x+\sqrt{x^2+1})$	$= \pm\ln(x+\sqrt{x^2-1})$	$= \frac{1}{2}\ln\frac{1+x}{1-x}$	$= \frac{1}{2}\ln\frac{x+1}{x-1}$
f 58 defined within	$-\infty < x < +\infty$	$1 \leqq x < +\infty$	$\lvert x\rvert < 1$	$\lvert x\rvert > 1$
f 59 primary value	$-\infty < y < +\infty$	$-\infty < y < +\infty$	$-\infty < y < +\infty$	$\lvert y\rvert > 0$

Ratios between inverse hyperbolic functions

	arsinh x =	arcosh x =	artanh x =	arcoth x =
f 60	$\pm$* arcosh $\sqrt{1+x^2}$	$\pm$* arsinh $\sqrt{x^2-1}$	arsinh $\frac{x}{\sqrt{1-x^2}}$	arsinh $\frac{1}{\sqrt{x^2-1}}$
f 61	artanh $\frac{x}{\sqrt{1+x^2}}$	$\pm$* artanh $\frac{\sqrt{x^2-1}}{x}$	$\pm$* arcosh $\frac{1}{\sqrt{1-x^2}}$	$\pm$* arcosh $\frac{x}{\sqrt{x^2-1}}$
f 62	arcoth $\frac{\sqrt{1+x^2}}{x}$	$\pm$* arcoth $\frac{x}{\sqrt{x^2-1}}$	arcoth $\frac{1}{x}$	artanh $\frac{1}{x}$

For the defined x values cf. f 58:

f 63 $\operatorname{arsinh}(-x) = -\operatorname{arsinh} x$

f 64 $\operatorname{artanh}(-x) = -\operatorname{artanh} x$ $\quad$ $\operatorname{arcoth}(-x) = -\operatorname{arcoth} x$

Addition theorems

f 65 $\operatorname{arsinh} a \pm \operatorname{arsinh} b = \operatorname{arsinh}\left(a\sqrt{b^2+1} \pm b\sqrt{a^2+1}\right)$

f 66 $\operatorname{arcosh} a \pm \operatorname{arcosh} b = \operatorname{arcosh}\left[ab \pm \sqrt{(a^2-1)(b^2-1)}\right]$

f 67 $\operatorname{artanh} a \pm \operatorname{artanh} b = \operatorname{artanh}\frac{a \pm b}{1 \pm ab}$

f 68 $\operatorname{arcoth} a \pm \operatorname{arcoth} b = \operatorname{arcoth}\frac{ab \pm 1}{a \pm b}$

* Sign + for $x > 0$; – for $x < 0$

Components, magnitude, direction cosines of vectors

Vector: Quantity with magnitude and direction

A: Coordinates of the origin of the vector $\vec{a}$: x_1, y_1, z_1
B: Coordinates of the end-point of the vector $\vec{a}$: x_2, y_2, z_2

Unit vectors along OX, OY, OZ: $\vec{i}, \vec{j}, \vec{k}$

Components with magnitude and direction

f 69 $a_x, \; a_y, \; a_z \gtreqless 0$

f 70 $a_x = x_2 - x_1$

f 71 $a_y = y_2 - y_1$

f 72 $a_z = z_2 - z_1$

f 73 $\vec{a} = \vec{a_x} + \vec{a_y} + \vec{a_z}$ *)

f 74 $\vec{a} = a_x \vec{i} + a_y \vec{j} + a_z \vec{k}$

*) vector equations

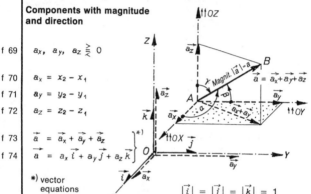

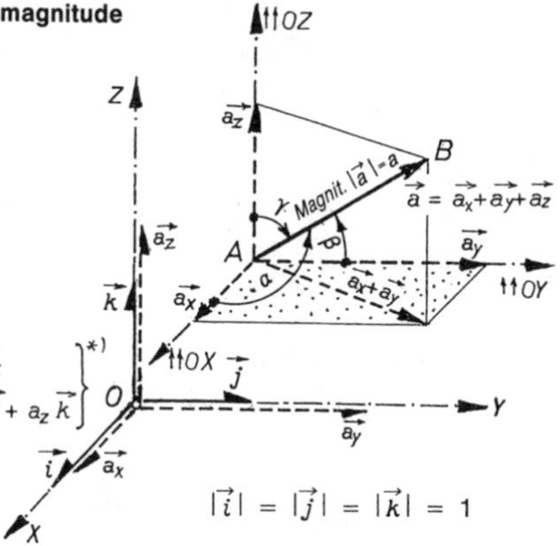

$|\vec{i}| = |\vec{j}| = |\vec{k}| = 1$

Magnitude or norm of the vector: $|\vec{a}|$ or a in engineering notation.

f 75 $|\vec{a}| = \sqrt{a_x^2 + a_y^2 + a_z^2}$ ($|\vec{a}|$ always ≥ 0)

Direction cosines of vectors: $\cos\alpha$, $\cos\beta$, $\cos\gamma$

α, β, γ, angles between the vector $\vec{a}$ and the axes OX, OY and OZ. ($\alpha, \beta, \gamma = 0° \ldots 180°$).

f 76 $\cos\alpha = \dfrac{a_x}{|\vec{a}|}; \quad \cos\beta = \dfrac{a_y}{|\vec{a}|}; \quad \cos\gamma = \dfrac{a_z}{|\vec{a}|}$

f 77 where $\cos^2\alpha + \cos^2\beta + \cos^2\gamma = 1$

Calculation of the components when $|\vec{a}|$ α, β, γ are known:

f 78 $a_x = |\vec{a}| \cos\alpha; \quad a_y = |\vec{a}| \cos\beta; \quad a_z = |\vec{a}| \cos\gamma$

Note: The components along OX, OY, OZ are used to determine the magnitude, direction cosines, sum of vectors and product of vectors.

Vector sum (difference)

Vector sum $\vec{s}$ of two vectors $\vec{a}$ and $\vec{b}$

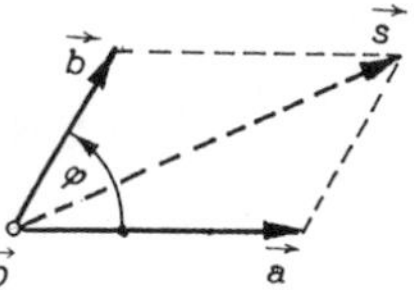

f 79 $\vec{s} = \vec{a} + \vec{b} = s_x\,\vec{i} + s_y\,\vec{j} + s_z\,\vec{k}$

f 80 $s_x = a_x + b_x \quad s_y = a_y + b_y \quad s_z = a_z + b_z$

f 81 $|\vec{s}| = \sqrt{s_x^2 + s_y^2 + s_z^2}$

Vector difference $\vec{s}$ of two vectors $\vec{a}$ and $\vec{b}$

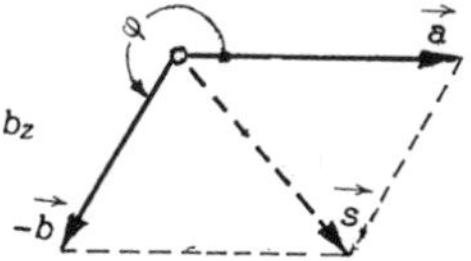

f 82 $\vec{s} = \vec{a} + (-\vec{b}) \qquad (-\vec{b}) \uparrow\downarrow \vec{b}$ *)

f 83 $s_x = a_x - b_x; \quad s_y = a_y - b_y; \quad s_z = a_z - b_z$

f 84 $|\vec{s}| = \sqrt{s_x^2 + s_y^2 + s_z^2}$

Special cases for $|\vec{s}|$

	φ	0°; 360°	90°	180°	270°
f 85	$\lvert\vec{a}\rvert \neq \lvert\vec{b}\rvert$	$\lvert\vec{a}\rvert + \lvert\vec{b}\rvert$	$\sqrt{\lvert\vec{a}\rvert^2 + \lvert\vec{b}\rvert^2}$	$\lvert\vec{a}\rvert - \lvert\vec{b}\rvert$	$\sqrt{\lvert\vec{a}\rvert^2 + \lvert\vec{b}\rvert^2}$
f 86	$\lvert\vec{a}\rvert = \lvert\vec{b}\rvert$	$2\lvert\vec{a}\rvert$	$\lvert\vec{a}\rvert\sqrt{2}$	0	$\lvert\vec{a}\rvert\sqrt{2}$

Vector sum $\vec{s}$ of vectors $\vec{a}$, $\vec{b}$, $-\vec{c}$, etc.:

f 87 $\vec{s} = \vec{a} + \vec{b} - \vec{c} + \ldots = s_x\,\vec{i} + s_y\,\vec{j} + s_z\,\vec{k}$ (Vector equations)

f 88 $s_x = a_x + b_x - c_x + \ldots; \quad s_y = a_y + b_y - c_y + \ldots; \quad s_z = a_z + b_z - c_z + \ldots$

f 89 $|\vec{s}| = \sqrt{s_x^2 + s_y^2 + s_z^2}$

Product of a scalar and a vector

Scalar: Quantity with magnitude only

Product of a scalar *k* and a vector $\vec{a}$ is the vector $\vec{c}$

f 90 $\vec{c} = k \times \vec{a} \qquad (k \gtreqless 0)$ (Vector equation)

f 91 $c_x = k \times a_x; \quad c_y = k \times a_y; \quad c_z = k \times a_z \qquad c = k\,|\vec{a}| \quad (c \lesseqgtr 0)$

If $k > 0$ then $\vec{c} \uparrow\uparrow \vec{a}$ ie

If $k < 0$ then $\vec{c} \uparrow\downarrow \vec{a}$ ie

Example: Force F_a = mass m times acceleration a

f 92 $m > 0; \quad \vec{F_a} \uparrow\uparrow \vec{a}; \quad \vec{F_a} = m \times \vec{a}; \quad F_a = m \times a$

*) The symbol $\uparrow\downarrow$ denotes that the vectors $(-\vec{b})$ and $(\vec{b})$ are parallel but opposite in direction.

Vector Products of 2 vectors

The scalar product of 2 vectors $\vec{a}$ and $\vec{b}$ is the scalar k.

Symbol for Scalar Product: Dot "·"

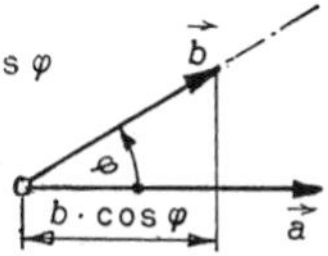

f 93 $k = \vec{a} \cdot \vec{b} = \vec{b} \cdot \vec{a} = a \cdot b \cdot \cos\varphi = |\vec{a}| \cdot |\vec{b}| \cdot \cos\varphi$

f 94 $k = a_x \cdot b_x + a_y \cdot b_y + a_z \cdot b_z \qquad (k \gtreqless 0)$

f 95 $\varphi = \arccos \dfrac{a_x \cdot b_x + a_y \cdot b_y + a_z \cdot b_z}{|\vec{a}| \cdot |\vec{b}|}$

Special cases	φ	0°; 360°;	90°	180°	270°
f 96	$\lvert\vec{a}\rvert \cdot \lvert\vec{b}\rvert \cdot \cos\varphi$	$+\lvert\vec{a}\rvert \cdot \lvert\vec{b}\rvert$	0	$-\lvert\vec{a}\rvert \cdot \lvert\vec{b}\rvert$	0

Example: Work done W, by a force F over distance s

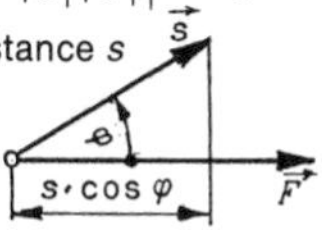

f 97 $W = \text{force} \cdot \text{distance} = \vec{F} \cdot \vec{s}$

f 98 $W = F \cdot s \cdot \cos\varphi \qquad (W \gtreqless 0; \quad F, s \geqq 0)$

Vector product of 2 vectors $\vec{a}$ and $\vec{b}$ is the vector $\vec{c}$

Symbol for Vector Product: Cross "x"

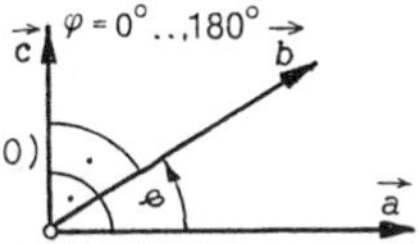

f 99 $\vec{c} = \vec{a} \times \vec{b} = -(\vec{b} \times \vec{a})$

f 100 $|\vec{c}| = a \cdot b \cdot \sin\varphi = |\vec{a}| \cdot |\vec{b}| \cdot \sin\varphi \quad (c \lesseqgtr 0)$

If $c \perp a$ and $\vec{c} \perp \vec{b}$

$\vec{a}$, $\vec{b}$, $\vec{c}$ form a basis

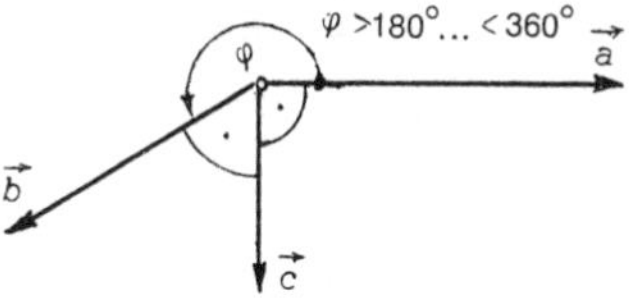

f 101 $c_x = a_y \cdot b_z - a_z \cdot b_y$

f 102 $c_y = a_z \cdot b_x - a_x \cdot b_z$

f 103 $c_z = a_x \cdot b_y - a_y \cdot b_x$

f 104 $|\vec{c}| = \sqrt{c_x^2 + c_y^2 + c_z^2}$

Special cases	φ	0°; 360°	90°	180°	270°
f 105	$\lvert\vec{a}\rvert \cdot \lvert\vec{b}\rvert \sin\varphi$	0	$+\lvert\vec{a}\rvert \cdot \lvert\vec{b}\rvert$	0	$-\lvert\vec{a}\rvert \cdot \lvert\vec{b}\rvert$

Example: Moment M of a force F
laround the point O

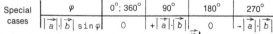

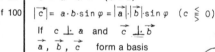

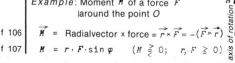

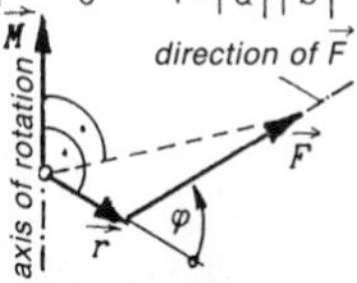

f 106 $\vec{M} = \text{Radialvector} \times \text{force} = \vec{r} \times \vec{F} = -(\vec{F} \times \vec{r})$

f 107 $M = r \cdot F \cdot \sin\varphi \qquad (M \gtreqless 0; \quad r, F \geqq 0)$

Theoretical probability $P(A)$

If E is the set of outcomes of an experiment all of which are assumed to be equally likely and an event A is satisfied by a subset

g 1 A of them, then $P(A) = n(A)/n(E)$.

Experimental probability $P(A)$

If an event A is satisfied by a certain outcome of an experiment and, when the experiment is repeated n times under exactly the same conditions, A occurs r times out of n, then

g 2 $P(A) = \lim_{n \to \infty} (r/n)$

Axioms to the probability

g 3 $P(A) \geqq 0$, event A has the probability $P(A)$

g 4 $h(A) = \dfrac{\text{number of events in which } A \text{ occurs}}{\text{number of possible events}}$

$= \text{relative frequency}$

g 5 $\sum_i P(A_i) = 1{\cdot}0$. The sum of the probabilities of all possible events A_i taking place must be 1·0.

g 6 $P(A \cap B)^{*)} = P(A) + P(B) - P(A \cap B)^{*)}$.

If A and B cannot take place at once, then

$= P(A) + P(B)$ and the events are said to be disjoint.

g 7 $P(A/B) = P(A \cap B)/P(B)^{*}$ is called the probability of A conditional on B (the probability of the event A, given that the event B has happened).

If the events are independent (if the knowledge that one event has occurred has no effect on the probability

g 8 of the other occurring) assuming $P(A)$ resp. $P(B) \neq 0$.

g 9 $P(A/B) = P(A)$, and $P(B/A) = P(B)$

g 10 $P(A \cap B) = P(A) \times P(B)$ if events are independent.

g 11 $P(A \cap \bar{A}) = P(A) \times P(\bar{A}) = 0$, as A and $\bar{A}$ are mutually exclusive.

*) **Venn Diagrams**

The rectangle	represents the sum of all events A
The large circle	represents the event A
The small circle	represents the event B

Hatched area shows the conjunction of the different cases.

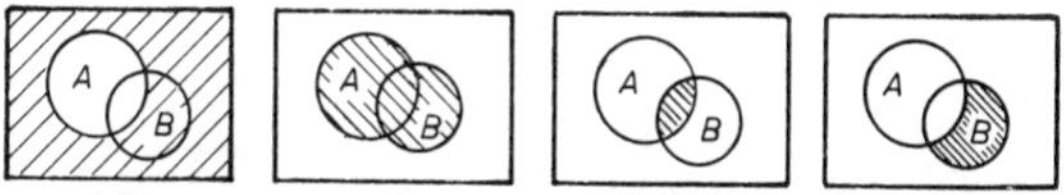

$\bar{A}$ ("not" A) $A \cup B$ (A "or" B) $A \cap B$ (A "and" B) $\bar{A} \cap B$ (B "but not" A)

The random variable A

The random variable A is a measurable quantity which can take any number x_i or a range of values with a given probability distribution.

The cumulative distribution function $F(x)$

The cumulative distribution function $F(x)$ shows the probability of the random variable being less than a specified value x.

g 12 $F(x)$ varies between 0 and 1·0.

g 13 $F(-\infty) = 0$ and $F(x)$ increases with x.

$F(x)$ for an experimental distribution

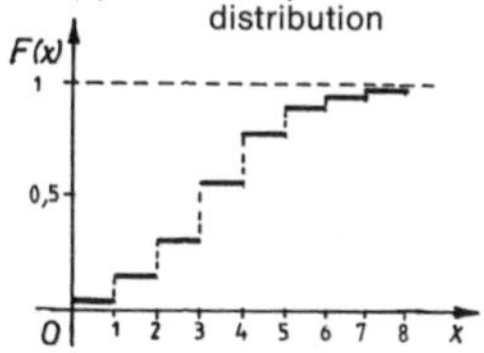

$F(x)$ for continuous functions or theoretical distribution

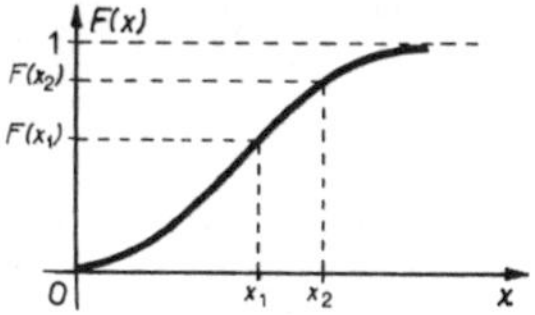

The probability density function $f(x)$

The probability density function $f(x)$ shows the number of times one particular value p_i or range of values $f(x)$ of the random variable A occurs.

14/15 $F(x) = \sum_{i<x} p_i$ $\qquad$ $F(x) = \int_{-\infty}^{x} f(x)\,dx$

p_i for an experimental distribution

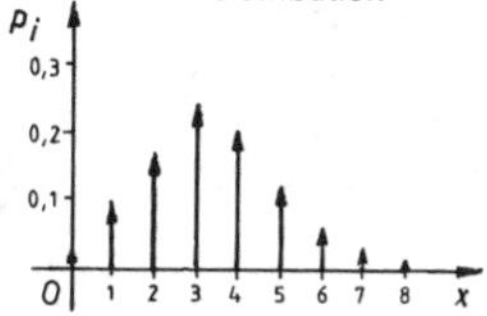

$f(x)$ for continuous functions or theoretical distribution

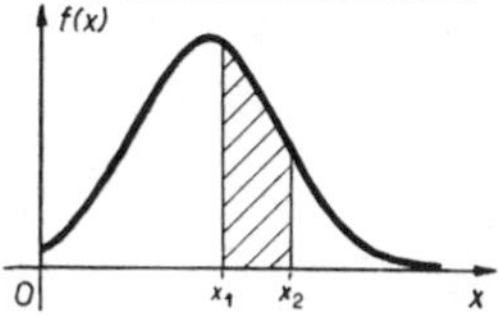

The hatched area under the probability density function curve shows the probability that the random variable A lies between x_1 and x_2.

g 16 $$P(x_1 \leqslant A < x_2) = \int_{x_1}^{x_2} f(x)\,dx$$

g 17 $$= F(x_2) - F(x_1) = P(A < x_2) - P(A < x_1)$$

STATISTICS

General terms

G 3

Mean value $\bar{x}$ **or expected mean** $E(x)$ or μ

	Random variable A discrete	Random variable A continuous
g 18	$\bar{x} = x_1 p_1 + x_2 p_2 + \ldots + x_n p_n$	
g 19		$\mu = \int_{-\infty}^{+\infty} x \; f(x) \; dx$
g 20	$= \sum_{i=1}^{n} x_i \; p_i$	

where p_i and $f(x)$ are probability densities.

Variance σ^2

	Random variable A discrete	Random variable A continuous
g 21	$\sigma^2 = (x_1 - \bar{x})^2 \times p_1 + (x_2 - \bar{x})^2 \; p^2 +$	
g 22	$+ \ldots + (x_n - \bar{x})^2 \times p_n$	$\sigma^2 = \int_{-\infty}^{+\infty} (x - \mu)^2 \times f(x) \; dx$
g 23	$= \sum_{i=1}^{n} (x_i - \bar{x})^2 \times p_i$	
g 24		$= \int_{-\infty}^{+\infty} x^2 \times f(x) \; dx - \mu^2$
g 25	$= \sum_{i=1}^{n} x_i^2 \times p_i - \bar{x}^2$	

where p_i and $f(x)$ are probability densities and σ is called the "Standard Deviation".

Central limit theorem (addition law)

When two or more random distributions A with expected values μ and variances σ^2 are combined

g 26 the random variable $A = \sum_{i=1}^{n} A_i$

g 27 the mean value $\mu = \sum_{i=1}^{n} \mu_i$ $\quad (\bar{x} = \sum_{i=1}^{n} \bar{x}_i)$

g 28 the variance $\sigma^2 = \sum_{i=1}^{n} \sigma_i^2 \;;$

If the random variables have normal distributions, then

g 29 $$P(A \leqslant x) = \Phi\left(\frac{x - \mu}{\sigma}\right)$$

where Φ is the cumulative distribution function for the standard normal distribution.

g 30 Example: If 10 batches of components, each batch having a standard deviation of 0·03 µm, are mixed together, the standard deviation of the whole σ_t, is given by:

$$\sigma_t^2 = 10\,\sigma^2 \;; \qquad \sigma_t = \pm\sigma\sqrt{10} \approx \pm 0{\cdot}095 \; \mu m$$

	Kind of distrib.	Probability density-function	Distribution function	Expected mean μ / mean val. $\bar{x}$	Variance σ^2	Form of the density function	Remarks, Application-field
g 31	Definition equat.	$f(x)$ continuous	$F(x)=\int_{-\infty}^{x} f(t)\,dt$	$\int_{-\infty}^{\infty} x\, f(x)\,dx$	$\int_{-\infty}^{\infty} x^2 f(x)\,dx-\mu^2$		k : numbers of errors n : number in random sample x_i : discrete value of a random variable p : error probability
g 32		p_i discrete	$F(x)=\sum_{i<x} p_i$	$\sum_{i=1}^{n} x_i\, p_i$	$\sum_{i=1}^{n} x_i^2\, p_i-\mu^2$		
g 33	hyper-geo-metric	$P(k)=\dfrac{\binom{pN}{k}\binom{N(1-p)}{n-k}}{\binom{N}{n}}$	$\sum_{k<x}\dfrac{\binom{pN}{k}\binom{N(1-p)}{n-k}}{\binom{N}{n}}$	$n\ p$	$n\ p\,\dfrac{N-n}{N-1}\,(1-p)$	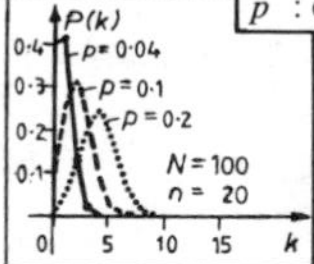	N : Size of the lot pN : Defect. parts within N Precise but expensive calculation. Without replacement.
		$P(k)$ means the probability that in n random samples from a lot N, exactly k are defective					
g 34	binomial	$P(k)=\binom{n}{k}p^k(1-p)^{n-k}$	$\sum_{k<x}\binom{n}{x}p^k(1-p)^{n-k}$	$n\ p$	$n\ p(1-p)$	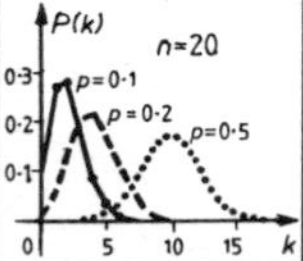	Conditions: Lot size, ∞
		$P(k)$ means the probability that in n random samples exactly k errors occur.					
g 35	Poisson	$P(k)=\dfrac{(np)^k}{k!}\,e^{-np}$	$\sum_{k<x}\dfrac{(np)^k}{k!}\,e^{-np}$	$n\ p$	$n\ p$	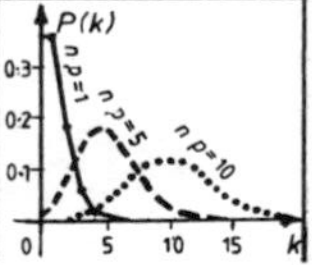	Conditions: Large value of random samples and small value of proportion defective. $n\cdot p$ = const. $n \to \infty;\ p \to 0$
		$P(k)$ means the probability that in n random samples k errors occur. Application: Curves for random sampling valuation (s. G 11)					

continued on G 5

	Kind of distrib.	Probability density-function	Distribution function	Expected mean μ / mean val. $\bar{x}$	Variance σ^2	Form of the density function	Remarks, Application-field
g 36	Definition equat.	$f(x)$ continuous	$F(x) = \int_{-\infty}^{x} f(t)\,dt$	$\int_{-\infty}^{\infty} x\,f(x)\,dx$	$\int_{-\infty}^{\infty} x^2 f(x)\,dx - \mu^2$		n: number in random sample x_i: discrete value of a random variable p: error probability
g 37		p_i discrete	$F(x) = \sum_{i<x} p_i$	$\sum_{i=1}^{n} x_i\ p_i$	$\sum_{i=1}^{n} x_i^2\ p_i - \mu^2$		
g 38	exponential	$f(x) = a\,e^{-ax}$ $a > 0$ $x \geqslant 0$	$1 - e^{-ax}$	$\frac{1}{a}$	$\frac{1}{a^2}$	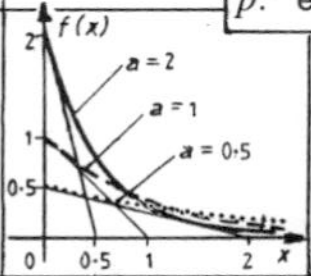	Special case of the Poisson distribution for $x = 0$. Gives the probability without error. $n \to \infty$; $p \to 0$.
		Used for realiability calculations. Replacement of $a\,x$ by failure rate λ multiplied by control time t (see G 12).					
g 39	normal Gaussian or error	$f(x) = \frac{1}{\sigma\sqrt{2\pi}}\,e^{\frac{-(x-\mu)^2}{2\sigma^2}}$	$\int_{-\infty}^{x} \frac{1}{\sigma\sqrt{2\pi}}\,e^{\frac{-(t-\mu)^2}{2\sigma^2}}\,dt$	μ	σ^2	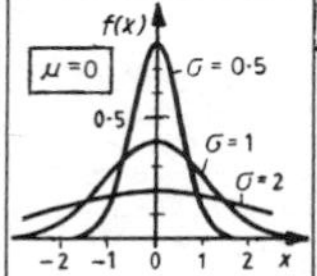	Special case of the Binomial distribution. $n \to \infty$ $p = 0{\cdot}5 = \text{const.}$
		Often obtained in practice, as measured values with bell-formed distribution occur around a mean value.					
g 40	uniform	$f(x) = \frac{1}{b-a}$ for $a \leqslant x \leqslant b$ $= 0$ for x outside	$F(x) = 0$ for $-\infty < x < a$ $= \frac{x-a}{b-a}$ for $a \leqslant x \leqslant b$	$\frac{a+b}{2}$	$\frac{(b-a)^2}{12}$	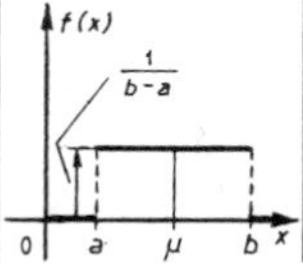	Random variable x is $\neq 0$ only within the interval a, b. There each value is of equal probability.
		Application when only maximum and minimum value and no other information about the distribution between is known.					

Determination of σ when discrete values are available

By calculation

Equation g 23 says:

g 41 $$\sigma^2 = \sum_{i=1}^{n} (x_i - \bar{x})^2 \, p_i \qquad \text{with} \qquad \bar{x} = \sum_{i=1}^{n} x_i \, p_i$$

g 42 $$= \sum_{i=1}^{n} x_i^2 \, p_i - \bar{x}^2$$

where x_i are measured values of the random variable A and
p_i are the frequencies of their occurrence.

By graphics

Standardise the distribution and choose four values of x_i spread across the range, say x_4, x_6, x_7 and x_9 shown in the drawing.

For each of these plot the cumulative frequency against the value of x_i e.g. 10% to value x_4, 38% to value x_6 and so on.

If a straight line can be drawn through these points, the distribution is proved to be normal. The values of the mean $\bar{x}$ and the standard deviation σ are obtained as shown in the diagram.

The mean value $\bar{x}$ is at 50%. The difference between the value A at 84% and the value A at 16% gives 2σ.

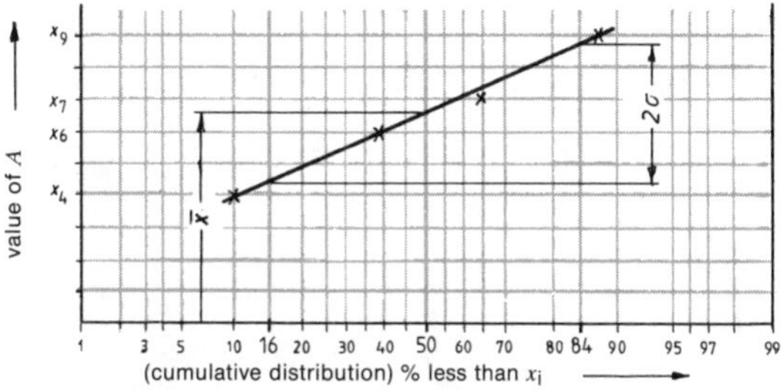

Normal curve for probability density $\varphi(\lambda)$

$\sigma^2 = 1$ and $\mu = 0$ in g 39 leads to the standardized probability density with mean value $\lambda = 0$.

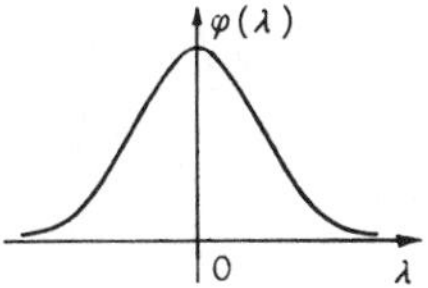

g 43 $$\varphi(\lambda) = \frac{1}{\sqrt{2\pi}}\, e^{\frac{-\lambda^2}{2}}$$

$\varphi(\lambda)$ is given in tables Z 26 and Z 27 for $0 \leq \lambda \leq 1{\cdot}99$, but can also be calculated from g 43.

The connection between standardized probability density $\varphi(\lambda)$ and the real probability density $f(x)$ for $\mu \neq 0$ and $\sigma^2 \neq 1$ is

g 44 $$f(x) = \frac{\varphi(\lambda)}{\sigma} = \frac{1}{\sigma\sqrt{2\pi}}\, e^{\frac{-(x-\mu)^2}{2\sigma^2}} \qquad \text{where} \qquad \lambda = \frac{x-\mu}{\sigma}$$

To use the table, first find the value of the standardized probability density $\varphi(\lambda)$ corresponding to λ. Divide by σ to get the real value of the probability density $f(x)$ for the value of x (see g 44).

Normal probability curve (Probability distribution function)

$\sigma^2 = 1$ and $\mu = 0$ in g 39 leads to the standardized normal distribution

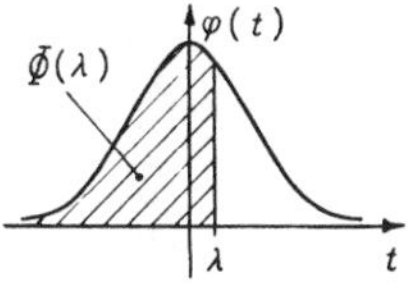

g 45 $$\Phi(\lambda) = \int_{-\infty}^{\lambda} \varphi(t)\, dt = \frac{1}{\sqrt{2\pi}} \int_{-\infty}^{\lambda} e^{\frac{-t^2}{2}}\, dt$$

As limit $\Phi(\lambda) = 1$ for $\lambda \to \infty$ and $\varphi(t)$ is a symmetrical function means that:

g 46 $$\Phi(-\lambda) = 1 - \Phi(\lambda)$$

The relation between the standardized $\Phi(\lambda)$ distribution function and the real distribution for $\mu \neq 0$ and $\sigma^2 \neq 1$ is

g 47/48 $$F(x) = \frac{\Phi(\lambda)}{\sigma} = \frac{1}{\sigma\sqrt{2\pi}} \int_{-\infty}^{x} e^{\frac{-(t-\mu)^2}{2\sigma^2}}\, dt \qquad \text{where} \qquad \lambda = \frac{t-\mu}{\sigma}$$

Gaussian or Error curve

The curve is based on the standardized normal distribution using g 45 for $\sigma^2 = 1$ and $\mu = 0$. The area under the curve gives the value of the distribution function between $-x$ and $+x$ of the symmetrical density function $\varphi(t)$.

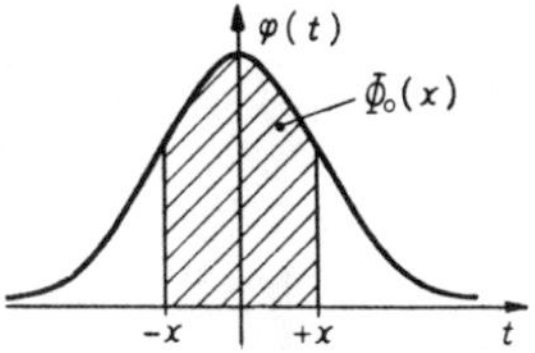

g 49 $$\Phi_0(x) = \frac{2}{\sqrt{2\pi}} \int_0^x e^{\frac{-t^2}{2}}\, dt$$

Values of $\Phi_0(x)$ between $0 \le x \le 1{\cdot}99$ are given in tables Z 26 and Z 27. For greater values of x look at the approximation in the next paragraph. The connection between $\Phi_0(x)$ and the

g 50 error function is $\Phi_0(x) = \operatorname{erf}(x/\sqrt{2})$.

Error function

g 51 $$\operatorname{erf}(x) = \Phi_0(x\sqrt{2}) = \frac{2}{\sqrt{\pi}} \int_0^x e^{-t^2}\, dt$$

g 52 $$= \frac{2}{\sqrt{\pi}} e^{-x^2} \times \sum_{n=0}^{\infty} \frac{2^n}{1 \times 3 \times \ldots \times (2n+1)} x^{2n+1}$$

Values of erf (x) between $0 \le x \le 1{\cdot}99$ are given in tables Z26 and Z27. For $x \ge 2$ values of erf(x) can be found approximately using

g 53 $$\operatorname{erf}(x) = 1 - \frac{a}{x\, e^{x^2}}$$

where $a = 0{\cdot}515$ for $2 \le x \le 3$
$a = 0{\cdot}535$ for $3 \le x \le 4$
$a = 0{\cdot}545$ for $4 \le x \le 7$
$a = 0{\cdot}56$ for $7 \le x \le \infty$

Area beneath the error curve when erf(x) is subtracted:

g 54 $$\operatorname{erfc}(x) = 1 - \operatorname{erf}(x) = \frac{2}{\sqrt{\pi}} \int_x^{\infty} e^{-t^2}\, dt$$

$\Phi_0(x)$ and $[1-\Phi_0(x)]$ in % relation to the whole area for special values

g 55

x	$\Phi_0(x)/\%$	$[1-\Phi_0(x)]/\%$
$\pm\sigma$	68·26	31·74
$\pm 2\sigma$	95·44	4·56
$\pm 2{\cdot}58\sigma$	99	1
$\pm 3\sigma$	99·73	0·27
$\pm 3{\cdot}29\sigma$	99·9	0·1

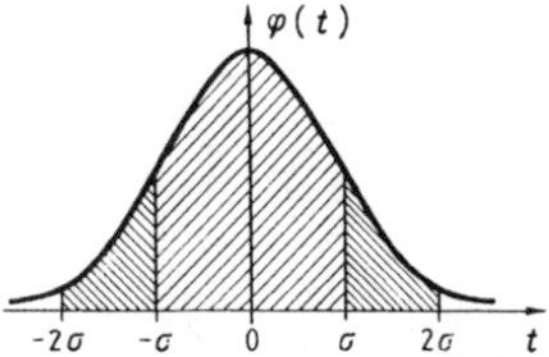

General: When testing each individual component is too expensive or not possible, test by random sampling is used. The samples must be chosen arbitrarily to give equal chances for all parts (i. e. good intermixing).

The aim of the test by random sampling is to predict the probability of the real failure rate of the whole lot on the basis of measured failure or error numbers in a sample.

Hypergeometric distribution: A hypergeometric distribution occurs when the sampling takes place without replacement. The probability $P(k)$ that in a lot of N using samples of n, without replacement, exactly k defective parts are found, when p is the assumed probability for a defective part (i. e. pN is the number of real defective parts in N and is a whole number).

g 56 $$P(k) = \frac{\binom{pN}{k}\binom{N(1-p)}{n-k}}{\binom{N}{n}}; \quad pN \text{ is a whole number}$$

The probability that not more than k defective parts are found is:

g 57 $$\sum_{x=0}^{k} P(k) = P(0) + P(1) + \dots + P(k)$$

g 58 $$= \sum_{x=0}^{k} \frac{\binom{pN}{x}\binom{N(1-p)}{n-x}}{\binom{N}{n}}; \quad pN \text{ is a whole number}$$

Example:

In a batch of 100 screws, a maximum of 3 can be defective ($N = 100$, $pN = 3$). Random samples of $n = 20$ are taken. How many defective parts are allowed in the sample? – The probabilities $\Sigma P(x)$ are

x	$P(x)$	$\sum_{x=0}^{k} P(x)$
0	0·508	0·508
1	0·391	0·899
2	0·094	0·993
3	0·007	1·000

The table shows that for ≤ 90% probability one part may be defective.

Further special distributions: Besides the hypergeometric distribution which takes much time for calculation there are other special distributions for defined assumptions and conditions. In tables G 4 and G 5 these are shown together with the hypergeometric distribution; special characteristics of these are explained.

STATISTICS

The confidence statement $P(x > k)$

From a lot N, a random sample of n is taken and k defective parts are found in it. If the probability of finding a defective part in the lot is p, the probability of finding more than k defective parts in the sample n can be derived from g 57.

g 59 $$P(x>k) = P(k+1) + P(k+2) + \ldots + P(n) = \sum_{x=k+1}^{n} P(x)$$

If N is large which is true for most manufacturing processes, and $p < 0{\cdot}1$ the Poisson distribution may be used:

g 60 $$P(x>k) = \sum_{x=k+1}^{x=n} \frac{(n\,p)^x}{x!}\,e^{-np} = 1 - \sum_{x=0}^{x=k} \frac{(n\,p)^x}{x!}\,e^{-np}$$

and if the size of the sample k is small, then

g 61 $$P(x>k) = 1 - \sum_{x=0}^{k} \frac{(np)^x}{x!}\,e^{-np} = 1 - e^{-np}\left[1 + \frac{np}{1!} + \frac{(np)^2}{2!} + \ldots + \frac{(np)^k}{k!}\right]$$

Using g 61, the confidence statement $P(x > k)$ for the proportion defective in a lot N can be determined when there are k defective parts found in the sample n, or g 61 may be used to find the size of the sample required if with an error probability of $p = k/n$, k defective parts are allowed for a given confidence statement $P(x > k)$.

The Operating Characteristic (OC)

A user needs to know whether a lot delivered by a producer meets his quality requirements. Assuming a proportion p defective in the whole batch $(p \leq p_0)$ he wants to know whether to accept or reject the whole lot if in a random sample of n parts, up to c are found to be defective. The probability that the lot will be accepted on the basis of the evidence of the sample is

g 62 $$L(p,c) \geqslant 1 - \alpha$$ where α is the producer risk or

from g 57:

g 63 $$L(p,c) = P(0) + P(1) + \ldots + P(k=c)$$

or using the Poisson distribution:

g 64 $$L(p,c) = \sum_{k=0}^{c} \frac{(np)^k}{k!}\,e^{-np} = e^{-np}\left[1 + np + \frac{(np)^2}{2!} + \ldots + \frac{(np)^c}{c!}\right]$$

continued on G11

continued from G 10

Using equation g 64, the operating characteristics $L(p,c)$ may be plotted in two ways:

Type A	Type B
n = constant; c: parameter	c = constant; n: parameter

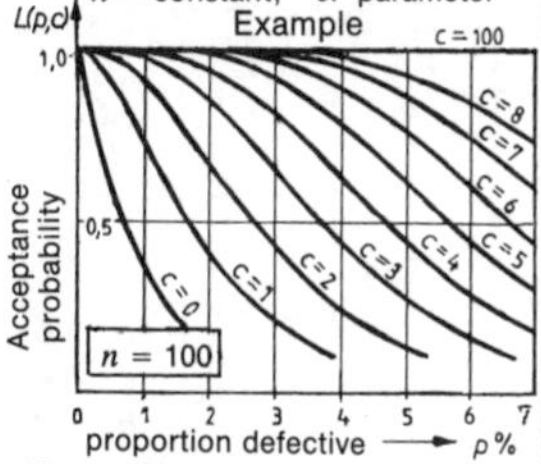

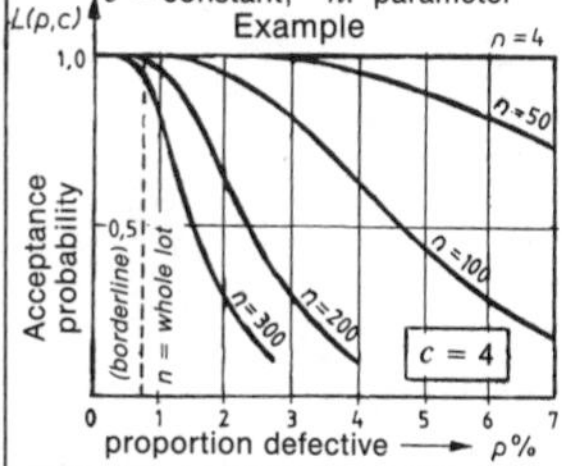

Note: The smaller the value of c becomes, the nearer the operating characteristic approaches $p = 0$.

c must be $\leq n$

Note: The bigger the value of n, the steeper is the operating characteristic; when $n = N$ the curve is parallel to the ordinate and every article is tested. The steeper the curve the more stringent is the control. n must be $\geq c$.

Acceptable Quality Level (A. Q. L.): Agreement between the producer and the user leads to the most important point on the operating characteristic, the AQL-value. The manufacturer needs to be assured that the method of sampling will accurately predict the quality of the lot. If this has a probability of 90%, then the producer risk from g 62:

$$L(p,c) \geq 1 - \alpha = 1 - 0{\cdot}9 = 10\%,$$

but the method of sampling may increase the producer risk. To overcome it, the producer may decide to hold his failure rate well below the agreed value of the AQL to say $p_0{}^{*)}$ which gives a permitted failure of c_1, in the sample as shown in the graph of $L(p,c)$ against p, which is less than c_2, the value originally required. As a result the probability of success in the lot rises to 99%. In practice, the AQL has a value of about 0·65.

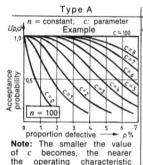

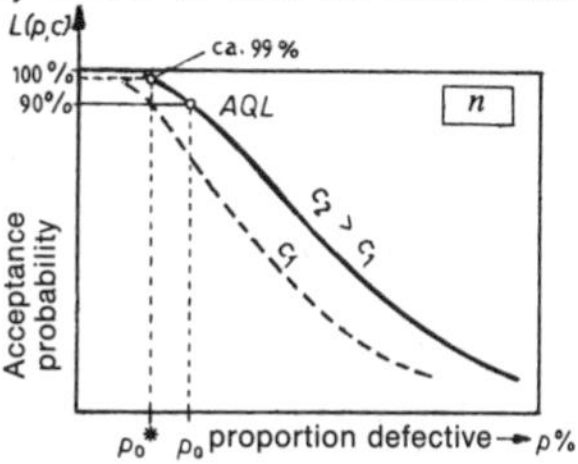

n : number in random samples

c : number of the maximum admissible defective parts

General definitions

g 65 Reliability $$R(t) = \frac{n(t)}{n_0} = e^{-\int_0^t \lambda(\tau)\, d\tau}$$

g 66 Probability to failure $$F(t) = 1 - R(t)$$

g 67 Failure density $$f(t) = -\frac{dR}{dt}$$
$$= \lambda(t)\, e^{-\int_0^t \lambda(\tau)\, d\tau}$$

g 68 Failure rate $$\lambda(t) = \frac{f(t)}{R(t)} = -\frac{1}{R(t)}\frac{dR}{dt}$$

MTTF (**m**ean **t**ime **t**o **f**ailure)

g 69 $$\text{MTTF} = \int_0^\infty f(t)\, t\, dt = \int_0^\infty R(t)\, dt$$

In systems which can be repaired, MTTF is replaced by the mean time between two errors, the mean failure distance m = MTBF (**m**ean **t**ime **b**etween **f**ailures). Values of MTTF and MTBF are equal.

g 70 $$\text{MTTF} = \text{MTBF} = m = \int_0^\infty R(t)\, dt$$

Product rule for the reliability R_s:

When $R_1 \ldots R_n$ are the reliabilities of the single elements 1 ... n, the reliability of the whole system becomes:

g 71 $$R_S = R_1\, R_2 \;\ldots\; R_n = \prod_{i=1}^{n} R_i$$

g 72 $$= e^{-\int_0^t [\lambda_1(\tau) + \lambda_2(\tau) \ldots \lambda_n(\tau)]\, d\tau}$$

Note

Expressions for the reliability functions $R(t)$ are the distribution functions $F(x)$ in tables G 4 and G 5 (for calculation use g 66). The exponential distribution, simple to calculate, usually fulfills the requirements (λ = const).

$n(t)$: number of elements at the time t
n_0 : number of elements at the beginning

Exponential distribution used as reliability function

g 73 Reliability $R(t) = e^{-\lambda t}$

g 74 Probability to failure $F(t) = 1 - e^{-\lambda t}$

g 75 Failure density $f(t) = \lambda \, e^{-\lambda t}$

g 76 Failure rate $\lambda(t) = \dfrac{f(t)}{R(t)} = \lambda = \text{const.}$

(Dimension: 1/time)

g 77 Failure distance (MTBF) $m = \int_0^\infty e^{-\lambda t} \, dt = \dfrac{1}{\lambda}$

Product rule for the reliability R_S:

g 78 $R_S = e^{-\lambda_1 t} \, e^{-\lambda_2 t} \cdot \ldots \, e^{-\lambda_n t}$

g 79 $= e^{-(\lambda_1 + \lambda_2 + \ldots + \lambda_n)t}$

g 80 Cumulative failure rate $\lambda_S = \lambda_1 + \lambda_2 + \ldots + \lambda_n = \dfrac{1}{\text{MTBF}}$

For small values the failure rate can be calculated approximately

g 81 $\lambda = \dfrac{\text{number of defectives}}{\text{number of elements at the beginning} \times \text{working time}}$

λ-values are mostly related to working hours

g 82 Unit: 1 fit = 1 failure/10^9 hours

Typical examples for failure rate λ in fit:

IC-digital bipolar (SSI)	15	Resistor-metal	1
IC-analog bipolar (OpAmp)	100	Resistor-wire wound	10
Transistor-Si-Universal	20	Small transformer	5
Transistor-Si-Power	200	HF-cool	1
Diode-Si	5	Quartz	10
Tantalum capacitor with liquid electrolyte	20	Light emitting diode (≙ luminous intensity is reduced to 50%)	500
Tantalum capacitor with solid electrolyte	5		
Alu-electrolytic capacitor	20	Soldered connection	0·5
Ceramic (multilayer) capacitor	10	Wrapped connection	0·0025
Paper capacitor	2	Crimped connection	0·26
Vulcanite capacitor	1	plug-in contact	0·3
Resistor-carbon ≥ 100 kΩ	5	plug-in socket per used contact	0·4
Resistor-carbon ≤ 100 kΩ	0·5	plug-in switch	5 . . . 30

Note: Specifications for reliability are DIN 29500, page 1, DIN 40040 and DIN 41611.

Differential coefficients (or derivatives)

Gradient of a curve

The gradient of a curve $y = f(x)$ varies from point to point. By the gradient of a curve at point P we mean the gradient of the tangent at the point. If x and y have equal dimensions – which is not the case in most technical diagrams – and are presented at equal scales, the gradient may be expressed by the tangent of angle α between the tangent at point P and the horizontal axis:

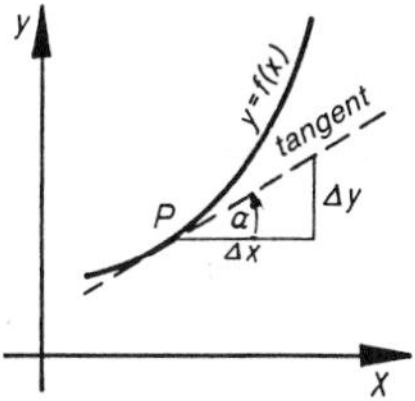

$$m = \tan \alpha$$

h 1 Always applicable is gradient: $m = \dfrac{\Delta y}{\Delta x}$

Difference coefficient

The difference coefficient or mean gradient of the function $y = f(x)$ between PP_1 is:

h 2 $$\frac{\Delta y}{\Delta x} = \frac{f(x+\Delta x)-f(x)}{\Delta x}$$

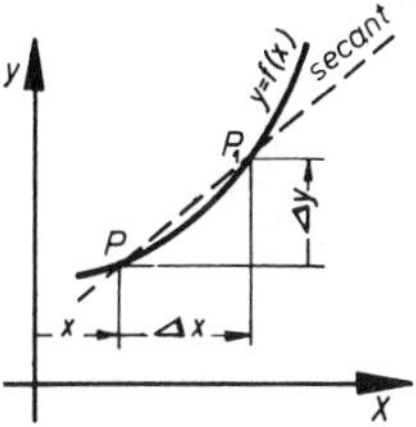

Differential coefficient

Where Δx is infinitely small, i.e. where Δx approaches zero, the slope at P becomes the limiting value of the slope of one of the secants. This slope is the "derivative" or "differential coefficient" of the function at P.

h 3 $$y' = \frac{dy}{dx} = f'(x)$$

$$y' = \lim_{\Delta x \to 0} \frac{\Delta y}{\Delta x} = \lim_{\Delta x \to 0} \frac{f(x+\Delta x)-f(x)}{\Delta x} = \frac{dy}{dx} = f'(x)$$

Geometric meaning of derivative

Gradient of a curve

If, for each point x of a curve, we plot its corresponding gradient as an ordinate y', we obtain the first gradient curve $y' = f'(x)$ or the first derivative of the original curve $y = f(x)$. If we now take the derivative of the first gradient $y' = f'(x)$ we obtain $y'' = f''(x)$ or the second derivative of the original curve $y = f(x)$ etc.

Example: $y = Ax^3 + Bx^2 + Cx + D$

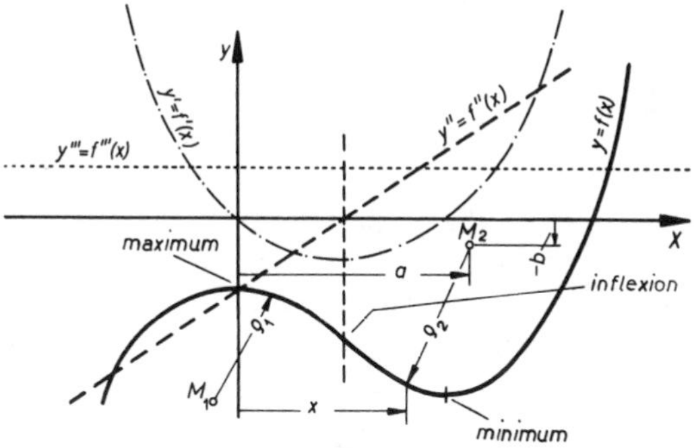

Radius of curvature ϱ at any point x

h 4 $$\varrho = \frac{\sqrt{(1 + y'^2)^3}}{y''}$$

M is below the curve where ϱ is $-$
M is above the curve where ϱ is $+$

Centre coordinates for radius ϱ

h 5 $$a = x - \frac{1 + y'^2}{y''}\,y'$$

h 6 $$b = y + \frac{1 + y'^2}{y''}$$

continued on H 3

Determination of minima, maxima and inflexions

Minima and maxima

The value $x = a$ obtained for $y' = 0$ is inserted in y''.

h 7 For $y''(a) > 0$ there is a minimum at $x = a$,
h 8 For $y''(a) < 0$ there is a maximum at $x = a$,
h 9 For $y''(a) = 0$ see h 19.

Inflexion

The value $x = a$ obtained for $y'' = 0$ is inserted in y'''.

h 10 For $y'''(a) \neq 0$ there is an inflexion at $x = a$.

Shape of the curve $y = f(x)$

Rise and fall

h 11 $y'(x) > 0$ $y(x)$ increases as x increases
h 12 $y'(x) < 0$ $y(x)$ decreases as x increases
h 13 $y'(x) = 0$ $y(x)$ is tangentially parallel the x-axis at x

Curve

h 14 $y''(x) < 0$ $y(x)$ is convex (viewed from above)
h 15 $y''(x) > 0$ $y(x)$ is concave (viewed from above)
h 16 $y''(x) = 0$ with / without a change of sign $y'(x)$ at x has a flexion / bottom point

Exceptional case

Where at a point $x = a$

h 17 $y'(a) = y''(a) = y'''(a) = \ldots y^{(n-1)}(a) = 0$, but
h 18 $y^{n}(a) \neq 0$, one of the 4 conditions is present:

n = even number		n = uneven number	
h 19 $y^{(n)}(a) > 0$	$y^{(n)}(a) < 0$	$y^{(n)}(a) > 0$	$y^{(n)}(a) < 0$

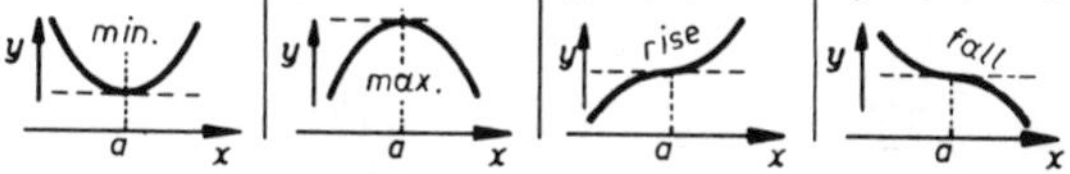

Derivatives

Basic rules

	function	derivative
h 21	$y = c\,x^n + C$	$y' = c\,n\,x^{n-1}$
h 22	$y = u(x) \pm v(x)$	$y' = u'(x) \pm v'(x)$
h 23	$y = u(x)\,v(x)$	$y' = u'v + u\,v'$
h 24	$y = \dfrac{u(x)}{v(x)}$	$y' = \dfrac{u'v - u\,v'}{v^2}$
h 25	$y = \sqrt{x}$	$y' = \dfrac{1}{2\sqrt{x}}$
h 26	$y = u(x)^{v(x)}$	$y' = u^v\left(\dfrac{u'v}{u} + v'\ln u\right)$

Derivative of a function of a function
(chain rule)

h 27 $y = f\left[u(x)\right]$ | $y' = f'(u)\,u'(x) = \dfrac{dy}{dx} = \dfrac{dy}{du}\,\dfrac{du}{dx}$

Parametric form of derivative

h 28 $y = f(x)$ $\quad \begin{cases} x = f(t) \\ y = f(t) \end{cases}$ | $y' = \dfrac{dy}{dt}\,\dfrac{dt}{dx} = \dfrac{\dot{y}}{\dot{x}}$

h 29 | $y'' = \dfrac{d^2y}{dx^2} = \dfrac{\dot{x}\ddot{y} - \dot{y}\ddot{x}}{\dot{x}^3}$

Derivative of inverse functions

The equation $y = f(x)$ solved for x gives the inverse function $x = \varphi(y)$.

h 30 $x = \varphi(y)$ | $f'(x) = \dfrac{1}{\varphi'(x)}$

Example:

h 31 $y = f(x) = \arccos x$

h 32 gives $x = \varphi(y) = \cos y$ | $f'(x) = \dfrac{1}{-\sin y} = -\dfrac{1}{\sqrt{1-x^2}}$

Derivatives

Exponential functions

	function	derivative
h 33	$y = e^x$	$y' = e^x = y'' = \dots$
h 34	$y = e^{-x}$	$y' = -e^{-x}$
h 35	$y = e^{ax}$	$y' = a\, e^{ax}$
h 36	$y = x\ e^x$	$y' = e^x\ (1 + x)$
h 37	$y = \sqrt{e^x}$	$y' = \frac{\sqrt{e^x}}{2}$
h 38	$y = a^x$	$y' = a^x \ln a$
h 39	$y = a^{nx}$	$y' = n\, a^{nx} \ln a$
h 40	$y = a^{x^2}$	$y' = a^{x^2}\ 2x \ln a$

Trigonometrical functions

	function	derivative
h 41	$y = \sin x$	$y' = \cos x$
h 42	$y = \cos x$	$y' = -\sin x$
h 43	$y = \tan x$	$y' = \frac{1}{\cos^2 x} = 1 + \tan^2 x = \sec^2 x$
h 44	$y = \cot x$	$y' = \frac{-1}{\sin^2 x} = -(1 + \cot^2 x) = -\mathrm{cosec}^2 x$
h 45	$y = a \sin(kx)$	$y' = a\ k \cos(kx)$
h 46	$y = a \cos(kx)$	$y' = -a\ k \sin(kx)$
h 47	$y = \sin^n x$	$y' = n \sin^{n-1} x \cos x$
h 48	$y = \cos^n x$	$y' = -n \cos^{n-1} x \sin x$
h 49	$y = \tan^n x$	$y' = n \tan^{n-1} x\ (1 + \tan^2 x)$
h 50	$y = \cot^n x$	$y' = -n \cot^{n-1} x\ (1 + \cot^2 x)$
h 51	$y = \frac{1}{\sin x}$	$y' = \frac{-\cos x}{\sin^2 x}$
h 52	$y = \frac{1}{\cos x}$	$y' = \frac{\sin x}{\cos^2 x}$

Derivatives

Logarithmic functions

	function	derivative
h 53	$y = \ln x$	$y' = \frac{1}{x}$
h 54	$y = \log_a x$	$y' = \frac{1}{x \times \ln a}$
h 55	$y = \ln (1 \pm x)$	$y' = \frac{\pm 1}{1 \pm x}$
h 56	$y = \ln x^n$	$y' = \frac{n}{x}$
h 57	$y = \ln \sqrt{x}$	$y' = \frac{1}{2x}$

Hyperbolic functions

	function	derivative
h 58	$y = \sinh x$	$y' = \cosh x$
h 59	$y = \cosh x$	$y' = \sinh x$
h 60	$y = \tanh x$	$y' = \frac{1}{\cosh^2 x}$
h 61	$y = \coth x$	$y' = \frac{-1}{\sinh^2 x}$

Inverse trigonometrical functions

	function	derivative
h 62	$y = \arcsin x$	$y' = \frac{1}{\sqrt{1 - x^2}}$
h 63	$y = \arccos x$	$y' = -\frac{1}{\sqrt{1 - x^2}}$
h 64	$y = \arctan x$	$y' = \frac{1}{1 + x^2}$
h 65	$y = \text{arccot}\ x$	$y' = -\frac{1}{1 + x^2}$
h 66	$y = \text{arsinh}\ x$	$y' = \frac{1}{\sqrt{x^2 + 1}}$
h 67	$y = \text{arcosh}\ x$	$y' = \frac{1}{\sqrt{x^2 - 1}}$
h 68	$y = \text{artanh}\ x$	$y' = \frac{1}{1 - x^2}$
h 69	$y = \text{arcoth}\ x$	$y' = \frac{1}{1 - x^2}$

Integration

Integration, inverse of differentiation

By integral calculus we mean the problem of finding a function $F(x)$ given $y = f(x)$, such that the derivate of $F(x)$ is equal to $f(x)$.

Thus

i 1 $$F'(x) = \frac{dF(x)}{dx} = f(x)$$

hence, we define

the indefinite integral

i 2 $$\int f(x)\,dx = F(x) + C$$

Here C is an unknown constant which disappears on differentiation, since the derivative of a constant equals zero.

Geometric interpretation of the indefinite integral

As this figure shows, there are an infinite number of curves $y = F(x)$ with gradient $y' = f(x)$. All $y = F(x)$ curves are the same shape, but intersect the x-axis at different points. the constant C, however, establishes a fixed curve. If the curve is to pass through the point x_0/y_0, then

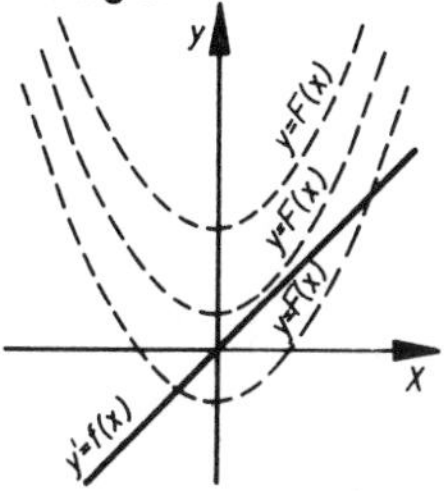

i 3 $$C = y_0 - F(x_0)$$

The definite integral

The definite integral is represented by

i 4 $$\int_a^b f(x)\,dx = F(x)\Big|_a^b = F(b) - F(a)$$

Here integration takes place between the limits a und b. The results obtained by substituting b and a are subtracted causing the constant C to disappear.

Integration

Basic formulae

i 5 $$\int x^n \, dx = \frac{x^{n+1}}{n+1} + C, \quad \text{here} \quad n \neq -1$$

i 6 $$\int \frac{dx}{x} = \ln|x| + C$$

i 7 $$\int [u(x) \pm v(x)] \, dx = \int u(x) \, dx \pm \int v(x) \, dx$$

i 8 $$\int \frac{u'(x)}{u(x)} \, dx = \ln |u(x)| + C$$

i 9 $$\int u(x) \cdot u'(x) \, dx = \frac{1}{2} [u(x)]^2 + C$$

Integration by parts

i10 $$\int u(x) \cdot v'(x) \, dx = u(x) \times v(x) - \int u'(x) \times v(x) \, dx$$

Integration by substitution

i11 $$\int f(x) \, dx = \int f[\varphi(z)] \, \varphi'(z) \, dz$$

here $x = \varphi(z)$ and $dx = \varphi'(z) \, dz$

Example:

i12 $$F(x) = \int \sqrt{3x - 5} \, dx.$$

Where $3x - 5 = z$, the derivative is $z' = \frac{dz}{dx} = 3.$

Thus $dx = \frac{dz}{3}$ expressed in terms of z, the integral becomes

$$F(x) = \frac{1}{3} \int \sqrt{z} \, dz = \frac{2}{9} z\sqrt{z} + C.$$ Insert value of z in above

expression: $$F(x) = \frac{1}{3} \int \sqrt{z} \, dz = \frac{2}{9} z \sqrt{z} + C.$$

Integrals

(omitting integral constant C)

i 13 $$\int \frac{1}{x^n}\,dx = -\frac{1}{n-1} \times \frac{1}{x^{n-1}} \qquad (n \neq 1)$$

i 14 $$\int a^{bx}\,dx = \frac{1}{b} \times \frac{a^{bx}}{\ln|a|}$$

i 15 $$\int \ln x\,dx = x\ln|x| - x$$

i 16 $$\int (\ln x)^2\,dx = x(\ln|x|)^2 - 2x\ln|x| + 2x$$

i 17 $$\int \frac{dx}{\ln x} = \ln|(\ln|x|)| + \ln|x| + \frac{(\ln|x|)^2}{2\times 2!} + \frac{(\ln|x|)^3}{3\times 3!} + \ldots$$

i 18 $$\int x\ln x\,dx = x^2\left[\frac{\ln|x|}{2} - \frac{1}{4}\right]$$

i 19 $$\int x^m \ln x\,dx = x^{m+1}\left[\frac{\ln|x|}{m+1} - \frac{1}{(m+1)^2}\right] \qquad (m \neq -1)$$

i 20 $$\int \frac{dx}{x\ln x} = \ln|(\ln|x|)|$$

i 21 $$\int e^{ax}\,dx = \frac{1}{a}e^{ax}$$

i 22 $$\int x e^{ax}\,dx = \frac{e^{ax}}{a^2}(ax - 1)$$

i 23 $$\int x^2 e^{ax}\,dx = e^{ax}\left(\frac{x^2}{a} - \frac{2x}{a^2} + \frac{2}{a^3}\right)$$

i 24 $$\int x^n e^{ax}\,dx = \frac{1}{a}x^n e^{ax} - \frac{n}{a}\int x^{n-1}e^{ax}\,dx$$

i 25 $$\int \frac{e^{ax}}{x}\,dx = \ln|x| + \frac{ax}{1\times 1!} + \frac{(ax)^2}{2\times 2!} + \frac{(ax)^3}{3\times 3!} + \ldots$$

i 26 $$\int \frac{e^{ax}}{x^n}\,dx = \frac{1}{n-1}\left(-\frac{e^{ax}}{x^{n-1}} + a\int \frac{e^{ax}}{x^{n-1}}\,dx\right) \qquad (n \neq 1)$$

i 27 $$\int \frac{dx}{1+e^{ax}} = \frac{1}{a}\ln\left|\frac{e^{ax}}{1+e^{ax}}\right|$$

i 28 $$\int \frac{dx}{b+c\,e^{ax}} = \frac{x}{b} - \frac{1}{ab}\ln|b+ce^{ax}|$$

Integrals
(omitting integral constant C)

i 29 $$\int \frac{e^{ax}\,dx}{b+c\,e^{ax}} = \frac{1}{ac}\ln\left|b+ce^{ax}\right|$$

i 30 $$\int e^{ax}\ln x\,dx = \frac{e^{ax}\ln|x|}{a} - \frac{1}{a}\int \frac{e^{ax}}{x}\,dx$$

i 31 $$\int e^{ax}\sin bx\,dx = \frac{e^{ax}}{a^2+b^2}(a\sin bx - b\cos bx)$$

i 32 $$\int e^{ax}\cos bx\,dx = \frac{e^{ax}}{a^2+b^2}(a\cos bx + b\sin bx)$$

---

i 33 $$\int \frac{dx}{ax+b} = \frac{1}{a}\ln|ax+b|$$

i 34 $$\int \frac{dx}{(ax+b)^n} = -\frac{1}{a(n-1)(ax+b)^{n-1}} \qquad (n \neq 1)$$

i 35 $$\int \frac{dx}{ax-b} = \frac{1}{a}\ln|ax-b|$$

i 36 $$\int \frac{dx}{(ax-b)^n} = -\frac{1}{a(n-1)(ax-b)^{n-1}} \qquad (n \neq 1)$$

i 37 $$\int \frac{dx}{(ax+b)(cx+d)} = \frac{1}{bc-ad}\ln\left|\frac{cx+d}{ax+b}\right| \qquad (bc-ad \neq 0)$$

i 38 $$\int \frac{dx}{(ax-b)(cx-d)} = \frac{1}{ad-bc}\ln\left|\frac{cx-d}{ax-b}\right| \qquad (ad-bc \neq 0)$$

---

i 39 $$\int \frac{x\,dx}{(ax+b)(cx+d)} = \frac{1}{bc-ad}\left[\frac{b}{a}\ln|ax+b| - \frac{d}{c}\ln|cx+d|\right] \qquad (bc-ad \neq 0)$$

i 40 $$\int \frac{x\,dx}{ax+b} = \frac{x}{a} - \frac{b}{a^2}\ln|ax+b|$$

i 41 $$\int \frac{x^2\,dx}{ax+b} = \frac{1}{a^3}\left[\frac{1}{2}(ax+b)^2 - 2b(ax+b) + b^2\ln|ax+b|\right]$$

i 42 $$\int \frac{x^3\,dx}{ax+b} = \frac{1}{a^4}\left[\frac{(ax+b)^3}{3} - \frac{3b(ax+b)^2}{2} + 3b^2(ax+b) - b^3\ln|ax+b|\right]$$

i 43 $$\int \frac{dx}{x(ax+b)} = -\frac{1}{b}\ln\left|a+\frac{b}{x}\right|$$

i 44 $$\int \frac{dx}{x^2(ax+b)} = -\frac{1}{bx} + \frac{a}{b^2}\ln\left|a+\frac{b}{x}\right|$$

Integrals

(omitting integral constant C)

i 45 $$\int \frac{dx}{x^3(ax+b)} = -\frac{1}{b^3}\left[a^2 \ln\left|\frac{ax+b}{x}\right| - \frac{2a(ax+b)}{x} + \frac{(ax+b)^2}{2x^2}\right]$$

i 46 $$\int \frac{x\,dx}{(ax+b)^2} = \frac{b}{a^2(ax+b)} + \frac{1}{a^2}\ln|ax+b|$$

i 47 $$\int \frac{x^2\,dx}{(ax+b)^2} = \frac{1}{a^3}\left[(ax+b) - 2b\ln|ax+b| - \frac{b^2}{ax+b}\right]$$

i 48 $$\int \frac{x^3\,dx}{(ax+b)^2} = \frac{1}{a^4}\left[\frac{(ax+b)^2}{2} - 3b(ax+b) + 3b^2\ln|ax+b| + \frac{b^3}{ax+b}\right]$$

i 49 $$\int \frac{x\,dx}{(ax+b)^3} = \frac{1}{a^2}\left[-\frac{1}{ax+b} + \frac{b}{2(ax+b)^2}\right]$$

i 50 $$\int \frac{x^2\,dx}{(ax+b)^3} = \frac{1}{a^3}\left[\ln|ax+b| + \frac{2b}{ax+b} - \frac{b^2}{2(ax+b)^2}\right]$$

i 51 $$\int \frac{x^3\,dx}{(ax+b)^3} = \frac{1}{a^4}\left[(ax+b) - 3b\ln|ax+b| - \frac{3b^2}{ax+b} + \frac{b^3}{2(ax+b)^2}\right]$$

i 52 $$\int \frac{dx}{x(ax+b)^2} = -\frac{1}{b^2}\left(\ln\left|\frac{ax+b}{x}\right| + \frac{ax}{ax+b}\right) = -\frac{1}{b^2}\left[\ln\left|a+\frac{b}{x}\right| + \frac{ax}{ax+b}\right]$$

i 53 $$\int \frac{dx}{x^2(ax+b)^2} = -a\left[\frac{1}{b^2(ax+b)} + \frac{1}{ab^2x} - \frac{2}{b^3}\ln\left|\frac{ax+b}{x}\right|\right]$$

i 54 $$\int \frac{dx}{x^3(ax+b)^2} = -\frac{1}{b^4}\left[3a^2\ln\left|\frac{ax+b}{x}\right| + \frac{a^3x}{ax+b} + \frac{(ax+b)^2}{2x^2} - \frac{3a(ax+b)}{x}\right]$$

i 55 $$\int \frac{dx}{a^2+x^2} = \frac{1}{a}\arctan\frac{x}{a}$$

i 56 $$\int \frac{x\,dx}{a^2+x^2} = \frac{1}{2}\ln|a^2+x^2|$$

i 57 $$\int \frac{x^2\,dx}{a^2+x^2} = x - a \times \arctan\frac{x}{a}$$

i 58 $$\int \frac{x^3\,dx}{a^2+x^2} = \frac{x^2}{2} - \frac{a^2}{2}\ln|a^2+x^2|$$

i 59 $$\int \frac{dx}{a^2-x^2} = -\int \frac{dx}{x^2-a^2} = \frac{1}{a}\frac{1}{2}\ln\left|\frac{a+x}{a-x}\right|$$

i 60 $$\int \frac{x\,dx}{a^2-x^2} = -\int \frac{x\,dx}{x^2-a^2} = -\frac{1}{2}\ln|a^2-x^2|$$

Integrals
(omitting integral constant C)

i 61 $\int \frac{x^2\,dx}{a^2 - x^2} = -\int \frac{x^2\,dx}{x^2 - a^2} = -x + a\frac{1}{2}\ln\left|\frac{a+x}{a-x}\right|$

i 62 $\int \frac{x^3\,dx}{a^2 - x^2} = -\int \frac{x^3\,dx}{x^2 - a^2} = -\frac{x^2}{2} - \frac{a^2}{2}\ln\left|a^2 - x^2\right|$

i 63 $\int \frac{dx}{(a^2 + x^2)^2} = \frac{x}{2a^2(a^2 + x^2)} + \frac{1}{2a^3}\arctan\frac{x}{a}$

i 64 $\int \frac{x\,dx}{(a^2 + x^2)^2} = -\frac{1}{2(a^2 + x^2)}$

i 65 $\int \frac{x^2\,dx}{(a^2 + x^2)^2} = -\frac{x}{2(a^2 + x^2)} + \frac{1}{2a}\arctan\frac{x}{a}$

i 66 $\int \frac{x^3\,dx}{(a^2 + x^2)^2} = \frac{a^2}{2(a^2 + x^2)} + \frac{1}{2}\ln\left|a^2 + x^2\right|$

i 67 $\int \frac{dx}{(a^2 + x^2)^n} = \frac{x}{2a^2(n-1)(a^2+x^2)^{n-1}} + \frac{2n-3}{2a^2(n-1)}\int \frac{dx}{(a^2+x^2)^{n-1}} \quad (n \neq 1)$

i 68 $\int \frac{dx}{(a^2 - x^2)^2} = \frac{x}{2a^2(a^2 - x^2)} + \frac{1}{2a^3}\,\frac{1}{2}\ln\left|\frac{a+x}{a-x}\right|$

i 69 $\int \frac{x\,dx}{(a^2 - x^2)^2} = \frac{1}{2(a^2 - x^2)}$

i 70 $\int \frac{x^2\,dx}{(a^2 - x^2)^2} = \frac{x}{2(a^2 - x^2)} - \frac{1}{2a}\,\frac{1}{2}\ln\left|\frac{a+x}{a-x}\right|$

i 71 $\int \frac{x^3\,dx}{(a^2 - x^2)^2} = \frac{a^2}{2(a^2 - x^2)} + \frac{1}{2}\ln\left|a^2 - x^2\right|$

i 72 $\int \sqrt{x}\,dx = \frac{2}{3}\sqrt{x^3}$

i 73 $\int \sqrt{ax + b}\,dx = \frac{2}{3a}\sqrt{(ax + b)^3}$

i 74 $\int x\sqrt{ax + b}\,dx = \frac{2(3ax - 2b)\sqrt{(ax + b)^3}}{15a^2}$

i 75 $\int x^2\sqrt{ax + b}\,dx = \frac{2(15a^2x^2 - 12abx + 8b^2)\sqrt{(ax + b)^3}}{105a^3}$

i 76 $\int \frac{dx}{\sqrt{x}} = 2\sqrt{x}$

Integrals

(omitting integral constant C)

i 77 $$\int \frac{dx}{\sqrt{ax+b}} = \frac{2\sqrt{(ax+b)}}{a}$$

i 78 $$\int \frac{x\,dx}{\sqrt{ax+b}} = \frac{2(ax-2b)}{3a^2}\sqrt{(ax+b)}$$

i 79 $$\int \frac{x^2\,dx}{\sqrt{ax+b}} = \frac{2(3a^2x^2-4abx+8b^2)\sqrt{(ax+b)}}{15a^3}$$

i 80 $$\int \sqrt{a^2+x^2}\,dx = \frac{x}{2}\sqrt{a^2+x^2} + \frac{a^2}{2}\operatorname{arsinh}\frac{x}{a}$$

i 81 $$\int x\sqrt{a^2+x^2}\,dx = \frac{1}{3}\sqrt{(a^2+x^2)^3}$$

i 82 $$\int x^2\sqrt{a^2+x^2}\,dx = \frac{x}{4}\sqrt{(a^2+x^2)^3} - \frac{a^2}{8}\left(x\sqrt{a^2+x^2} + a^2\operatorname{arsinh}\frac{x}{a}\right)$$

i 83 $$\int x^3\sqrt{a^2+x^2}\,dx = \frac{\sqrt{(a^2+x^2)^5}}{5} - \frac{a^2\sqrt{(a^2+x^2)^3}}{3}$$

i 84 $$\int \frac{\sqrt{a^2+x^2}}{x}\,dx = \sqrt{a^2+x^2} - a\ln\left|\frac{a+\sqrt{a^2+x^2}}{x}\right|$$

i 85 $$\int \frac{\sqrt{a^2+x^2}}{x^2}\,dx = -\frac{\sqrt{a^2+x^2}}{x} + \operatorname{arsinh}\frac{x}{a}$$

i 86 $$\int \frac{\sqrt{a^2+x^2}}{x^3}\,dx = -\frac{\sqrt{a^2+x^2}}{2x^2} - \frac{1}{2a}\ln\left|\frac{a+\sqrt{a^2+x^2}}{x}\right|$$

i 87 $$\int \frac{dx}{\sqrt{a^2+x^2}} = \operatorname{arsinh}\frac{x}{a}$$

i 88 $$\int \frac{x\,dx}{\sqrt{a^2+x^2}} = \sqrt{a^2+x^2}$$

i 89 $$\int \frac{x^2\,dx}{\sqrt{a^2+x^2}} = \frac{x}{2}\sqrt{a^2+x^2} - \frac{a^2}{2}\operatorname{arsinh}\frac{x}{a}$$

i 90 $$\int \frac{x^3\,dx}{\sqrt{a^2+x^2}} = \frac{\sqrt{(a^2+x^2)^3}}{3} - a^2\sqrt{x^2+a^2}$$

i 91 $$\int \frac{dx}{x\sqrt{a^2+x^2}} = -\frac{1}{a}\ln\left|\frac{a+\sqrt{a^2+x^2}}{x}\right|$$

i 92 $$\int \frac{dx}{x^2\sqrt{a^2+x^2}} = -\frac{\sqrt{x^2+a^2}}{a^2x}$$

Integrals

(omitting integral constant C)

i 93 $$\int \frac{dx}{x^3\sqrt{a^2+x^2}} = -\frac{\sqrt{a^2+x^2}}{2x^2a^2} + \frac{1}{2a^3}\ln\left|\frac{a+\sqrt{a^2+x^2}}{x}\right|$$

i 94 $$\int \sqrt{a^2-x^2}\,dx = \frac{1}{2}\left[x\sqrt{a^2-x^2} + a^2\arcsin\frac{x}{a}\right]$$

i 95 $$\int x\sqrt{a^2-x^2}\,dx = -\frac{1}{3}\sqrt{(a^2-x^2)^3}$$

i 96 $$\int x^2\sqrt{a^2-x^2}\,dx = -\frac{x}{4}\sqrt{(a^2-x^2)^3} + \frac{a^2}{8}\left(x\sqrt{a^2-x^2} + a^2\arcsin\frac{x}{a}\right)$$

i 97 $$\int x^3\sqrt{a^2-x^2}\,dx = \frac{\sqrt{(a^2-x^2)^5}}{5} - a^2\frac{\sqrt{(a^2-x^2)^3}}{3}$$

i 98 $$\int \frac{dx}{\sqrt{a^2-x^2}} = \arcsin\frac{x}{a}$$

i 99 $$\int \frac{x\,dx}{\sqrt{a^2-x^2}} = -\sqrt{a^2-x^2}$$

i 100 $$\int \frac{x^2\,dx}{\sqrt{a^2-x^2}} = -\frac{x}{2}\sqrt{a^2-x^2} + \frac{a^2}{2}\arcsin\frac{x}{a}$$

i 101 $$\int \frac{x^3\,dx}{\sqrt{a^2-x^2}} = \frac{\sqrt{(a^2-x^2)^3}}{3} - a^2\sqrt{a^2-x^2}$$

i 102 $$\int \frac{dx}{x\sqrt{a^2-x^2}} = -\frac{1}{a}\ln\left|\frac{a+\sqrt{a^2-x^2}}{x}\right|$$

i 103 $$\int \frac{dx}{x^2\sqrt{a^2-x^2}} = -\frac{\sqrt{a^2-x^2}}{a^2x}$$

i 104 $$\int \frac{dx}{x^3\sqrt{a^2-x^2}} = -\frac{\sqrt{a^2-x^2}}{2a^2x^2} - \frac{1}{2a^3}\ln\left|\frac{a+\sqrt{a^2-x^2}}{x}\right|$$

i 105 $$\int \sqrt{x^2-a^2}\,dx = \frac{1}{2}\left(x\sqrt{x^2-a^2} - a^2\operatorname{arcosh}\frac{x}{a}\right)$$

i 106 $$\int x\sqrt{x^2-a^2}\,dx = \frac{1}{3}\sqrt{(x^2-a^2)^3}$$

i 107 $$\int x^2\sqrt{x^2-a^2}\,dx = \frac{x}{4}\sqrt{(x^2-a^2)^3} + \frac{a^2}{8}\left(x\sqrt{x^2-a^2} - a^2\operatorname{arcosh}\frac{x}{a}\right)$$

i 108 $$\int x^3\sqrt{x^2-a^2}\,dx = \frac{\sqrt{(x^2-a^2)^5}}{5} + \frac{a^2\sqrt{(x^2-a^2)^3}}{3}$$

Integrals
(omitting integral constant C)

109 $\displaystyle\int \frac{\sqrt{x^2-a^2}}{x}\,dx = \sqrt{x^2-a^2} - a\arccos\frac{a}{x}$

110 $\displaystyle\int \frac{\sqrt{x^2-a^2}}{x^2}\,dx = -\frac{\sqrt{x^2-a^2}}{x} + \operatorname{arcosh}\frac{x}{a}$

111 $\displaystyle\int \frac{\sqrt{x^2-a^2}}{x^3}\,dx = -\frac{\sqrt{x^2-a^2}}{2x^2} + \frac{1}{2a}\arccos\frac{a}{x}$

112 $\displaystyle\int \sin ax\,dx = -\frac{1}{a}\cos ax$

113 $\displaystyle\int \sin^2 ax\,dx = \frac{x}{2} - \frac{1}{4a}\sin 2ax$

114 $\displaystyle\int \sin^3 ax\,dx = -\frac{1}{a}\cos ax + \frac{1}{3a}\cos^3 ax$

115 $\displaystyle\int \sin^n ax\,dx = -\frac{1}{na}\cos ax\,\sin^{n-1} ax + \frac{n-1}{n}\int \sin^{n-2} ax\,dx$

(n is an integer > 0)

116 $\displaystyle\int x\sin ax\,dx = \frac{\sin ax}{a^2} - \frac{x\cos ax}{a}$

117 $\displaystyle\int x^2\sin ax\,dx = \frac{2x}{a^2}\sin ax - \left(\frac{x^2}{a} - \frac{2}{a^3}\right)\cos ax$

118 $\displaystyle\int x^3\sin ax\,dx = \left(\frac{3x^2}{a^2} - \frac{6}{a^4}\right)\sin ax - \left(\frac{x^3}{a} - \frac{6x}{a^3}\right)\cos ax$

119 $\displaystyle\int \frac{\sin ax}{x}\,dx = ax - \frac{(ax)^3}{3\times 3!} + \frac{(ax)^5}{5\times 5!} - \frac{(ax)^7}{7\times 7!} + \ldots$

120 $\displaystyle\int \frac{\sin ax}{x^2}\,dx = -\frac{\sin ax}{x} + a\int \frac{\cos ax}{x}\,dx$

121 $\displaystyle\int \frac{\sin ax}{x^n}\,dx = -\frac{1}{n-1}\times\frac{\sin ax}{x^{n-1}} + \frac{a}{n-1}\int \frac{\cos ax}{x^{n-1}}\,dx$

122 $\displaystyle\int \cos ax\,dx = \frac{1}{a}\sin ax$

123 $\displaystyle\int \cos^2 ax\,dx = \frac{x}{2} + \frac{1}{4a}\sin 2ax$

124 $\displaystyle\int \cos^3 ax\,dx = \frac{1}{a}\sin ax - \frac{1}{3a}\sin^3 ax$

Integrals

(omitting integral constant C)

i 125 $$\int \cos^n ax \; dx = \frac{1}{na} \sin ax \times \cos^{n-1} ax + \frac{n-1}{n} \int \cos^{n-2} ax \; dx$$

i 126 $$\int x \cos ax \; dx = \frac{\cos ax}{a^2} + \frac{x \sin ax}{a}$$

i 127 $$\int x^2 \cos ax \; dx = \frac{2x}{a^2} \cos ax + \left(\frac{x^2}{a} - \frac{2}{a^3}\right) \sin ax$$

i 128 $$\int x^3 \cos ax \; dx = \left(\frac{3x^2}{a^2} - \frac{6}{a^4}\right) \cos ax + \left(\frac{x^3}{a} - \frac{6x}{a^3}\right) \sin ax$$

i 129 $$\int \frac{\cos ax}{x} \; dx = \ln|ax| - \frac{(ax)^2}{2 \times 2!} + \frac{(ax)^4}{4 \times 4!} - \frac{(ax)^6}{6 \times 6!} + \dots$$

i 130 $$\int \frac{\cos ax}{x^2} \; dx = -\frac{\cos ax}{x} - a \int \frac{\sin ax \; dx}{x}$$

i 131 $$\int \frac{\cos ax}{x^n} \; dx = -\frac{\cos ax}{(n-1)x^{n-1}} - \frac{a}{n-1} \int \frac{\sin ax \; dx}{x^{n-1}} \qquad (n \neq 1)$$

i 132 $$\int \tan ax \; dx = -\frac{1}{a} \ln|\cos ax|$$

i 133 $$\int \tan^2 ax \; dx = \frac{1}{a} \tan ax - x$$

i 134 $$\int \tan^n ax \; dx = \frac{\tan^{n-1} ax}{a(n-1)} - \int \tan^{n-2} ax \; dx \qquad (n \neq 1)$$

i 135 $$\int \cot ax \; dx = \frac{1}{a} \ln|\sin ax|$$

i 136 $$\int \cot^2 ax \; dx = -x - \frac{1}{a} \cot ax$$

i 137 $$\int \cot^n ax \; dx = \frac{\cot^{n-1} ax}{a(n-1)} - \int \cot^{n-2} ax \; dx \qquad (n \neq 1)$$

i 138 $$\int \frac{dx}{\sin ax} = \frac{1}{a} \ln\left|\tan \frac{ax}{2}\right|$$

i 139 $$\int \frac{dx}{\sin^2 ax} = -\frac{1}{a} \cot ax$$

i 140 $$\int \frac{dx}{\sin^n ax} = -\frac{1}{a(n-1)} \times \frac{\cos ax}{\sin^{n-1} ax} + \frac{n-2}{n-1} \int \frac{dx}{\sin^{n-2} ax} \qquad (n > 1)$$

Integrals

(omitting integral constant C)

i 141 $$\int \frac{x\,dx}{\sin^2 ax} = -\frac{x}{a}\cot ax + \frac{1}{a^2}\ln|\sin ax|$$

i 142 $$\int \frac{dx}{\cos ax} = \frac{1}{a}\ln\left|\tan\left(\frac{ax}{2}+\frac{\pi}{4}\right)\right|$$

i 143 $$\int \frac{dx}{\cos^2 ax} = \frac{1}{a}\tan ax$$

i 144 $$\int \frac{dx}{\cos^n ax} = \frac{1}{a(n-1)} \times \frac{\sin ax}{\cos^{n-1} ax} + \frac{n-2}{n-1} \cdot \int \frac{dx}{\cos^{n-2} ax} \qquad (n > 1)$$

i 145 $$\int \frac{x\,dx}{\cos^2 ax} = \frac{x}{a}\tan ax + \frac{1}{a^2}\ln|\cos ax|$$

i 146 $$\int \frac{dx}{1+\sin ax} = \frac{1}{a}\tan\left(\frac{ax}{2}-\frac{\pi}{4}\right)$$

i 147 $$\int \frac{dx}{1+\cos ax} = \frac{1}{a}\tan\frac{ax}{2}$$

i 148 $$\int \frac{dx}{1-\sin ax} = -\frac{1}{a}\cot\left(\frac{ax}{2}-\frac{\pi}{4}\right) = \frac{1}{a}\tan\left(\frac{\pi}{4}+\frac{ax}{2}\right)$$

i 149 $$\int \frac{dx}{1-\cos ax} = -\frac{1}{a}\cot\frac{ax}{2}$$

i 150 $$\int \sin ax \times \sin bx\,dx = -\frac{\sin(ax+bx)}{2(a+b)} + \frac{\sin(ax-bx)}{2(a-b)} \qquad (|a| \neq |b|)$$

i 151 $$\int \sin ax \times \cos bx\,dx = -\frac{\cos(ax+bx)}{2(a+b)} - \frac{\cos(ax-bx)}{2(a-b)} \qquad (|a| \neq |b|)$$

i 152 $$\int \cos ax \times \cos bx\,dx = \frac{\sin(ax+bx)}{2(a+b)} + \frac{\sin(ax-bx)}{2(a-b)} \qquad (|a| \neq |b|)$$

i 153 $$\int x^n \sin ax\,dx = -\frac{x^n}{a}\cos ax + \frac{n}{a}\int x^{n-1}\cos ax\,dx$$

i 154 $$\int x^n \cos ax\,dx = \frac{x^n}{a}\sin ax + \frac{n}{a}\int x^{n-1}\sin ax\,dx$$

i 155 $$\int \frac{dx}{\sin ax \cdot \cos ax} = \frac{1}{a}\ln|\tan ax|$$

i 156 $$\int \frac{dx}{\sin^2 ax \cdot \cos ax} = \frac{1}{a}\left[\ln\left|\tan\left(\frac{\pi}{4}+\frac{ax}{2}\right)\right| - \frac{1}{\sin ax}\right]$$

Integrals

(omitting integral constant C)

i 157 $$\int \frac{dx}{\sin^3 ax \cos ax} = \frac{1}{a}\left(\ln|\tan ax| - \frac{1}{2\sin^2 ax}\right)$$

i 158 $$\int \frac{dx}{\cos^2 ax \sin ax} = \frac{1}{a}\left(\ln\left|\tan\frac{ax}{2}\right| + \frac{1}{\cos ax}\right)$$

i 159 $$\int \frac{dx}{\cos^3 ax \sin ax} = \frac{1}{a}\left(\ln|\tan ax| + \frac{1}{2\cos^2 ax}\right)$$

i 160 $$\int \frac{dx}{\sin^2 ax \cos^2 ax} = -\frac{2}{a}\cot 2ax$$

i 161 $$\int \sin^m ax \cos^n ax \, dx = \frac{1}{a(m+n)}\sin^{m+1} ax \cdot \cos^{n-1} ax + \frac{n-1}{m+n}\int \sin^m ax \cdot \cos^{n-2} ax \, dx$$

If n is an odd number, solution for the Remainder-integral:

i 162 $$\int \sin^m ax \cos ax \, dx = \frac{\sin^{m+1} ax}{a(m+1)} \qquad (m \neq -1)$$

i 163 $$\int \arcsin x \, dx = x \arcsin x + \sqrt{1-x^2}$$

i 164 $$\int \arccos x \, dx = x \arccos x - \sqrt{1-x^2}$$

i 165 $$\int \arctan x \, dx = x \arctan x - \frac{1}{2}\ln\left|1+x^2\right|$$

i 166 $$\int \operatorname{arccot} x \, dx = x \operatorname{arccot} x + \frac{1}{2}\ln\left|1+x^2\right|$$

i 167 $$\int \sinh(ax) \, dx = \frac{1}{a}\cosh(ax)$$

i 168 $$\int \sinh^2 x \, dx = \frac{1}{4}\sinh(2x) - \frac{x}{2}$$

i 169 $$\int \sinh^n x \, dx = \frac{1}{n}\cosh x \times \sinh^{n-1} x - \frac{n-1}{n}\int \sinh^{n-2} x \, dx \qquad (n > 0)$$

i 170 $$\int \cosh(ax) \, dx = \frac{1}{a}\sinh(ax)$$

Integrals

(omitting integral constant C)

171 $\int \cosh^2 x \, dx = \frac{1}{4} \sinh(2x) + \frac{x}{2}$

172 $\int \cosh^n x \, dx = \frac{1}{n} \sinh x \cosh^{n-1} x + \frac{n-1}{n} \int \cosh^{n-2} x \, dx \quad (n > 0)$

173 $\int \tanh(ax) \, dx = \frac{1}{a} \ln|\cosh(ax)|$

174 $\int \tanh^2 x \, dx = x - \tanh x$

175 $\int \tanh^n x \, dx = -\frac{1}{n-1} \tanh^{n-1} x + \int \tanh^{n-2} x \, dx \quad (n \neq 1)$

176 $\int \coth(ax) \, dx = \frac{1}{a} \ln|\sinh(ax)|$

177 $\int \coth^2 x \, dx = x - \coth x$

178 $\int \coth^n x \, dx = -\frac{1}{n-1} \coth^{n-1} x + \int \coth^{n-2} x \, dx \quad (n \neq 1)$

179 $\int \frac{dx}{\sinh ax} = \frac{1}{a} \ln\left|\tanh \frac{ax}{2}\right|$

180 $\int \frac{dx}{\sinh^2 x} = -\coth x$

181 $\int \frac{dx}{\cosh ax} = \frac{2}{a} \arctan e^{ax}$

182 $\int \frac{dx}{\cosh^2 x} = \tanh x$

183 $\int \operatorname{arsinh} x \, dx = x \operatorname{arsinh} x - \sqrt{x^2 + 1}$

184 $\int \operatorname{arcosh} x \, dx = x \operatorname{arcosh} x - \sqrt{x^2 - 1}$

185 $\int \operatorname{artanh} x \, dx = x \operatorname{artanh} x + \frac{1}{2} \ln|(1 - x^2)|$

186 $\int \operatorname{arcoth} x \, dx = x \operatorname{arcoth} x + \frac{1}{2} \ln|(x^2 - 1)|$

Arc differential $ds = \sqrt{dx^2 + dy^2} = \sqrt{1 + \left(\frac{dy}{dx}\right)^2}\, dx$

	arc length	surface area where the curve rotates around the x-axis
i 189	$s = \int_a^b \sqrt{1 + y'^2}\; dx$	$A_m = 2\pi \int_a^b y \sqrt{1 + y'^2}\; dx$

static moment of a curve

	x-axis	y-axis
i 190	$M_x = \int_a^b y\sqrt{1+y'^2}\,dx$	$M_y = \int_a^b x\sqrt{1+y'^2}\,dx$

coordinates of centre of gravity

i 191	$x_S = \frac{M_y}{s}$	$y_S = \frac{M_x}{s}$

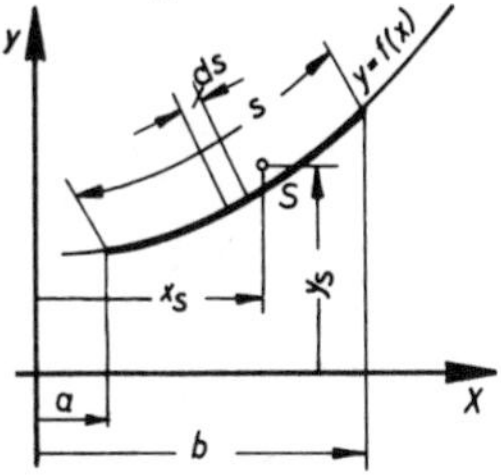

	area	volume of a rotating body where area A rotates around the x-axis	volume of a body, the cross section A_1 of which is a function of x
i 192	$A = \int_a^b y\; dx$	$V = \pi \int_a^b y^2\; dx$	$V = \int_a^b A_1\,(x)\; dx$

static moment of a curve in relation to the

	x-axis	y-axis
i 193	$H_x = \int_a^b \frac{y^2}{2}\,dx$	$H_y = \int_a^b xy\; dx$

coordinates of centre of gravity

i 194	$x_S = \frac{H_y}{A}$	$y_S = \frac{H_x}{A}$

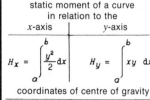

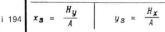

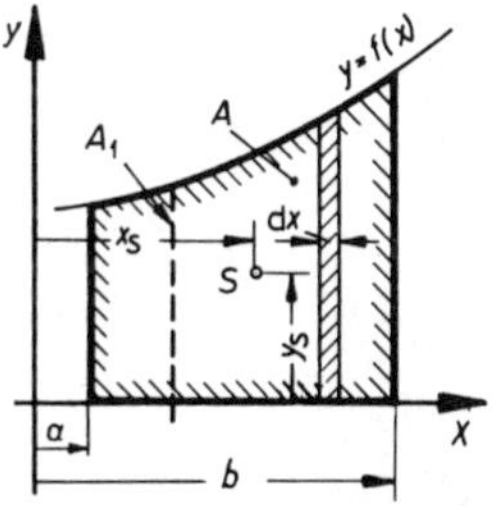

Static moment of a body
(in relation to the y–z plane)

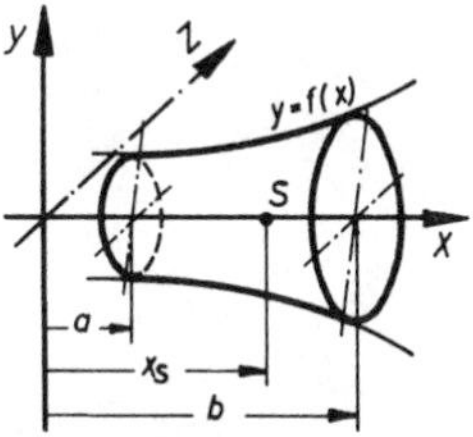

i 195 $$M_{yz} = \pi \int_a^b x \; y^2 \; dx$$

Coordinates of centre of gravity

i 196 $$x_S = \frac{M_{yz}}{V}$$

Pappus theorems

Surface area of a revolving body

i 197 A_m = arc length s times the distance covered by the centre of gravity

$= 2\pi s y_S$ (see also formulae i 189 and i 191)

Volume of a revolving body

i 198 V = area A times the distance covered by the centre of gravity

$= 2\pi A y_S$ (see also formulae i 192 and i 194)

Numerical integration

Division of area into an even number n of strips of equal

i 199 width $h = \frac{b_1 - b_0}{n}$.

Then, according to the

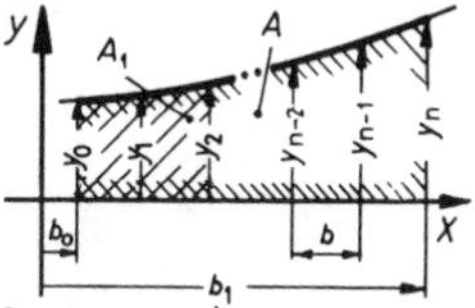

Trapezium rule

i 200 $$A = \frac{b}{2}\,(y_0 + 2y_1 + 2y_2 + \ldots + y_n)$$

Simpson's rule for three ordinates:

i 201 $$A_1 = \frac{b}{3}\,(y_0 + 4y_1 + y_2)$$

Simpson's rule for more than three ordinates:

i 202 $$A = \frac{b}{3}\left[y_0 + y_n + 2(y_2 + y_4 + \ldots + y_{n-2}) + 4(y_1 + y_3 + \ldots + y_{n-1})\right]$$

INTEGRAL CALCULUS

Moment of inertia

(Second moment of area)

General

By moment of inertia about an axis x or a point O, we mean the sum of the products of line-, area-, volume- or mass-elements and the squares of their distances from the x-axis or point O.

Moment of inertia		Second moment of area	
i 203 $I = \int x^2 \, dm$	kg m²	$I = \int x^2 \, dA$	m⁴

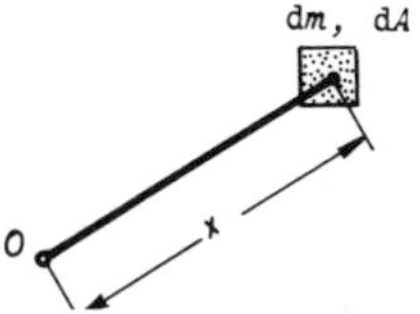

Steiner's theorem (Parallel axis theorem)

For every mass moment of inertia, both axial and polar, the following equation will apply:

i 204 $I_{xx} = I_{yy} + m\, l_s^2$ kg m²

Similar equations will apply for line, area and volume moments of inertia:

$I_{xx} = I_{yy} + A\, l_s^2$ m⁴

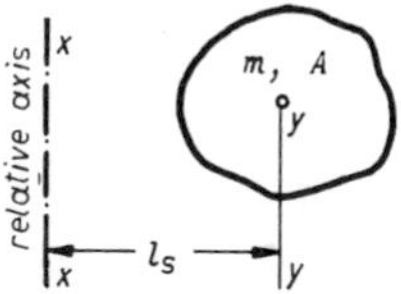

Moment of inertia of plane curves

in relation to the x-axis	y-axis
i 205 $I_{ox} = \int_a^b y^2 \sqrt{1+y'^2} \, dx$	$I_{oy} = \int_a^b x^2 \sqrt{1+y'^2} \, dx$

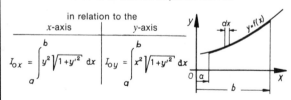

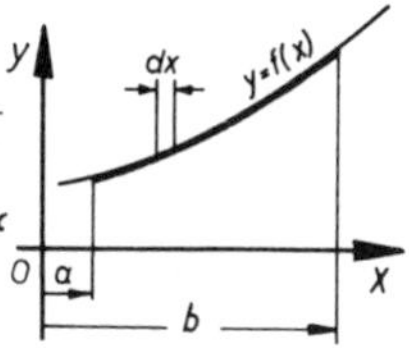

I_{xx} : moment of inertia about a general axis xx

I : moment of inertia about the centre of gravity

m, A : total length, area, volume, or mass

l_s : distance of centre of gravity from axis or point

Moments of inertia, centrif. moments of plane surfaces

By **axial second moment of area** of a plane surface in relation to an axis x or y within the plane we mean the sum of the products of the area-elements $\mathrm{d}A$ and the squares of their distances from axis x or y, respectively:

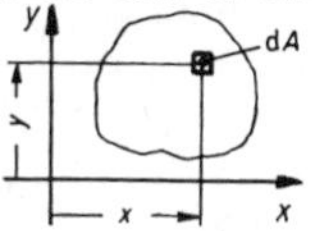

i 206 $$I_x = \int y^2 \,\mathrm{d}A \;; \quad I_y = \int x^2 \,\mathrm{d}A$$

A given function $y = f(x)$ yields:

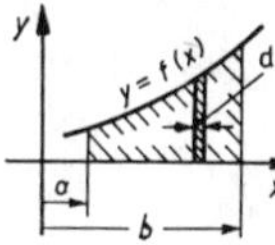

x-axis	y-axis
i 207 $I_x = \int_a^b \frac{y^3}{3}\,\mathrm{d}x$	$I_y = \int_a^b x^2 y \,\mathrm{d}x$

By **polar second moment of area** of a plane surface in relation to a point O within the plane we mean the sum of the products of the area-elements $\mathrm{d}A$ and the squares of their distances r from point O.

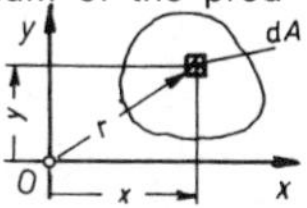

i 208 $$I_p = \int r^2 \,\mathrm{d}A$$

Where the relative axes of I_x and I_y are perpendicular to each other, the polar second moment of area in relation to the pole (intersection O of axis x and y) is:

i 209 $$I_p = \int r^2 \,\mathrm{d}A = \int (y^2 + x^2)\,\mathrm{d}A = I_x + I_y$$

By **centrifugal moment** (product of inertia) of a plane surface in relation to 2 axes within the plane we mean the sum of the products of the area-elements $\mathrm{d}A$ and the products of their distances x and y from the two axes:

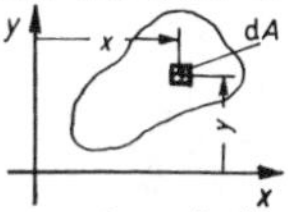

i 210 $$I_{xy} = \int x\,y \,\mathrm{d}A \gtreqless 0$$

One of the relative axes being an axis of symmetry of the plane surface results in $I_{xy} = 0$.

Conversion to an inclined axis x': Where moments I_x, I_y, and I_{xy} in relation to two perpendicular axes x and y are known, the second moment of area I_α in relation to an axis inclined x' by an angle α with respect to the x-axis can be calculated by:

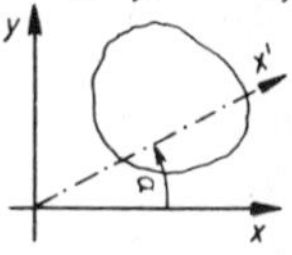

i 211 $$I_\alpha = I_x \cos^2\alpha + I_y \sin^2\alpha - I_{xy} \sin 2\alpha$$

Examples in conjunction to second moments of area on page I 17

Rectangle

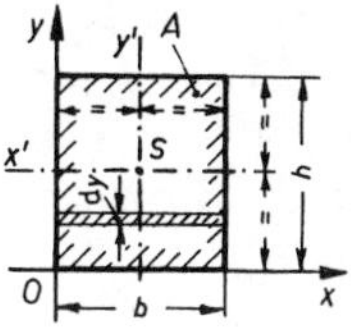

i 212 $$I_x = \int_0^h y^2 b \, dy = b\left[\frac{y^3}{3}\right]_0^h = \frac{b\,h^3}{3}$$

i 213 $$I_{x'} = I_x - A\left(\frac{h}{2}\right)^2 = \frac{b\,h^3}{12}$$

i 214 $$I_y = \frac{b^3 h}{3}; \quad I_{y'} = \frac{b^3 h}{12}$$

i 215 $$I_{po} = I_x + I_y = \frac{b\,h^3}{3} + \frac{b^3 h}{3} = \frac{b\,h}{3}(b^2+h^2); \quad I_{ps} = \frac{b\,h}{12}(b^2+h^2)$$

i 216 $$I_{xy} = I_{x'y'} + \frac{b}{2}\times\frac{h}{2}A.$$

As x' and/or y' are axes of symmetry, $I_{x'y'}$ is zero. Hence:

i 217 $$I_{xy} = \frac{b}{2}\times\frac{h}{2}(b\,h) = \left(\frac{b\,h}{2}\right)^2$$

Circle

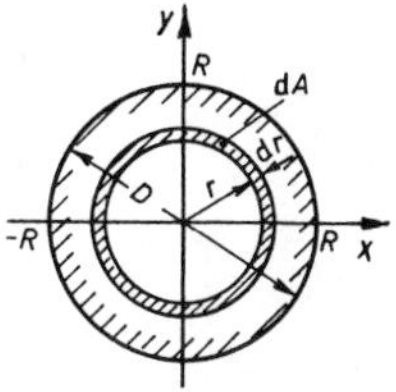

i 218 $$I_p = \int_0^R r^2 \, dA = \int_0^R r^2 \, 2\pi r \, dr$$

i 219 $$= 2\pi\left[\frac{r^4}{4}\right]_0^R = \frac{\pi R^4}{2}$$

i 220 $$I_x = I_y = \frac{I_p}{2} = \frac{\pi R^4}{4} = \frac{\pi D^4}{64}$$

i 221 $I_{xy} = 0$, as x and y are axes of symmetry.

Semicircle

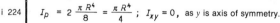

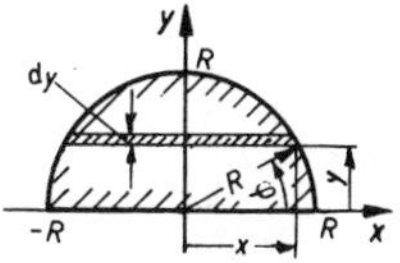

i 222 $$I_x = \int_0^R y^2 \, dA = \int_0^R y^2 \, 2\,x \, dy$$

i 223 $$= 2\int_0^R y^2 \sqrt{R^2 - y^2} \, dy = \frac{\pi R^4}{8} = I_y$$

i 224 $$I_p = 2\,\frac{\pi R^4}{8} = \frac{\pi R^4}{4}; \quad I_{xy} = 0,$$ as y is axis of symmetry.

Regular polygon

i 225 $$I_x = I_y = \frac{I_p}{2} = \frac{n\,a\,r}{2\times 48}(12r^2+a^2) = \frac{n\,a\sqrt{R^2 - \frac{a^2}{4}}}{48}(6R^2 - a^2)$$

$$I_{xy} = 0$$

r : radius of inscribed circle
R : radius of circumscribed circle
a : length of side
n : number of sides

Second moment of volume of a solid

'Moment of inertia' of a cuboid

Where $\left(\frac{b\,h^3}{12} + \frac{b^3\,h}{12}\right)$ is the polar moment of inertia of a rectangle (see I 18), the equation for the z-axis:

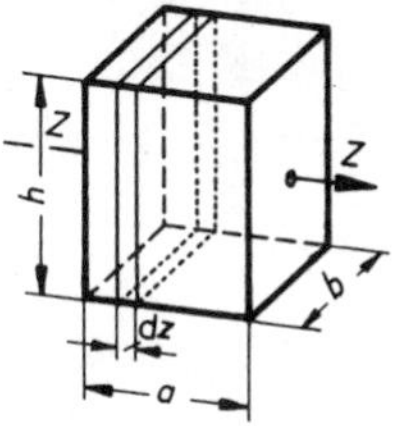

i 226 $$I_{v,zz} = \int_0^a \left(\frac{bh^3}{12} + \frac{b^3 h}{12}\right) dz = \frac{abh}{12}(b^2 + h^2)$$

'Moment of inertia' of a circular cylinder

for the axis zz:

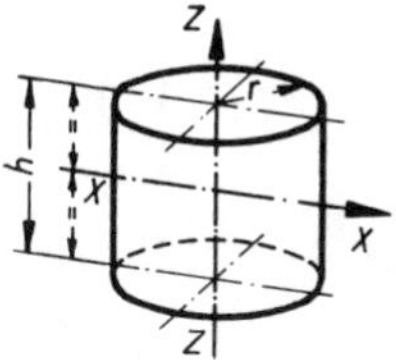

i 227 $$I_{v,zz} = \int_{-\frac{h}{2}}^{+\frac{h}{2}} \frac{\pi\ r^4}{2}\ dz = \frac{\pi\ r^4\ h}{2}$$

for the axis xx:

i 228 $$I_{v,xx} = \int_{-\frac{h}{2}}^{+\frac{h}{2}} \left(\frac{\pi\ r^4}{4} + \pi\ r^2 z^2\right) dz = \frac{\pi\ r^2 h}{12}(3r^2 + h^2)$$

Dynamic moment of inertia (mass moment of inertia)

The mass moment of inertia I about a particular axis is the product of the second moment of volume I_v about that axis and the density ϱ.

i 229 $$I = I_v \times \varrho \qquad \text{kg m}^2,\ \text{kgf m s}^2,\ \text{lb ft}^2$$

i 230 where $$\varrho = \frac{m}{V} \qquad \text{kg m}^{-3},\ \text{kg dm}^{-3},\ \text{lb ft}^{-3}$$

e.g. for a cylinder about the axis zz:

i 231 $$I_{zz} = I_{v,zz} \times \frac{m}{V} = \frac{\pi\ r^4\ h}{2} \times \frac{m}{r^2\ \pi\ h} = \frac{m\ r^2}{2}$$

For other mass moments of inertia see M 3

DIFFERENTIAL EQUATIONS

General terms

Definition of the Differential Equation (DE)

A DE is an equation of unknown functions which contains derivatives (partial derivatives) of the unknown functions and independent variables. The different kinds are:

Ordinary Differential Equation (ODE): the unknown functions depend only on one independent variable, e. g.:

j 1 $$y'' + 2x^2y = \sin x \qquad y = f(x)$$

Partial Differential Equation (PDE): the unknown functions depend on a number of independent variables, e. g.:

j 2 $$\frac{\partial^2 x}{\partial u\, \partial v} = x^2\, v\, w\, \frac{\partial x}{\partial u} \times \frac{\partial x}{\partial v} \qquad x = f(u, v, w)$$

Partial Differential Equations will not be specially considered here, as methods for Ordinary Differential Equations can be applied.

Ordinary Differential Equations

j 3 **Form:** $F\left(x, y(x), y'(x), \ldots y^{(n)}(x)\right) = 0.$

Where $y(x)$ is the unknown function. $y' \ldots y^{(n)}$ are the 1st to n^{th} derivatives; x is the independent variable.

j 4 *Example:* $y'''(x) + m(x)\, y'(x) + n(x) y^2(x) + p(x) y = q(x).$

j 5 Order: the highest derivative occuring in the ODE 3th order in the above example.

j 6 Degree: the highest exponent of the unknown function and its derivatives; 2nd degree in the above example.

j 7 Linear: ODE means, that the nighest exponent of the required function is one; i. e. an ODE of degree 1.

j 8 Homogeneous ODE implies the forcing function, $q(x) = 0$.

j 9 Inhomogeneous ODE implies the forcing function, $q(x) = 0$.

j 10 Solution: $y = y(x)$ of an ODE means, that this function and its derivatives satisfy the ODE.

j 11 Integration of the ODE yields the solution.

j 12 The general solution of an nth order ODE contains n constants $C_1, C_2, \ldots, C_n$. These constant are uniquely determined from n boundary conditions.

j 13 $$y'(x_0) = y_0' \quad \cdots \quad y^{(n-1)}(x_0) = y_0^{(n-1)}$$

The particular integral of the ODE is a special solution.

Methods to solve an ODE

1. Transform the ODE into one of the standard forms listed in J 6, J 8 ... J 12.

2. Application of a special method (cf. J 8).
   Using this method ODE can often be reduced to a standard ODE of lower order or degree (cf. J 9 ... J 12).

3. Use of transformations, particularly of the Laplace-Transform cf. D 18 ... D 20.

Linear Differential Equations

j 15 **Form:** $y^{(n)} + p_1(x)\, y^{(n-1)} + \ldots + p_{n-1}(x)\, y' + p_n(x)\, y = q(x).$

j 16 Here $y = y(x)$ is the required function. $y' \ldots y^{(n)}$ the 1st
j 17 to n^{th} derivative of $y(x)$ and $P_1(x) \ldots P_n(x)$ are functions of x.

General solution of a linear inhomogeneous ODE.

j 18 $$y = y_{hom} + y_{part}$$

Solution of the homogeneous ODE y_{hom}

y_{hom} is determined by setting the forcing function $q(x) = 0$. Each linear homogeneous n-th order ODE has n linear inde-
j 19 pendent solutions $y_1, y_2 \ldots y_n$ with n independent constants $C_1 \ldots C_n$.

j 20 $$y_{hom} = C_1 y_1(x) + C_2 y_2(x) + \ldots + C_n y_n(x)$$

J 9 ... J 12 give solutions for 1st and 2nd order Linear Differential Equations.

j 21 Particular solution of the inhomogeneous ODE y_{part}

y_{part} is determined for $q(x) \neq 0$. J 3, J 6 and J 7 suggest how to find solutions. J 9 and J 12 give solutions for 1st and 2nd order Linear Differential Equations.

Particular Solution

Determination using "Variation of constants" when y_{hom} of a linear n^{th} order ODE is well known (cf. J 2, j 20), the following formulation always leads to a particular solution:

j 23 $$y_{part} = C_1(x)\, y_1 + C_2(x)\, y_2 + \ldots + C_n(x)\, y_n.$$

Method to determine $C_1(x), C_2(x) \ldots C_n(x)$:

j 24 Form the simultaneous equations

$$\begin{aligned} C_1'(x)\, y_1 + C_2'(x)\, y_2 + \ldots + C_n'(x)\, y_n &= 0 \\ C_1'(x)\, y_1 + C_2'(x)\, y_2 + \ldots + C_n'(x)\, y_n' &= 0 \\ &\vdots \\ C_1'(x)\, y_1^{(n-2)} + C_2'(x)\, y_2^{(n-2)} + \ldots + C_n'(x)\, y_n^{(n-2)} &= 0 \\ C_1'(x)\, y_1^{(n-1)} + C_2'(x)\, y_2^{(n-1)} + \ldots + C_n'(x)\, y_n^{(n-1)} &= q(x) \end{aligned}$$

j 25 Determine $C_1'(x)$ for i = 1, 2… n using the above equation system.

j 26 Integration of $C_1'(x)$ for i = 1, 2… n yields the values of $C_i(x)$ for the solution.

Example: Solution for y_{part} of the ODE:

j 27 $$y'' + \frac{1}{x} y' = 2x.$$

j 28 Acc. to j 121: $y_{hom} = \int C_1 e^{-\int \frac{1}{x} dx}\, dx + C_2 = C_1 \ln|x| + C_2$

j 29 $= C_1\, y_1(x) + C_2\, y_2(x)$

using $y_1(x) = \ln|x|$ and $y_2(x) = 1$

j 30 let $y_{part} = C_1(x)\, y_1 + C_2(x)\, y_2$

j 31 using j 24 $$\begin{cases} C_1'(x) \ln|x| + C_2'(x) \times 1 = 0 \\ C_1'(x) \frac{1}{x} + C_2'(x) \times 0 = 2x \end{cases}$$

j 32 hence $C_1'(x) = 2x^2;\quad C_2'(x) = -2x^2 \ln|x|$

Integration of $C_1(x)$ and $C_2(x)$ gives:

j 33 $$C_1(x) = \frac{2}{3}x^3;\quad C_2(x) = -\frac{2}{3}x^3 \left[\ln|x| - \frac{1}{3}\right]$$

j 34 So $$y_{part} = \frac{2}{3}x^3 \times \ln|x| - \frac{2}{3}x^3 \left(\ln|x| - \frac{1}{3}\right) \times 1 = \frac{2}{9}x^3$$

General solution:

j 35 $$y = y_{hom} + y_{part} = C_1 \times \ln|x| + C_2 + \frac{2}{9}x^3.$$

Check: $y' = \frac{C_1}{x} + \frac{2}{3}x^2 \qquad y'' = -\frac{C_1}{x^2} + \frac{4}{3}x$

$$y'' + \frac{y'}{x} = -\frac{C_1}{x^2} + \frac{4}{3}x + \frac{C_1}{x^2} + \frac{2}{3}x = 2x$$

1st Order Linear ODE

j 36 **Form:** $y' + p(x)y = q(x)$.

The form corresponds to J 2, j 15 for $n = 1$; the highest derivative here is y'. Solutions for y, y_{hom} and y_{part} are given in J 2 and J 9.

j 37 *Example:* $y' + \frac{y}{x} = \sin x \qquad y = y_{hom} + y_{part}$

j 38 from j 110 $\quad p(x) = \frac{1}{x} \qquad q(x) = \sin x.$

from j 109 the homogeneous solution is:

j 39 $$y_{hom} = C_1\, e^{-\int \frac{1}{x}\,dx} = C_1\, e^{-\ln|x|} = \frac{C_1}{x} \quad \text{with} \quad C_1 \gtrless 0.$$

from j 110 the particular solution is

j 40 $$y_{part} = \int \sin x\, e^{\int \frac{1}{x}\,dx}\, dx\;\; e^{-\int \frac{1}{x}\,dx}$$
$$= \int (\sin x\, e^{\ln|x|})\, dx\, e^{-\ln|x|} = \int (\sin x \times x)\,dx\, \frac{1}{x}$$
$$= \frac{1}{x}\sin x - \cos x$$

j 41 $$y = y_{hom} + y_{part} = \frac{1}{x}(C_1 + \sin x) - \cos x.$$

Check: $$y' = -\frac{C_1}{x^2} + \frac{x\cos x - \sin x}{x^2} + \sin x$$
$$y' + \frac{y}{x} = \sin x$$

j 42 $C_1 \gtrless 0$; C_1 is determined using the boundary condition e. g.

j 43 $y(x_o) = 1$ for $x_o = \pi/2$

j 44 Then: $$1 = \frac{1}{\pi/2}\left(C_1 + \sin\frac{\pi}{2}\right) - \cos\frac{\pi}{2}$$

j 45 Gives: $$C_1 = \frac{\pi}{2} - 1.$$

2nd Order Linear ODE

j 46 **Form:** $y'' + p_1(x)\, y' + p_2(x)\, y = q(x)$

The form corresponds to J 2, j 15, for $n = 2$; the highest derivative is y''. Solutions for y, y_{hom} and y_{part} are given in J 11 and J 12.

Linear 2nd order ODE with constant coefficients

j 47 Due to the great importance of this ODE-type for oscillation-problems, special cases are considered.

j 48 **Form:** $y'' + 2ay' + b^2 y = q(x).$

j 49 a and b are constants $\neq 0$,
$q(x)$ is a forcing function

General solution, according to J 2, j 15 :

j 50 $$y = y_{\text{hom}} + y_{\text{part}}$$

j 51 Overdamped solution: $k^2 = a^2 - b^2 > 0$

j 52 $$y_{\text{hom}} = C_1\, e^{(-a+k)x} + C_2\, e^{(-a-k)x}$$

j 53 $$y_{\text{part}} = \frac{e^{(-a+k)x}}{2k} \int e^{(a-k)x}\, q(x)\, dx - \frac{e^{(-a-k)x}}{2k} \int e^{(a+k)x}\, q(x)\, dx \quad {}^{*)}$$

j 54 Critically damped solution: $k^2 = a^2 - b^2 = 0$

j 55 $$y_{\text{hom}} = C_1\, e^{-ax} + C_2\, x\, e^{-ax}$$

j 56 $$y_{\text{part}} = -e^{-ax} \int x\, e^{ax}\, q(x)\, dx + x\, e^{-ax} \int e^{ax}\, q(x)\, dx \quad {}^{*)}$$

j 57 Underdamped solution: $k^2 = a^2 - b^2 < 0$

j 58 $$y_{\text{hom}} = e^{-ax}[C_1 \sin(\omega x) + C_2 \cos(\omega x)]$$

with $\omega = \sqrt{b^2 - a^2}$

j 59 $$y_{\text{part}} = \frac{e^{-ax} \sin(\omega x)}{\omega} \int e^{ax} \cos(\omega x)\, q(x)\, dx - \frac{e^{-ax} \cos(\omega x)}{\omega} \int e^{ax} \sin(\omega x)\, q(x)\, dx \quad {}^{*)}$$

*) Note: For the special case $q(x) = A_o \sin(\omega_o x)$ is:

j 60 $$y_{\text{part}} = A \sin(\omega_o x - \gamma),$$

j 61 where: $$A = \frac{A_o}{\sqrt{(b^2 - \omega_o^2)^2 + 4a^2 \omega_o^2}}$$

j 62 and: $$\gamma = \operatorname{arccot} \frac{b^2 - \omega_o^2}{2a\, \omega_o}$$

Linear n^{th} order ODE with constant-coefficients

j 63 **Form:** $a_n \cdot y^{(n)} + a_{n-1} \cdot y^{(n-1)} + \ldots + a_1 y' + a_o y = q(x).$

Solution of the homogeneous n^{th} order ODE with constant coeffi-

j 64 cients ($q(x) = 0$).

j 65 **Let** $y = e^{rx}; \quad y' = r \cdot e^{rx}; \quad \ldots y^{(n)} = r^n \cdot e^{rx}$

Substitution in the homogeneous ODE of j 63 leads to the algebraic equation:

j 66 $$a_n r^n + a_{n-1} r^{n-1} + \ldots + a_1 r + a_o = 0.$$

The roots $r_1, r_2 \ldots r_n$ can be determined. Depending on the type of the roots, different solutions for y_{hom} are found.

Case a): $r_1, r_2 \ldots r_n$ are all real and different:

j 67 $$y_{hom} = C_1 \cdot e^{r_1 \cdot x} + C_2 \cdot e^{r_2 \cdot x} + \ldots + C_n \cdot e^{r_n \cdot x} \quad *)$$

Case b): There are real single and multiple roots:

$$r_1 = r_2 = \ldots = r_m; \quad r_{m+1}, r_{m+2}, \ldots r_n.$$

j 68 $$y_{hom} = C_1 \cdot e^{r_1 \cdot x} + C_2 \cdot x \cdot e^{r_1 \cdot x} + C_3 x^2 \cdot e^{r_1 \cdot x} + \ldots + C_m \cdot x^{m-1} \cdot e^{r_1 \cdot x} + C_{m+1} \cdot e^{r_{m+1}} + \ldots + C_n \cdot e^{r_n \cdot x} \quad *)$$

j 69 $$= e^{r_1 \cdot x}(C_1 + C_2 \cdot x + \ldots + C_m \cdot x^{m-1}) + C_{m+1} \cdot e^{r_{m+1} \cdot x} + \ldots + C_n \cdot e^{r_n \cdot x}.$$

Case c): There are conjugate complex roots:

$$r_1 = \alpha + i\beta; \quad r_2 = \alpha - i\beta = \overline{r_1}.$$

j 70 $$y_{hom} = C_1 \cdot e^{r_1 \cdot x} + C_2 \cdot e^{r_2 \cdot x} \quad *)$$
$$= e^{\alpha x} \cdot (A \cdot \cos \beta x + B \cdot \sin \beta x)$$

$$A = C_1 + C_2; \quad B = i(C_1 - C_2)$$

Particular solution of the inhomogeneous n^{th} order ODE with constant-coefficients

j 71 $$y_{part} = g_1(x) + g_2(x) + \ldots + g_k(x).$$

The form of the particular solution depends on $q(x)$. Some examples are given in J 7.

Using an appropriate form for y_{part}, the derivatives y'_{part}, y''_{part} etc. are found and substituted in the ODE. By comparison of the coefficients, the unknowns α_v and β can be determined (cf. example on J 7).

*) $C_1, C_2 \ldots C_n$ are arbitrary constants

Linear n^{th} order ODE with constant coefficients

	for $q(x)$	Form of $y_{part} =$
j 72	A	α
j 73	x^m	$\alpha_o + \alpha_1 x + \alpha_2 x^2 + \ldots + \alpha_m x^m$
j 74	$A_o + A_1 x + A_2 x^2 + \ldots + A_m x^m$	$\alpha_o + \alpha_1 \ + \alpha_2 x^2 + \ldots + \alpha_m x^m$
j 75	$A \cdot e^{\lambda x}$	$\alpha \cdot e^{\lambda x}$
j 76	$A \cos mx$	$\alpha \cos mx + \beta \sin mx$
j 77	$B \sin mx$	" + "
j 78	$A \cos mx + B \sin mx$	" + "
j79	$A \cosh mx$	$\alpha \cosh mx + \beta \sinh mx$
j80	$B \sinh mx$	" + "
j81	$A \cosh mx + B \sinh mx$	" + "
j82	$A\, e^{\lambda x} \cos mx$	$\alpha\, e^{\lambda x} \cos mx + \beta e^{\lambda x} \sin mx$
j83	$B\, e^{\lambda x} \sin mx$	" + "
j 84	$A\, e^{\lambda x} \cos mx + B\, e^{\lambda x} \sin mx$	" + "

j 85 *Example*: $y'' - y = \cos 2x$; according to form of J 6, j 65 let:

j 86 $y = e^{rx}$; $\quad y' = r\, e^{rx}$; $\quad y'' = r^2 e^{rx}$

Substitution in the ODE of example j 85 gives

j 87 $r^2 - 1 = 0; \quad r^2 = 1; \quad r_1 = 1; \quad r_2 = -1$

j 88 $y_{hom} = C_1\, e^{r_1 x} + C_2\, e^{r_2 x} = C_1\, e^x + C_2\, e^{-x}.$

Form of:

j 89 $y_{part} = \alpha \cos 2x + \beta \sin 2x$

j 90 $y'_{part} = -2\alpha \sin 2x + 2\beta \cos 2x$

j 91 $y''_{part} = -4\alpha \cos 2x - 4\beta \sin 2x$.

Equations j 89 and j 91 used in ODE (line j 85) yield:

j 92 $-5\alpha \cos 2x - 5\beta \sin 2x = \cos 2x.$

Comparison of terms yields:

j 93 $\beta = 0\,; \ \alpha = -\frac{1}{5}$ and therefore $y_{part} = -\frac{1}{5} \cos 2x$

General solution:

j 94 $y = y_{hom} + y_{part} = C_1\, e^x + C_2\, e^{-x} - \frac{1}{5} \cos 2x$

Check: $y' = C_1 \ e^x - C_2 \ e^{-x} + \frac{1}{5} \sin 2x \times 2$

$$y'' = C_1 \ e^x + C_2 \ e^{-x} + \frac{1}{5} \ 4 \ \cos 2x$$

$$y'' - y = C_1 \ e^x + C_2 \ e^{-x} + \frac{1}{5} \ 4 \ \cos 2x - C_1 \ e^x -$$

$$- C_2 \ e^{-x} + \frac{1}{5} \cos 2x = \cos 2x$$

Reduction of order by variable-substitution to solve an nth order ODE

	Form of the ODE	assumption	substitution	comment
j 95	$y^{(n)} = f(y, y', \dots y^{(n-1)})$ *(cf. example A)*	x not explicitly available	$y' = p = \frac{dy}{dx}$ $y'' = p' = \frac{dp}{dy}\, p$	Reduction from order n to order $n-1$
j 96	$y^{(n)} = f(x, y', \dots y^{(n-1)})$	y not explicitly available	$y' = p$ $y'' = \frac{dp}{dx}$	Reduction from order n to order $n-1$
j 97	$y^{(n)} = f(x, y^{(k+1)}, \dots y^{(n-1)})$ *(cf. example B)*	1st to k^{th} derivative not available	$y^{(k+1)} = p$ $y^{(k+2)} = p' = \frac{dp}{dx}$	Reduction from order n to order $n-k$

Example A:

j 98 $y\, y'' - y'^2 = 0;$

j 99 Substitution: $y' = p; \quad y'' = p\, \frac{dp}{dy}$

j 100 $y\, p\, \frac{dp}{dy} - p^2 = 0; \quad \frac{dp}{p} = \frac{dy}{y}$

j 101 $\ln|p| = \ln|y| + \ln C$

j 102 $\ln|p| = \ln C\, y = \frac{dy}{dx} = y'$

j 103 $p = C\, y = y'$

j 104 $y' - C\, y = 0$

j 105 $y = C_1\, e^{-\int C\, dx} = C_1\, e^{-C x}$

Check: $y' = -C_1 \quad e^{-Cx} \quad C$

$y'' = C_1 \quad C^2 \quad e^{-Cx}$

$y\, y'' - y'^2 = C_1\, e^{-Cx} \quad C_1 \quad C^2 \quad e^{-Cx} -$

$- C_1^2 \quad C^2 \quad e^{-Cx\, 2} = 0$

Example B:

$y''' + 2y'' - 4x = 0.$

Substitution: $y'' = p; \quad y''' = \frac{dp}{dx} = p'$

$p' + 2p - 4x = 0,$ using j 110:

$p = C_1\, e^{-2x} + 2x - 1 = \frac{d}{dx}\, y' = y''$

$y' = \int (C_1\, e^{-2x} + 2x - 1)\, dx + C_2$

$y' = -\frac{C_1}{2}\, e^{-2x} + x^2 - x + C_2 = \frac{dy}{dx}$

$y = \int (-\frac{C_1}{2}\, e^{-2x} + x^2 - x + C_2) dx + C_3$

$y = \frac{1}{4} C_1\, e^{-2x} + C_2\, x + \frac{x^3}{3} - \frac{x^2}{2} + C_3$

Check: $y''' + 2y'' - 4x =$

$-2C_1\, e^{-2x} + 2 + 2C_1\, e^{-2x} + 4x - 2 - 4x = 0$

	Kind	Form	Substitution	Solution	Comment
j 106	Separable ODE	$y' = \frac{dy}{dx} = \frac{f(x)}{g(y)}$		$\int g(y)\,dy = \int f(x)\,dx + C$	The variables x and y can be separated into the left and right side of the equation.
j 107	Not-directly separable ODE	$y' = f(\alpha x + \beta y + \gamma)$	$\alpha x + \beta y + \gamma = u$ $\frac{du}{dx} = \alpha + \beta y'$ $y' = \frac{1}{\beta}(\frac{du}{dx} - \alpha)$	$\int dx = \int \frac{du}{\beta f(u) + \alpha} + C$	Substitution after integration
j 108	Similarity ODE	$y' = f(\frac{y}{x})$	$\frac{y}{x} = u$ $y' = u + x\frac{du}{dx}$	$\int \frac{dx}{x} = \int \frac{du}{f(u) - u} + C$	Check if it is possible to transform to $f(y/x)$
j 109	Homogen-linear 1st order ODE	$y' + p(x)\,y = 0$		$y = C\,e^{-\int p(x)\,dx} = y_{hom}$	$y = y_{hom}$
j 110	Inhomogeneous linear 1st order ODE	$y' + p(x)\,y = q(x)$	$y = y_{hom} + y_{part}$ $y_p = C(x)\,e^{-\int p(x)dx}$ $y_p' = C'(x)\,e^{-\int p(x)dx} - C(x)\,p(x) \times$ $\times\, e^{-\int p(x)dx}$	$y = e^{-\int p(x)dx}\left[C + \int q(x)\,e^{\int p(x)dx}\,dx\right]$ where $y_p = \int q(x)\,e^{\int p(x)dx}\,dx\;e^{-\int p(x)dx}$	y_{hom} cf. j 109 Determination of the particulary solution using variation of constants cf. J 2, J 3
j 111	Implicit 1st order ODE no x term	$y = f(y')$	$y' = p$	$x = \int \frac{f'(p)\,dp}{p} + C$ $y = f(p)$	By elimination of p the solution is found in parametric form

	Kind	Form	Substitution	Solution	Comment
j 112	Implicit 1st order ODE, no y-term	$x = f(y')$	$y' = p$	$x = f(p)$ $y = \int p\, f'(p)\, \mathrm{d}p + C$	Elimination of p leads to solution in parametric form
j 113	Implicit 1st order d'Alembert's ODE	$y = x\, g(y') + f(y')$	$y' = p$	$\frac{\mathrm{d}x}{\mathrm{d}p} = \frac{g'(p)}{p-g(p)} x + \frac{f'(p)}{p-g(p)}$	
j 114	Clairaut's ODE	$y = x\, y' + f(y')$	$y' = p$ $f(y') = f(p)$	with $C_1 = y'$: $y = x\, C_1 + f(C_1)$ (group of lines, general integral) $x = -f'(p)$ $y = -p\, f'(p) + f(p)$	Parameter description x and y. Singular integral (envelope curve) by elimination of p
j 115	Bernoulli-ODE 1st order, nth degree ODE	$y' + p(x)y + q(x)y^n = 0$ with $n \neq 0$; $n \neq 1$	$z = y^{1-n}$ $y = z^{\frac{1}{1-n}}$ $y' = \frac{1}{1-n}\, z^{\frac{n}{1-n}}\, z'$	$\frac{1}{1-n} z' + p(x)\, z = -q(x)$ $z = \mathrm{e}^{-\int (1-n)p(x)\,\mathrm{d}x}\Big[C - (1-n)\int q(x) \times \mathrm{e}^{\int (1-n)p(x)\,\mathrm{d}x}\mathrm{d}x\Big]$ $y = z^{\frac{1}{1-n}} = \frac{1}{\sqrt[n-1]{z}}$	Reduction to 1st order ODE as in J 9, j 110
j 116	Ricatti ODE 1st order 2nd degree ODE	$y' + p(x)y + q(x)y^2 = r(x)$	$y(x) = u(x) + y_1(x)$ where $y_1(x)$ is a standard particular solution $z(x) = \frac{1}{u(x)}$	$z' - [p(x) + 2q(x)\, y_1(x)]\, z = q(x)$ $y(x) = y_1(x) + \frac{1}{\mathrm{e}^{+\int [p(x)+2q(x)\, y_1(x)]\mathrm{d}x}} \times$ $\times \frac{1}{[C + \int q(x)\, \mathrm{e}^{-\int [p(x)+2q(x)\, y_1(x)]\mathrm{d}x}\, \mathrm{d}x]}$	Inhomog. ODE in z; solution as in J 9. At least one particular solution must be known.

	Kind	Form	Substitution	Solution	Comment
j 117	y and y' terms are missing	$y'' = f(x)$		$y = C_1 + C_2 x + \int[\int f(x)\,dx]\,dx$	Start the calculation with inner integral
j 118	Homogen. linear 2nd order ODE with constant coefficients	$y'' + a_1 y' + a_o y = 0$	$y = e^{rx}$ $y' = r\,e^{rx}$ $y'' = r^2 e^{rx}$	$y(x) = C_1\,e^{r_1 x} + C_2\,e^{r_2 x} = y_{hom}$ with $r_{1,2} = -\frac{a_1}{2} \pm \sqrt{\frac{a_1^2}{4} - a_o}$ or $y(x) = e^{\alpha x}(A\cos\beta x + B\sin\beta x)$ with $r_1 = \alpha + i\beta$; $r_2 = \alpha - i\beta = \bar{r}_1$	C_1 and C_2 are arbitrary constants $A = C_1 + C_2$ $B = i\,(C_1 - C_2)$
j 119	Inhomogen. linear 2nd order ODE with constant coeffic.	$y'' + a_1 y' + a_o y = q(x)$	$y = y_{hom} + y_{part}$	$y(x) = C_1 \cdot e^{r_1 x} + C_2 \cdot e^{r_2 x} + y_{part}$ (for $r_1 \neq r_2$ cf. j 120) or $y(x) = e^{\alpha x}(A\cos\beta x + B\sin\beta x) + y_{part}$ (for $r_1 = \alpha + i\beta$; $r_2 = \alpha - i\beta = \overline{r_1}$)	y_{part} dependent on $q(x)$; calculation cf. J 6, J 7. Comment cf. j 120
j 120	Homogen. linear 2nd order ODE Euler's Equation	$x^2 \cdot y'' + b_1 x y' + b_o y = 0$	$y = x^r$ $y' = r\,x^{r-1}$ $y'' = r(r-1)x^{r-2}$	$y(x) = C_1\,x^{r_1} + C_2\,x^{r_2}$; $r_1 \neq r_2$ with $r_{1,2} = \frac{1-b_1}{2} \pm \sqrt{\frac{(b_1-1)^2}{4} - b_o}$ or $y(x) = x^{\alpha}[A\cos(\beta\ln\lvert x\rvert) +$ $+ B\sin(\beta\ln\lvert x\rvert)]$ for $r_1 = \alpha + i\beta$ and $r_2 = \alpha - i\beta$	C_1 and C_2 are arbitrary constants $A = C_1 + C_2$ $B = i(C_1 - C_2)$
j 121	Homogen. linear 2nd order ODE y-term is missing	$y'' + p_1(x)\,y' = 0$	$y' = u$ $y'' = \frac{du}{dx}$	$y = \int C_1\,e^{-\int p_1(x)dx}\,dx + C_2 = y_{hom}$	By substitution, first reduce to 1st order ODE, then solve.

	Kind	Form	Substitution	Solution	Comment
j 122	Inhom. lin. 2nd order ODE y-term is missing	$y''+p_1(x)y'=q(x)$	$y'=u$ $y''=\frac{du}{dx}$ $y=y_{hom}+y_{part}$	$y=\int\left[e^{-\int p_1(x)dx}\left(C_1+\int q(x)\, e^{\int p_1(x)dx}\, dx\right)\right]dx+C_2$ $y_{part}=\left[e^{-\int p_1(x)dx}\left(\int q(x)^{\int p_1(x)dx}\, dx\right)\right]dx$	Start calculation with inner integrals
j 123	Homog. lin. 2nd order ODE y-term is missing	$y''+p_1(x)\ f(y')=0$	$y'=u$; $y''=\frac{du}{dx}$ $f(y')=f(u)$	$\int\frac{du}{f(u)}=-\int p_1(x)\, dx+C_1$ $y=\int u\, dx+C_2$	
j 124	2nd order ODE: y-term is missing	$y''=f(y)$	$y'=u(y)$ $y''=u(y)\ \frac{du}{dy}$	$x=\pm\int\frac{dy}{\sqrt{2\int f(y)\, dy+C_1}}+C_2$	
j 125	2nd order ODE: y'-term is missing	$y''=f(x,\ y')$	$y'=u(x)$	$\frac{du}{dx}=f(x,u)$; $y=\int u(x)\, dx+C$	Often not solvable
j 126	2nd order ODE: x-term is missing	$y''=f(y')$	$y'=u$ $y''=u'$ $f(y')=f(u)$	$x=\int\frac{du}{f(u)}+C_1$; $y=\int\frac{u\, du}{f(u)}+C_2$	After elimination of u, a solution can be found
j 127	2nd order ODE x-term is missing	$y''=f(y,y')$	$y'=u$ $y''=\frac{du}{dx}=f(y,u)=u\frac{du}{dy}$ $u=u(y)$ $y=y(x)$	$u\frac{du}{dy}=f(y,u)$ $x=\int\frac{dy}{u(y)}+C$	Finally substitute $y'=dy/dx$ for u
j 128	Homogen linear ODE 2nd order	$y''+p_1(x)\ y'+$ $+p_2(x)\ y=0$	$v(x)=\frac{y}{y_1(x)}$ $v'(x)=w$ $=\frac{d}{dx}\left(\frac{y}{y_1(x)}\right)$	after transformation to: $y_1(x)\ w'+[2y_1'(x)+p_1(x)\ y_1(x)]w=0$ $y=y_1\ v(x)$ $y=y_1(x)\left[\int C_1\frac{1}{y_1^2(x)}e^{-\int p_1(x)\cdot dx}\, dx+C_2\right]$	$y_1(x)$ must be a standard particular solution. Then transf. to a lin. homog. 1st order ODE $y_1(x)$ cf. J 9

STATICS

General terms

General

Statics deals with the theory of equilibrium and with the determination of external forces acting on stationary solid bodies (e.g. support reactions). The contents of page K1 ... K14 are applicable only to forces acting in one plane.

The most important quantities of statics and their units

Length l

Is a base quantity, see preface.

Units: m; cm; km

Force F (see explanation on M 1)

Being a vector a force is defined by its magnitude, direction, and point of application *(P)*.

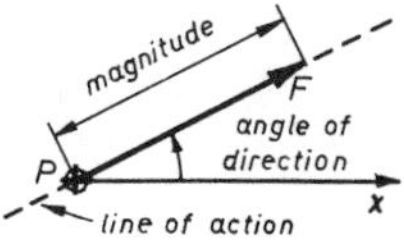

Gravitational force G

definition: force of earth's attraction
point of application: centre of grav. S
line of action: vertical line intersecting centre of gravity
direction: downwards (towards earth's centre)
magnitude: determined by spring balance.

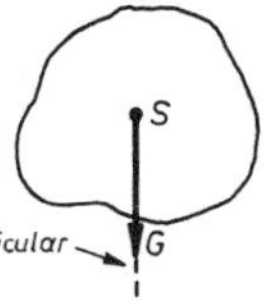

Support reaction F_A

force applied to body by support A.

Resultant force F_R

force representing the total action of several external forces.

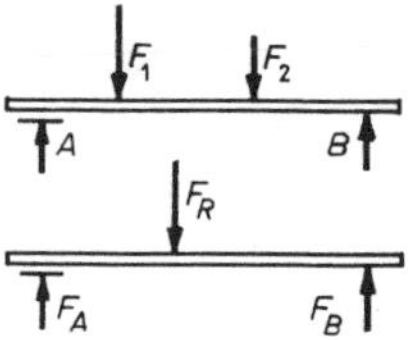

Moment M **of a force** F **about a point** O

The perpendicular distance from point O to the line of action of force F is called lever arm l.

F_1 and F_2 form a pair of forces. The moment may be represented by a vector.

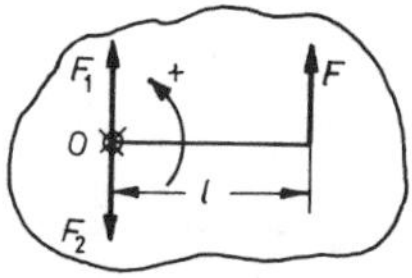

k 1 $F_1 = F$; $F_2 = -F_1$

k 2 moment $M = \pm F\, l$

Moment theorem: The moment of the resultant force is equal to the sum of the moments of the individual forces.

STATICS

Forces composition

K 2

Graphical composition of forces

diagram of forces	force polygon

Two forces

k 4

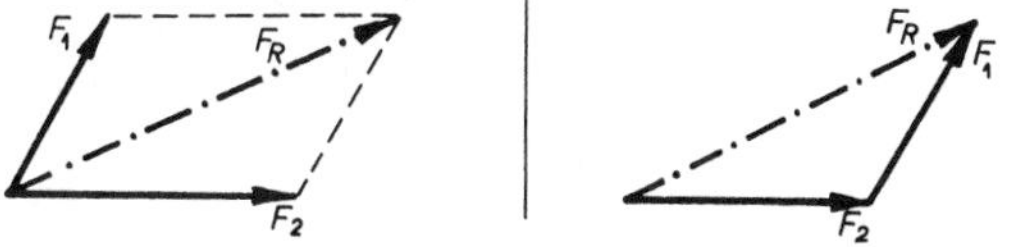

A number of forces and a common point of application

k 5

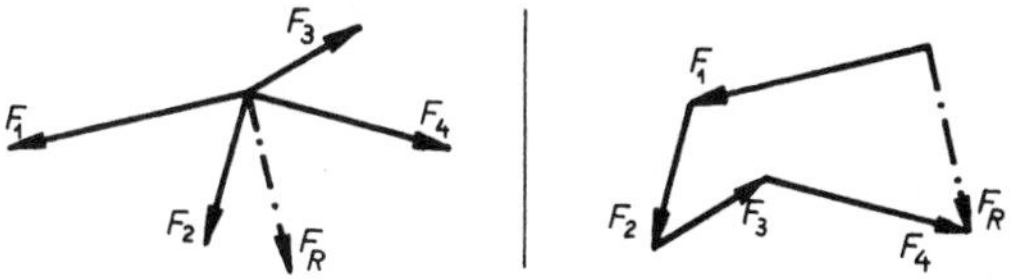

Parallel forces

k 6

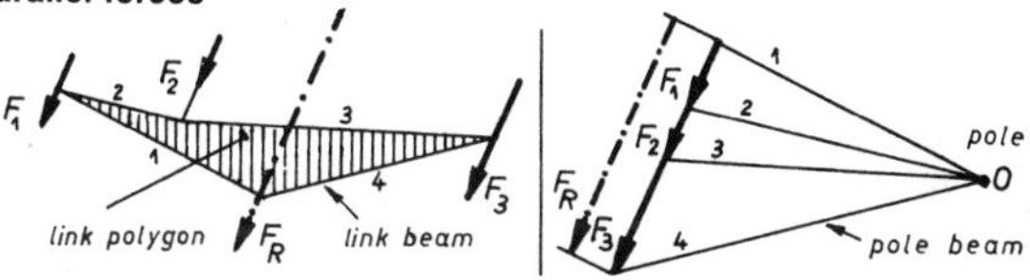

A number of forces and any random point of application

k 7

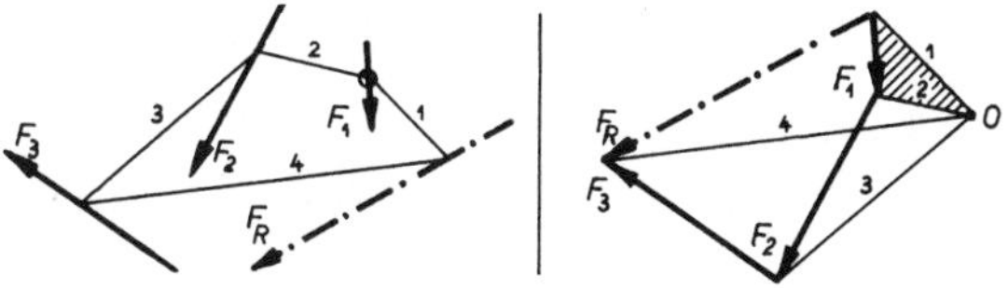

Construction of the link polygon

Draw force polygon and determine pole O so as to avoid any link rays running parallel. Draw pole rays. Construct link polygon such that link rays run parallel to corresponding pole rays. Thereby each point of intersection in the link polygon corresponds to a triangle in the force polygon (e.g. triangle F_1-1-2 of force polygon corresponds to point of intersection F_1-1-2 of link polygon).

Mathematical composition of forces

Resolution of a force

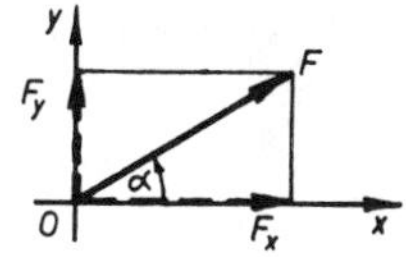

k 8 $F_x = F \times \cos\alpha$ | $F_y = F \times \sin\alpha$

k 9 $F = +\sqrt{F_x^2 + F_y^2}$ | $\tan\alpha = \dfrac{F_y}{F_x}$

(for signs of trigonometrical functions of α see table k 16 to k 19)

Moment M_O of a force about a point O

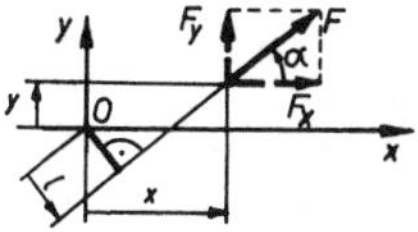

k 10 $M_O = +F \times l = F_y\,x - F_x\,y$

(for signs of trigonometrical functions of α see table k 16 to k 19)

Resultant force F_R of any random given forces

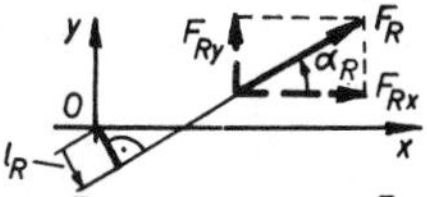

k 11 components $F_{Rx} = \Sigma F_x$ | $F_{Ry} = \Sigma F_y$

k 12 magnitude $F_R = +\sqrt{F_{Rx}^2 + F_{Ry}^2}$

k 13 angle of direction α_{FR} $\tan\alpha_R = \dfrac{F_{Ry}}{F_{Rx}}$; $\sin\alpha_R = \dfrac{F_{Ry}}{F_R}$; $\cos\alpha_R = \dfrac{F_{Rx}}{F_R}$

k 14 distance $l_R = \dfrac{|\Sigma M_O|}{|F_R|}$ (moment theorem)

k 15 sign of $F_R \times l_R$ = sign of ΣM_O

Signs of trigonometrical functions of x, y; F_x, F_y; F_{Rx}, F_{Ry}

	quadrant	α, α_R	$\cos\alpha$	$\sin\alpha$	$\tan\alpha$	x, F_x, F_{Rx}	y, F_y, F_{Ry}
k 16	I	0 ... 90°	+	+	+	+	+
k 17	II	90 ... 180°	−	+	−	−	+
k 18	III	180 ... 270°	−	−	+	−	−
k 19	IV	270 ... 360°	+	−	−	+	−

F_x, F_y : components of F parallel to x-axis and y-axis
F_{Rx}, F_{Ry} : components of F_R parallel to x-axis and y-axis
x, y : coordinates of F
α, α_R : angles of F and F_R
l, l_R : distances of F and F_R from reference point

Conditions of equilibrium

A body is said to be in equilibrium when both the resultant force and the sum of the moments of all external forces about any random point are equal to zero.

	forces	graphical	mathematical
k 20	with common point of application	closed force polygon	$\Sigma F_x = 0;\ \Sigma F_y = 0;$
k 21	parallel to vertical axis	force polygon and link polygon closed	$\Sigma F_y = 0;\ \Sigma M = 0$
k 22	arbitrary		$\Sigma F_x = 0;\ \Sigma F_y = 0;\ \Sigma M = 0$

Simply supported beam with point loads W_1 and W_2

Find reactions R_A and R_B:

Graph. solution:

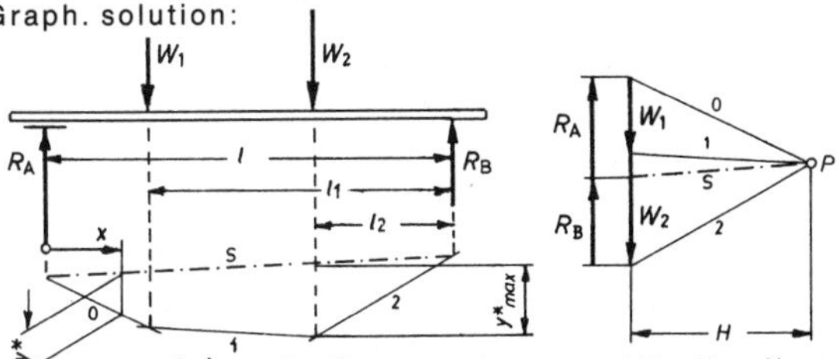

$M(x) = y^* m_L H m_F$ kNm, Ncm, Nmm

k 23 m_L : scale of lengths = true length/diagram length

k 24 m_F : scale of forces = force/diagram length

H : pole spacing y^*: vertical distance between closing line s and link polygon.

k 25 Mathem. solution: $R_A = W_1\ l_1/l + W_2\ l_2/l$; $R_B = (W_1 + W_2) - R_A$

Distributed loads are divided into small sections and considered as corresponding point forces acting through the centres of mass of the sections.

Wall mounted crane (3 forces): Find reactions F_A, F_B.

problem

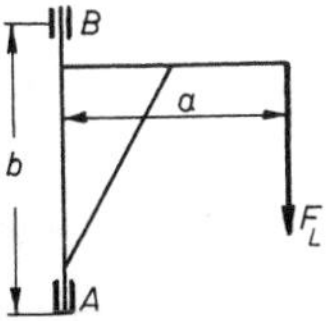

solution

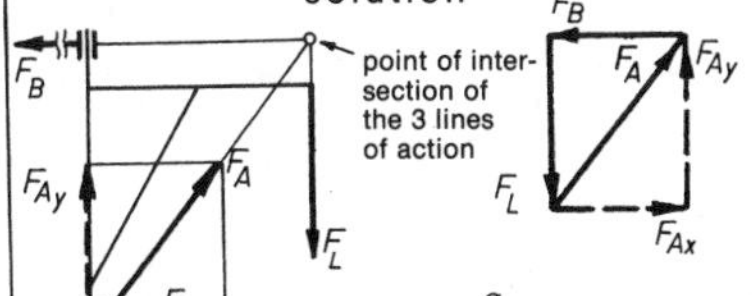

$F_{Ax} = F_B = \frac{a}{b} F_L$; $F_{Ay} = F_L$

Mathematical determination of member loads

(Ritter method – Method of Sections)

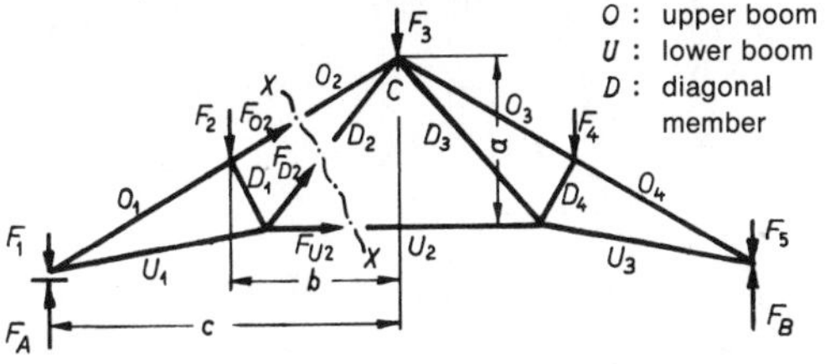

k 26

Determine the support reactions from K 4 (girder on two supports). Draw a line $X \ldots X$ through the framework to the bar in question, but intersecting no more than 3 bars. Take tensile forces as positive, so that compressive forces are negative.

Establish the equation of moments $\Sigma M = 0$ with moments of external and internal forces taken about the point of intersection of two unknown forces.

Rule for moment signs

Where turning moment is counter-clockwise the sign is positive.
Where turning moment is clockwise the sign is negative.

Example (from the above girder)

problem: to find force F_{U2} in bar U_2.
solution:
Draw a line $X \ldots X$ through $O_2 - D_2 - U_2$. Since the lines O_2 and D_2 meet at C, this is the relative point of intersection selected so that the moment O_2 and D_2 may equal zero.

Proceed as follows: $$\Sigma M_C = 0$$

$$+ a\,F_{U2} + b\,F_2 - c(F_A - F_1) = 0$$

$$F_{U2} = \frac{-b\,F_2 + c(F_A - F_1)}{a}$$

Graphical determination of forces in members

(Cremona method – Bow diagramm)

k 27

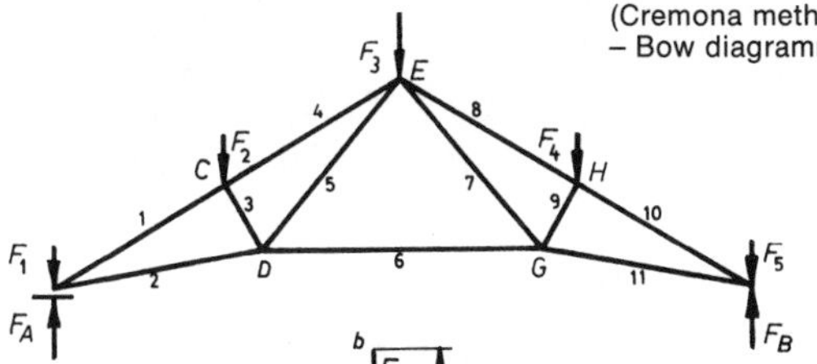

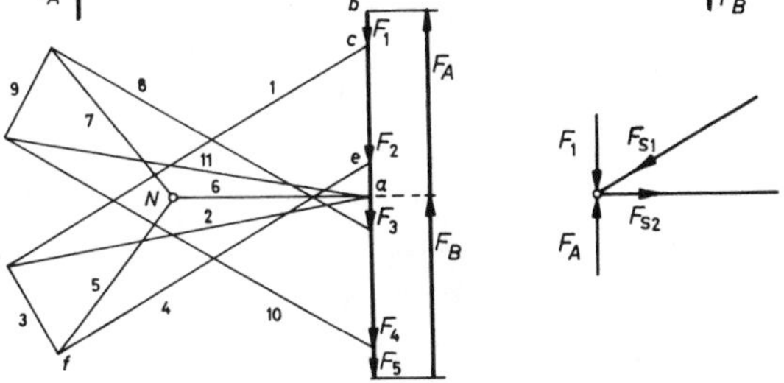

Basic principles

Each bar is confined by two adjacent joints. The external forces only act through the joints.

Procedure

Establish a scale of forces and determine the support reactions. Since each force polygon must not contain more than two unknown forces, start at joint A. Establish identical order of forces (cw or ccw) for all joints(e. g. $F_A - F_1 - F_{S1} - F_{S2}$).

Joint A: Force polygon $a - b - c - d - a$. Keep a record of forces being tensile or compressional.

Joint C: Force polygon $d - c - e - f - d$.

etc.

Check

Forces acting through a single joint in the framework form a polygon in the Bow diagram.

Forces acting through a single point in the diagram of forces form a triangle in the framework.

Arc of circle

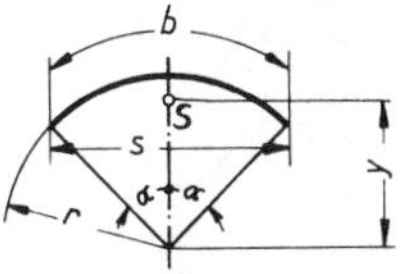

k 28 $y = \dfrac{r \sin \alpha \, 180^\circ}{\pi \, \alpha} = \dfrac{r \, s}{b}$

k 29 $y = 0{\cdot}6366 \, r$ at $2\alpha = 180^\circ$

k 30 $y = 0{\cdot}9003 \, r$ at $2\alpha = 90^\circ$

k 31 $y = 0{\cdot}9549 \, r$ at $2\alpha = 60^\circ$

Triangle

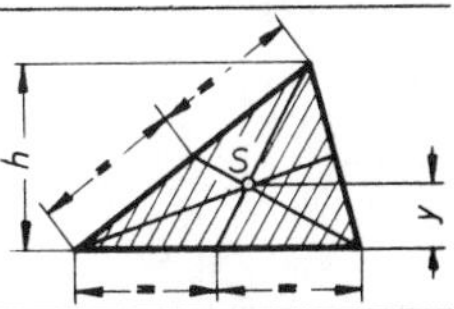

k 31 $y = \dfrac{1}{3} h$

S is the point of intersection of the medians

Sector of a circle

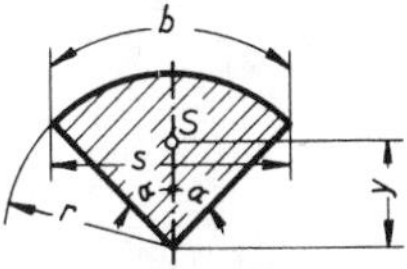

k 32 $y = \dfrac{2r \sin \alpha \, 180^\circ}{3 \, \pi \, \alpha} = \dfrac{2r \, s}{3b}$

k 33 $y = 0{\cdot}4244 \, r$ at $2\alpha = 180^\circ$

k 34 $y = 0{\cdot}6002 \, r$ at $2\alpha = 90^\circ$

k 35 $y = 0{\cdot}6366 \, r$ at $2\alpha = 60^\circ$

Trapezium

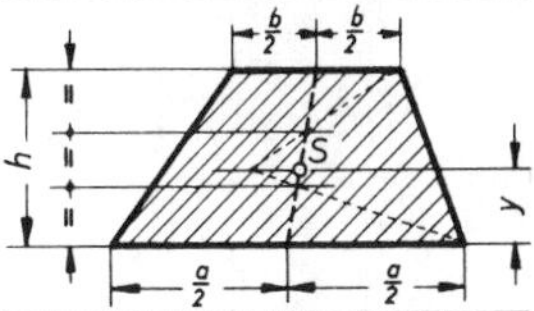

k 36 $y = \dfrac{h}{3} \times \dfrac{a + 2b}{a + b}$

Sector of an annulus

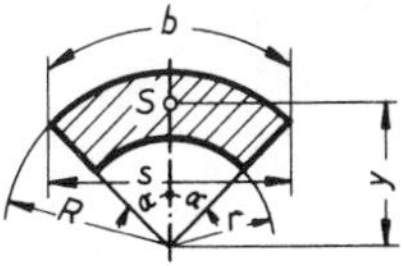

k 37 $y = \dfrac{2}{3} \times \dfrac{R^3 - r^3}{R^2 - r^2} \times \dfrac{\sin \alpha}{\operatorname{arc} \alpha}$

k 38 $ = \dfrac{2}{3} \times \dfrac{R^3 - r^3}{R^2 - r^2} \times \dfrac{s}{b}$

Segment of a circle

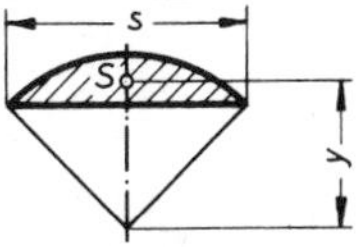

k 39 $y = \dfrac{s^3}{12 \, A}$

for area *A* see B 3

For determination of centre of gravity *S*, see also I 14

Determination of centre of gravity of any random surface area

Graphical solution

Subdivide the total area A into partial areas A_1, A_2 ... A_n the centres of gravity of which are known. The size of each partial area is represented as a force applied to the centre of area of each partial area. Use the force polygon (see K 2) to determine the mean forces A_{Rx} and A_{Ry} operating in any two directions (preferably at right angles). The point of intersection of the lines of application will indicate the position of the centre of area A.

k 40

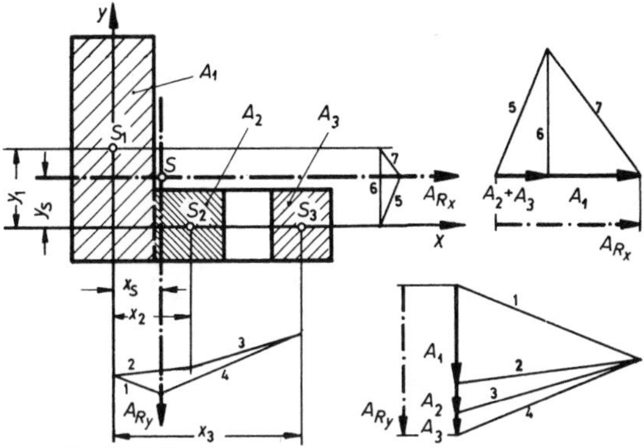

Mathematical solution

Subdivide the above total area A into partial areas A_1, A_2 ... A_n; we now get

	distance	in general	in the above example
k 41	$x_S =$	$\dfrac{\sum_{i=1}^{n} A_i \; x_i}{A}$	$\dfrac{A_1 \; x_1 + A_2 \; x_2 + A_3 \; x_3}{A}$
k 42	$y_S =$	$\dfrac{\sum_{i=1}^{n} A_i \; y_i}{A}$	$\dfrac{A_1 \; y_1 + A_2 \; y_2 + A_3 \; y_3}{A}$

Note: In the above example the distances x_1, y_2 and y_3 each equal zero.

Force acting parallel to a sliding plane

	static friction	limiting friction	sliding friction
k 43	$\upsilon = 0$	$\upsilon = 0$	$\longrightarrow \upsilon > 0$

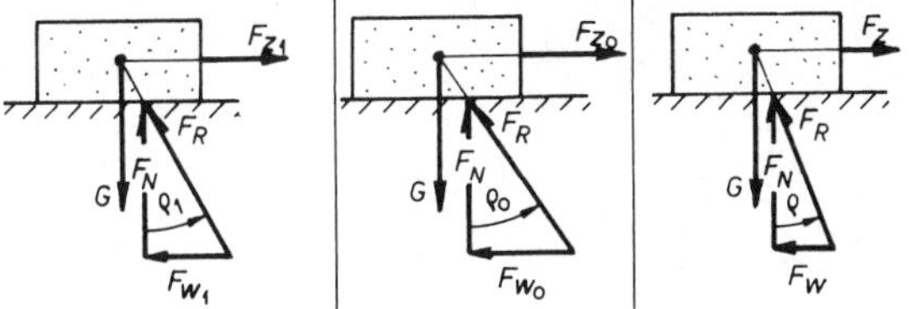

	static friction	limiting friction	sliding friction
k 44	$F_{W1} = -F_{Z1} = G \tan \rho_1$	$F_{W0} = -F_{Z0} = G \tan \rho_0$	$F_W = -F_Z = G \tan \rho$
k 45	$F_N = -G$	$F_N = -G$	$F_N = -G$
k 46		$\mu_0 = \tan \rho_0 > \mu$	$\mu = \tan \rho < \mu_0$
k 47	$0 < \rho_1$ (variable) $< \rho_0$	$\rho_0 = \text{const.} > \rho$	$\rho = \text{const.} < \rho_0$

A force F_{Z1} gradually increasing from zero is compensated by an increasing F_{W1} without causing the body to move, until F_{Z1} reaches the value

k 48 $$F_{Z0} = G\mu_0 .$$

As soon as this happens, the body starts sliding, whereby F_Z drops to $G\mu$. Any excessive force will now accelerate the body.

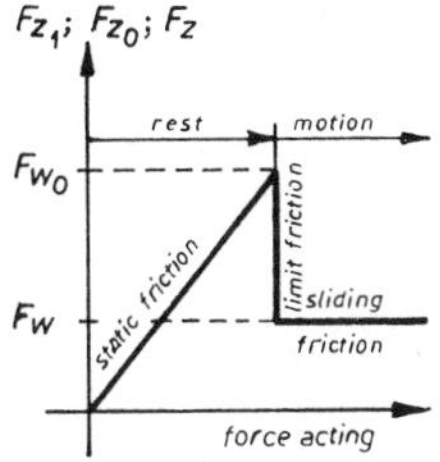

Force applied obliquely

The force F needed to set in motion a body, weight G:

k 49 $$F = G\frac{\mu_0}{\sin\alpha - \mu_0 \cos\alpha} = G\frac{\sin\rho_0}{\sin(\alpha - \rho_0)}$$

The force needed to maintain the motion is ascertained by replacing μ_0 by μ. No motion possible when result of F is negative.

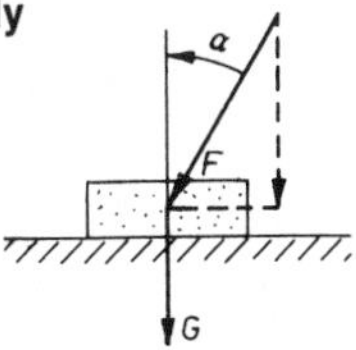

F_{W1}, F_{W0}, F_W: friction force
F_{Z1}, F_{Z0}, F_Z: traction force
$-, \mu_0, \mu$: friction coefficient } see
ρ_1, ρ_0, ρ: angle of friction } Z 7

Inclined plane

General

The angle α at which a body will move easily down an inclined plane is the angle of friction ϱ.

k 50 $$\tan\alpha = \tan\varrho = \mu$$

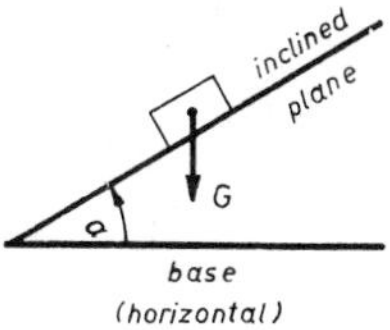

Application in the experimental determination of the angle of friction or the friction coefficient:

$$\mu = \tan\varrho$$

k 51 **Condition of automatic locking:** $\alpha < \varrho$

Friction properties

motion	constant velocity maintained by tractive force F parallel to inclined plane	constant velocity maintained by tractive force F parallel to base
upwards $0 < \alpha < \alpha^*$	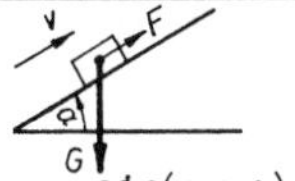 k 52 $F = G\,\dfrac{\sin(\alpha+\varrho)}{\cos\varrho}$	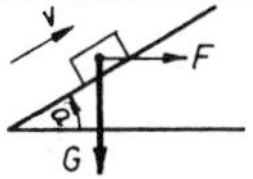 $F = G\,\tan(\alpha+\varrho)$
downwards $0 < \alpha < \varrho$	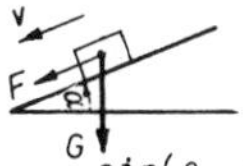 k 53 $F = G\,\dfrac{\sin(\varrho-\alpha)}{\cos\varrho}$	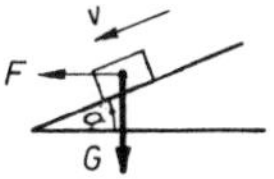 $F = G\,\tan(\varrho-\alpha)$
downwards $\varrho < \alpha < \alpha^*$	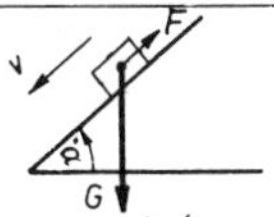 k 54 $F = G\,\dfrac{\sin(\alpha-\varrho)}{\cos\varrho}$	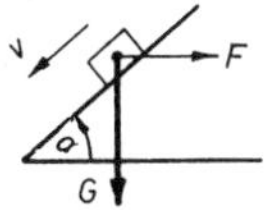 $F = G\,\tan(\alpha-\varrho)$

Note: For static friction replace μ by μ_0 and ϱ by μ_0.

α^*: Tilting angle of body

Wedges

		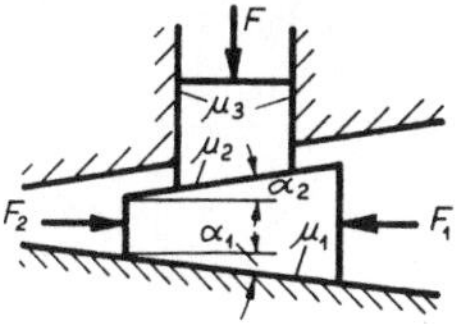	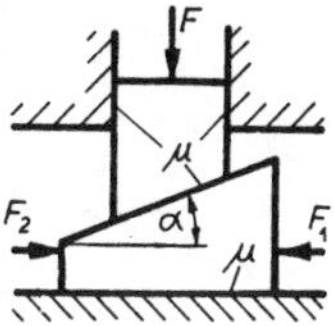
k 55	driving in	$F_1 = F \dfrac{\tan(\alpha_1+\rho_1)+\tan(\alpha_2+\rho_2)}{1-\tan\rho_3\times\tan(\alpha_2+\rho_2)}$	$F_1 = F\times\tan(\alpha+2\rho)$
k 56	loosening	$F_2 = F \dfrac{\tan(\alpha_1-\rho_1)+\tan(\alpha_2-\rho_2)}{1+\tan\rho_3\times\tan(\alpha_2-\rho_2)}$	$F_2 = F\times\tan(\alpha-2\rho)$
k 57	automatic locking	$\alpha_1 + \alpha_2 \leqq \rho_{0_1} + \rho_{0_2}$	$\alpha \leqq 2\rho_0$

Screws

			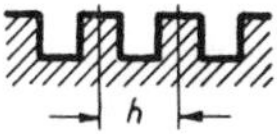	
k 58	turning moment when	raising	$M_1 = F\times r\times\tan(\alpha+\rho)$	$M_1 = F\times r\times\tan(\alpha+\rho')$
k 59		lowering	$M_2 = F\times r\times\tan(\alpha-\rho)$	$M_2 = F\times r\times\tan(\alpha-\rho')$
k 60	conditions of automatic locking when lowering		$\alpha < \rho$	$\alpha < \rho'$
k 61	efficiency of screw when	raised	$\eta = \dfrac{\tan\alpha}{\tan(\alpha+\rho)}$	$\eta = \dfrac{\tan\alpha}{\tan(\alpha+\rho')}$
k 62		lowered	$\eta = \dfrac{\tan(\alpha-\rho)}{\tan\alpha}$	$\eta = \dfrac{\tan(\alpha-\rho')}{\tan\alpha}$

M_1 : raising moment — N m, [kgf m]

M_2 : lowering moment — N m, [kgf m]

k 63 α : lead of thread $\left(\tan\alpha = \dfrac{h}{2\pi r}\right)$

k 64 ρ : angle of friction $(\tan\rho = \mu)$

ρ' : angle of friction $\left(\tan\rho' = \dfrac{\mu}{\cos\beta/2}\right)$

k 65 r : mean radius of thread — m, mm

Bearing friction

	radial bearing	longitudinal bearing
	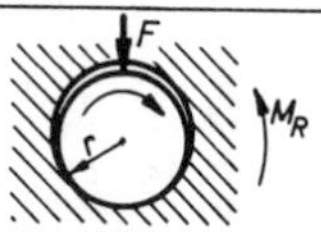	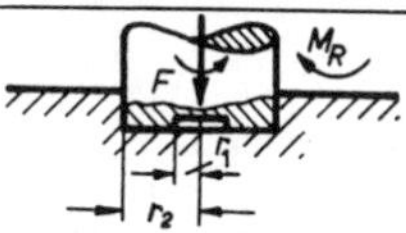
k 66	$M_R = \mu_q \, r \, F$	$M_R = \mu_l \, \frac{r_1 + r_2}{2} \, F$

M_R : moment of friction

μ_q :	coefficient of	radial	bearing (not
μ_l :	friction of a	longitudinal	constant values)

Note: μ_q and μ_l are determined experimentally as a function of bearing condition, bearing clearance, and lubrication. For run

k 67 in condition: $\mu_o \approx \mu_l \approx \mu_q$. Always use $r_1 > 0$ to allow for lu-

k 68 brication.

Rolling resistance

Rolling of a cylinder

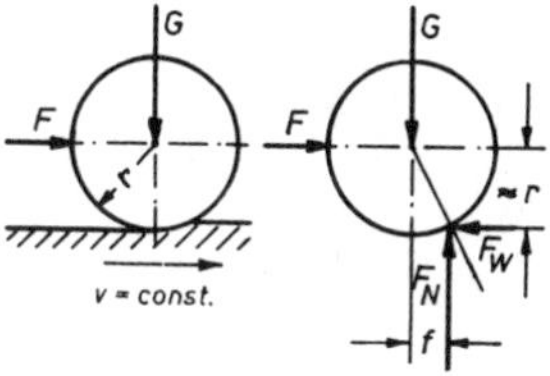

k 69 $F = \frac{f}{r} F_N \approx \frac{f}{r} G$

k 70 Rolling condition: $F_W < \mu_o F_N$

F_W: force of rolling resistance

f : lever arm of rolling resistance – value on Z 7 (caused by deformation of cylinder and support)

μ_o: coefficient of static friction between cylinder and support

Displacement of a plate supported by cylinders

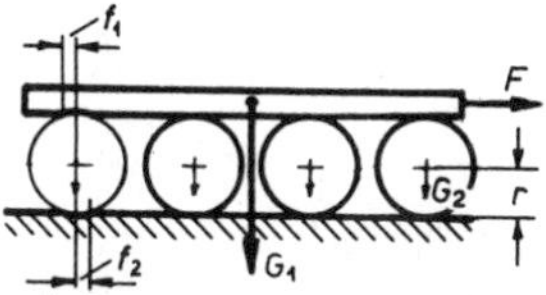

k 71 $F = \frac{(f_1 + f_2)\, G_1 + n \, f_2 \, G_2}{2\, r}$

k 72 where $f_1 = f_2 = f$ and $n\, G_2 < G_1$:

k 73 $F = \frac{f}{r} G_1$

G_1 , G_2 : weight of plate and one cylinder

f_1 , f_2 : lever arms of force of rolling resistance

F : tractive force

r : radius of cylinder

n : number of cylinders

Rope friction

	traction force and friction force for raising load	lowering load
k 75	$F_1 = e^{\mu\hat{\alpha}} G$	$F_2 = e^{-\mu\hat{\alpha}} G$
k 76	$F_R = (e^{\mu\hat{\alpha}} - 1) G$	$F_R = (1 - e^{-\mu\hat{\alpha}}) G$

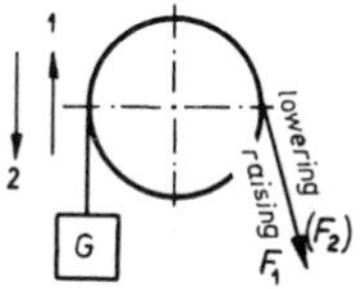

Formulae apply where cylinder is stationary and rope is moving at constant velocity, or where rope is stationary and cylinder is rotating at constant angular velocity.

k 77 **Condition of equilibrium:** $F_2 < F < F_1$ | $G\, e^{-\mu\hat{\alpha}} < F < G\, e^{\mu\hat{\alpha}}$

(F : force without friction)

Belt drive

k 78 $F_U = \dfrac{M_a}{r}$

k 79 $F_U = F_R$

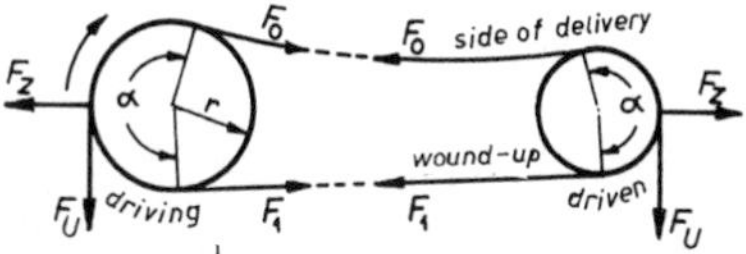

	forces	in motion	at rest
k 80	F_o	$F_o = \dfrac{F_U}{e^{\mu\hat{\alpha}} - 1}$	$F_o = F_1 = \dfrac{F_z}{2} \dfrac{(e^{\mu\hat{\alpha}} + 1)}{(e^{\mu\hat{\alpha}} - 1)}$
k 81	F_1	$F_1 = F_U \dfrac{e^{\mu\hat{\alpha}}}{e^{\mu\hat{\alpha}} - 1}$	
k 82	F_z	$F_z = F_U \dfrac{e^{\mu\hat{\alpha}} + 1}{e^{\mu\hat{\alpha}} - 1}$	

F_U : tangential force of driving wheel
F_R : frictional force of rope
M_a : driving torque
$\hat{\alpha}$: angle of contact (radians). Always introduce lowest value into formula
μ : coefficient of sliding friction (value of experience for leather
k 83 belt on steel drum: $\mu = 0{\cdot}22 + 0{\cdot}012\, v$ s/m
v : belt velocity
e = 2·718 281 83 . . . (base of natural logs)

Rope operated machines

The following figures deal solely with rope rigidity, disregarding bearing friction.

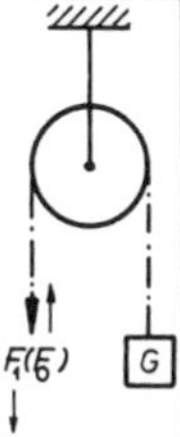

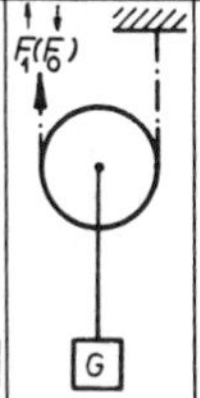

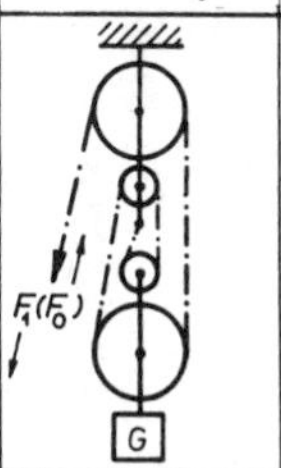

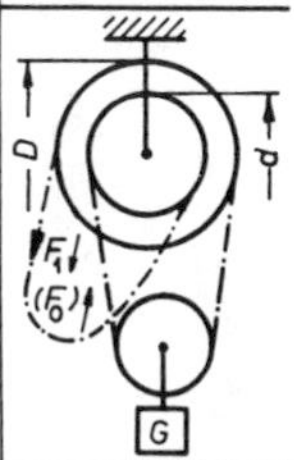

	unknown quantity	fixed sheave	free sheave	pulley block ordinary	pulley block differential
k 84	$F_1 =$	$\varepsilon\, G$	$\dfrac{\varepsilon}{1+\varepsilon}\, G$	$\dfrac{\varepsilon^n(\varepsilon-1)}{\varepsilon^n-1}\, G$	$\dfrac{\varepsilon^2-\dfrac{d}{D}}{\varepsilon+1}\, G$
k 85	$F_0 =$	$\dfrac{1}{\varepsilon}\, G$	$\dfrac{1}{1+\varepsilon}\, G$	$\dfrac{\dfrac{1}{\varepsilon^n}\left(\dfrac{1}{\varepsilon}-1\right)}{\dfrac{1}{\varepsilon^n}-1}\, G$	$\dfrac{\varepsilon}{1+\varepsilon}\left(\dfrac{1}{\varepsilon^2}-\dfrac{d}{D}\right) G$
k 86	$F =$	G	$\dfrac{1}{2}\, G$	$\dfrac{1}{n}\, G$	$\dfrac{1}{2}\left(1-\dfrac{d}{D}\right) G$
k 87	$s =$	h	$2\, h$	$n\, h$	$\dfrac{2}{1-\dfrac{d}{D}}\, h$

k 88 mechanical advantage $= \dfrac{\text{force}}{\text{effort}} = \dfrac{F}{G} = \dfrac{h}{s}$

F_1 : force required to raise load, disregarding bearing friction
F_0 : force required to lower load, disregarding bearing friction
F : force, disregarding both rope rigidity and bearing friction

k 89 $\varepsilon = \dfrac{1}{\eta}$: loss factor for rope rigidity (for wire ropes and chains ≈ 1·05)

η : efficiency
n : number of sheaves
h : path of load
s : path of force

General

Kinematics deals with motions of bodies as a function of time.

The most important quantities of kinematics and their units

Length l, see K 1

Units: m; km

Rotational angle φ

Unit: rad

Time t

Is a base quantity, see preface.

Units: s; min; h

Frequency f

The frequency of a harmonic or sinusoidal oscillation is the ratio of the number of periods (full cycles) and the corresponding time.

I 1 $$f = \frac{\text{number of oscillations}}{\text{corresponding time}}$$

Units: Hz (Hertz) = 1/s = cycle/s; 1/min

Period T

The period T is the time required for one full cycle. It is the reciprocal of the frequency f.

I 2 $$T = \frac{1}{f}$$

Units: s; min; h

Rotational speed n

Where an oscillation is tightly coupled with the rotation of a shaft, and one revolution of the shaft corresponds exactly to one full cycle of the oscillation, the rotational speed n of the shaft is equal to the frequency f of the oscillation.

I 3 $$n = f \; ; \quad f = \frac{n \text{ min}}{60 \text{ s}}$$

Units: revolutions/second (1/s)
rev./min. r.p.m. (1/min)

continued on L 2

continued from L 1

Velocity v

The velocity v is the first derivative of the distance s with respect to the time t:

I 4 $$v = \frac{ds}{dt} = \dot{s}$$

Where the velocity is constant, the following relation applies:

I 5 $$v = \frac{s}{t}$$

Units: m/s; km/h

Angular velocity ω, angular frequency ω

The angular velocity ω is the first derivative of the angle turned through φ, with respect to the time t:

I 6 $$\omega = \frac{d\varphi}{dt} = \dot{\varphi}$$

Hence, for constant angular velocity:

I 7 $$\omega = \frac{\varphi}{t}$$

Where $f = n$ (see I 3), the angular velocity ω is equal to the angular frequency ω.

I 8 $$\omega = 2\pi f = 2\pi n = \dot{\varphi}$$

Units: 1/s; rad/s; 1°/s

Acceleration a

The acceleration a is the first derivative of the velocity v with respect to the time t:

I 9 $$a = \frac{dv}{dt} = \dot{v} = \frac{d^2s}{dt^2} = \ddot{s}$$

Units: m/s^2; km/h^2

Angular acceleration α

The angular acceleration α is the first derivative of the angular velocity ω with respect to the time t:

I 10 $$\alpha = \frac{d\omega}{dt} = \dot{\omega} = \frac{d^2\varphi}{dt^2} = \ddot{\varphi}$$

Units: $1/s^2$; rad/s^2; $1°/s^2$

Distance, velocity, and acceleration of mass point in motion

Distance-time curve

An s–t curve is recorded for the motion. The first derivative of this curve is the instantaneous velocity v:

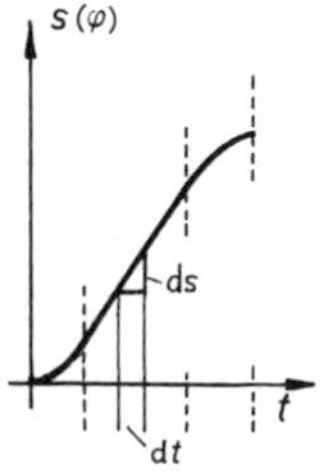

I 11 $$v = \frac{ds}{dt} = \dot{s}$$

It is the slope d*t* the tangent to the s–t curve.

Velocity-time curve

The velocity-time history is expressed as a v–t curve. The first derivative of this curve is the instantaneous acceleration a. Hence the acceleration is the second derivative of the distance-time curve.

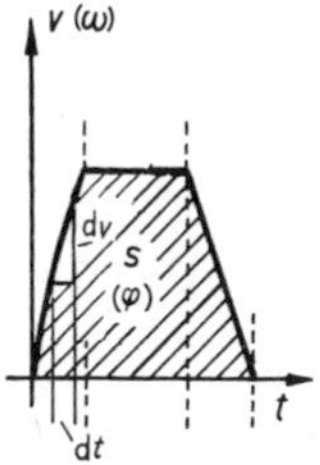

I 12 $$a = \frac{dv}{dt} = \dot{v} = \ddot{s}$$

It is the slope d*t* the tangent to the v–t curve.

The shaded area represents the distance travelled $s\ (t)$.

Acceleration-time curve

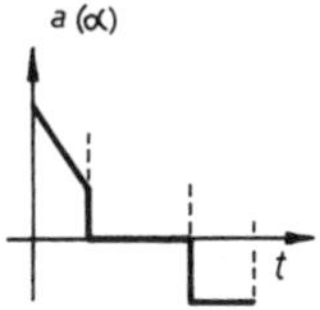

The acceleration-time history is shown as an a–t curve, which enables peak accelerations to be determined.

I 13 $a > 0$: acceleration (increasing velocity)

I 14 $a < 0$: retardation (decreasing velocity)

Note to diagrams

The letters in brackets apply to rotations (for explanation see L 5 and L 6).

KINEMATICS

The most important kinds of motion

Linear motion

Paths are straight lines. All points of a body cover congruent paths.

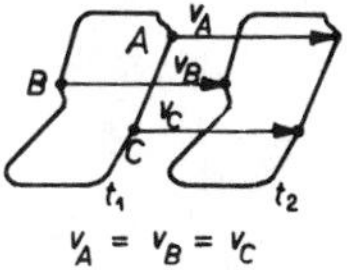

Special linear motions

uniform	uniform accelerated
motion	
I 15 $\upsilon = \upsilon_o =$ const.	$a = a_o =$ constant

Rotational motion

Paths are circles about the axis. Angle turned through $\widehat{\varphi}$, angular velocity ω, and angular acceleration α are identical for all points of the body.

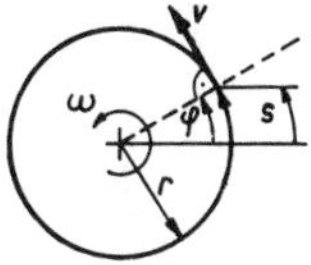

Special rotational motions

uniform	uniform accelerated
motion	
I 16 $\omega = \omega_o =$ const.	$\alpha = \alpha_o =$ constant

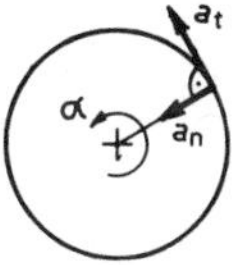

Distance s, velocity υ, and tangential acceleration a_t are proportional to the radius:

I 17 $s = r\widehat{\varphi}\ ; \quad \upsilon = r\omega \ ; \quad a = r\alpha = a_t$

I 18 centripetal acceleration $a_n = \omega^2 r = \dfrac{\upsilon^2}{r}$

Harmonic oscillation

Paths are straight lines or circles. The body moves back and forth about a position of rest. The maximum deflection from this position is called "amplitude".

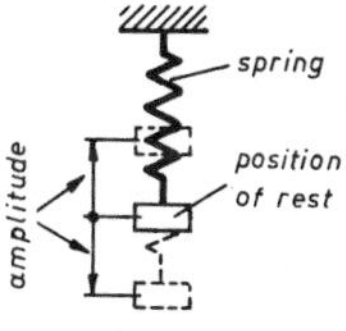

Instantaneous position, velocity, and acceleration are harmonic functions of time.

Uniform and uniform accelerated linear motion

	unknown parameter	uniform $a = 0$ $v =$ const.	uniform { accelerated ($a > 0$) / retarded ($a < 0$) $a =$ constant $v_0 = 0$	 $v_0 > 0$	EU
		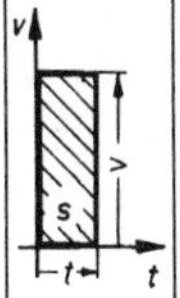	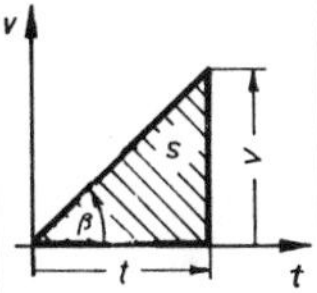	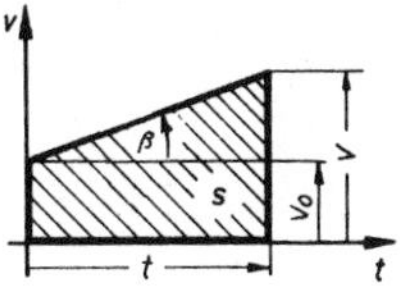	
I 19	$s =$	$v\,t$	$\frac{v\,t}{2} = \frac{a\,t^2}{2} = \frac{v^2}{2\,a}$	$\frac{t}{2}(v_0 + v) = v_0 t + \frac{1}{2} a\,t^2$	m cm km
I 20	$v =$	$\frac{s}{t}$	$\sqrt{2as} = \frac{2\,s}{t} = a\,t$	$v_0 + a\,t = \sqrt{v_0^2 + 2as}$	m/s cm/s km/h
I 21	$v_0 =$	constant	0	$v - a\,t = \sqrt{v^2 - 2as}$	
I 22	$a =$	0	$\frac{v}{t} = \frac{2s}{t^2} = \frac{v^2}{2s}$	$\frac{v - v_0}{t} = \frac{v^2 - v_0^2}{2s}$	m/s^2 cm/h^2 km/h^2
I 23	$t =$	$\frac{s}{v}$	$\sqrt{\frac{2\,s}{a}} = \frac{v}{a} = \frac{2\,s}{v}$	$\frac{v - v_0}{a} = \frac{2s}{v_0 + v}$	s min h

Note

The shaded areas represent the distance s covered during the time period t.

The tangent β represents the linear acceleration a

Uniform and uniform accelerated rotation about a fixed axis

	unknown parameter	uniform $\alpha = 0$, $\omega =$ const.	uniform { accelerated ($\alpha > 0$) / retarded ($\alpha < 0$) }, $\alpha =$ constant: $\omega_0 = 0$	$\omega_0 > 0$	EU
		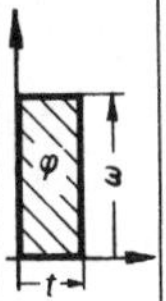	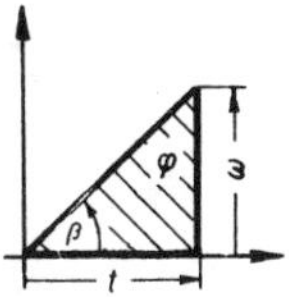	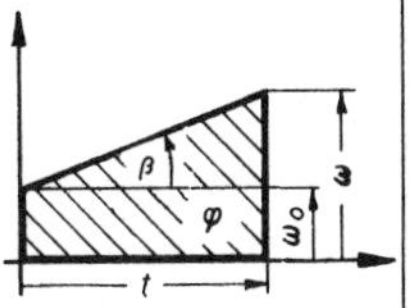	
I 24	$\varphi =$	ωt	$\frac{\omega t}{2} = \frac{\alpha t^2}{2} = \frac{\omega^2}{2\alpha}$	$\frac{t}{2}(\omega_0 + \omega) = \omega_0 t + \frac{1}{2}\alpha t^2$	– rad
I 25	$\omega =$	$\frac{\varphi}{t}$	$\sqrt{2\alpha\varphi} = \frac{2\varphi}{t} = \alpha t$	$\omega_0 + \alpha t = \sqrt{\omega_0^2 + 2\alpha\varphi}$	1/s m/ms rad/s
I 26	$\omega_0 =$	constant	0	$\omega - \alpha t = \sqrt{\omega^2 - 2\alpha\varphi}$	
I 27	$\alpha =$	0	$\frac{\omega}{t} = \frac{2\varphi}{t^2} = \frac{\omega^2}{2\varphi}$	$\frac{\omega - \omega_0}{t} = \frac{\omega^2 - \omega_0^2}{2\varphi}$	$1/s^2$ m/ms^2 rad/s^2
I 28	$t =$	$\frac{\varphi}{\omega}$	$\sqrt{\frac{2\varphi}{\alpha}} = \frac{\omega}{\alpha} = \frac{2\varphi}{\omega}$	$\frac{\omega - \omega_0}{\alpha} = \frac{2\varphi}{\omega_0 + \omega}$	s min h

Note

The shaded areas represent the angle of rotation φ covered during a time period t.
(Angle of rotation $\varphi = 2\pi \times$ number of rotations respective $360^\circ \times$ number of rotations).

The tangent β represents the angular acceleration α.

Linear simple harmonic motion

A body supported by a spring performs a linear harmonic oscillation. For this kind of motion, quantities s, v, and a as functions of time are equal to the projections $\underline{s}$, $\underline{v}$, and $\underline{a_n}$ of a uniform rotation of a point.

uniform rotation	path (project.)	simple harmonic motion

position — **position-time curve**

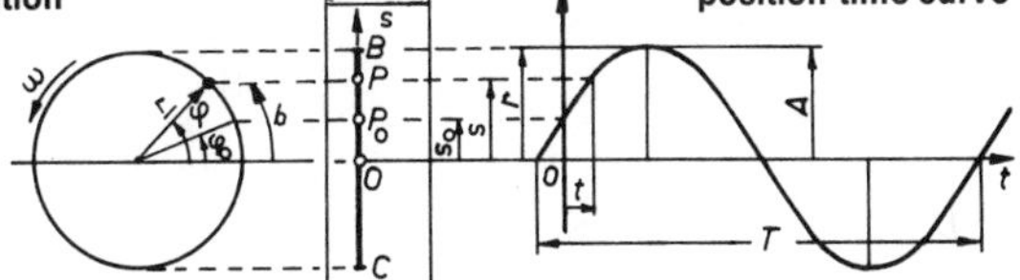

I 29 $\varphi = \omega t + \varphi_0$; $b = r(\omega t + \varphi_0)$ $\qquad$ $s = A \sin(\omega t + \varphi_0)$

velocity — **velocity-time curve**

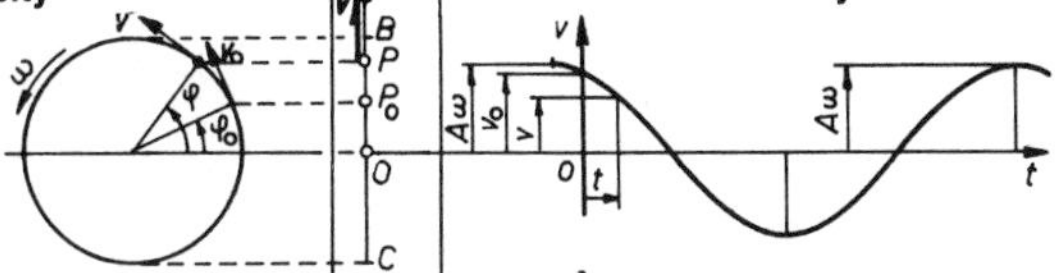

I 30 $v = r\,\omega$ $\qquad$ $v = \dfrac{ds}{dt} = A\omega \cos(\omega t + \varphi_0)$

acceleration — **acceleration-time curve**

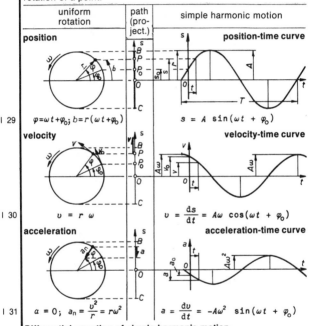

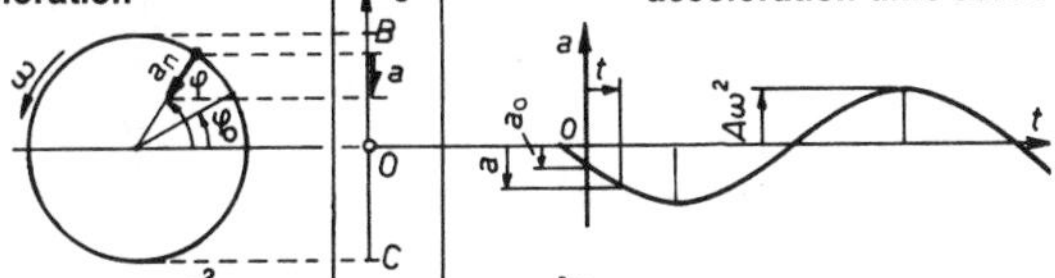

I 31 $a = 0$; $a_n = \dfrac{v^2}{r} = r\omega^2$ $\qquad$ $a = \dfrac{dv}{dt} = -A\omega^2 \sin(\omega t + \varphi_0)$

Differential equation of simple harmonic motion

I 32 $$a = \frac{d^2 s}{dt^2} = -\omega^2 s$$

φ_0	angular position	at time $t = 0$	s : displacement
φ		at time t	A : amplitude (max. displacement)

a_n : centripetal acceleration $\qquad$ r : radius of circle

$\underline{r}$: radius vector (origin: centre of circle; head: position of body)

B, C : extreme positions of oscillating point

KINEMATICS

Free fall and projection

L 8

Free fall and vertical projection

	un-known param.	free fall $v_0 = 0$	vertical project. } upwards ($v_0 > 0$) / downwards ($v_0 < 0$)	EU
		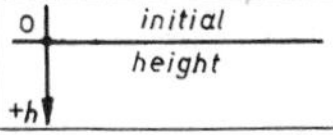	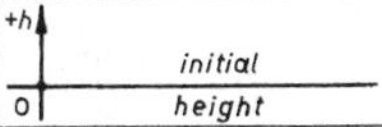	
I 33	$h =$	$\frac{g}{2}t^2 = \frac{v}{2}t = \frac{v^2}{2g}$	$v_0 t - \frac{g}{2}t^2 = \frac{v_0 + v}{2}t$	m cm
I 34	$v =$	$g\,t = \frac{2h}{t} = \sqrt{2gh}$	$v_0 - g\,t = \sqrt{v_0^2 - 2gh}$	m/s km/h
I 35	$t =$	$\frac{v}{g} = \frac{2h}{v} = \sqrt{\frac{2h}{g}}$	$\frac{v_0 - v}{g} = \frac{2h}{v_0 + v}$	s min

Horizontal and angled projection

	un-known param.	horizontal projection $v_0 > 0$	angled project. } upwards ($\alpha > 0$) / downwards ($\alpha < 0$) $v_0 > 0$	EU
		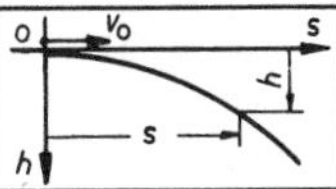	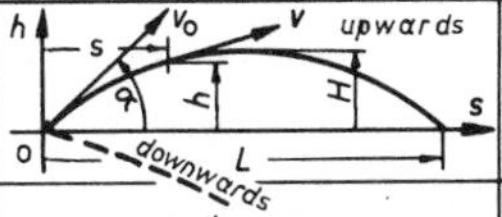	
I 36	$s =$	$v_0 t = v_0\sqrt{\frac{2h}{g}}$	$v_0 t \cos\alpha$	m cm
I 37	$h =$	$\frac{g}{2}t^2$	$v_0 t \sin\alpha - \frac{g}{2}t^2$	m cm
I 38	$v =$	$\sqrt{v_0^2 + g^2 t^2}$	$\sqrt{v_0^2 - 2gh}$	m/s km/s

Range L and max. height H for an angled upwards projection

I 39	general	$L = \frac{v_0^2}{g}\sin 2\alpha$	$H = \frac{v_0^2}{2g}\sin^2\alpha$	m cm
I 40		$t_L = \frac{2v_0}{g}\sin\alpha$	$t_H = \frac{v_0}{g}\sin\alpha$	s min
	maxima	at $\alpha = 45^\circ$	at $\alpha = 90^\circ$	
I 41		$L_{max} = \frac{v_0^2}{g}$	$H_{max} = \frac{v_0^2}{2g}$	m cm
I 42		$t_{L_{max}} = \frac{v_0\sqrt{2}}{g}$	$t_{H_{max}} = \frac{v_0}{g}$	s min

α : angle of projection with respect to horizontal plane

t_H : time for height H | t_L : time for distance L

Motion on an inclined plane

Sliding motion on an inclined plane

	unknown param.	excluding friction $\mu = 0$	including friction $\mu > 0$
I 43	$a =$	$g \sin\alpha$	$g(\sin\alpha - \mu\cos\alpha)$
I 44			otherwise $g \dfrac{\sin(\alpha - \varrho)}{\cos\varrho}$
I 45	$v =$	$a\,t = \dfrac{2s}{t} = \sqrt{2as}$	
I 46	$s =$	$\dfrac{t^2}{2} = \dfrac{vt}{2} = \dfrac{v^2}{2a}$	
	α	$0 \ldots \alpha^*$	$\varrho_0 \ldots \alpha^*$

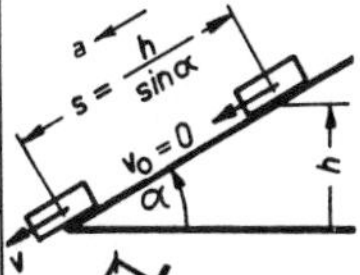

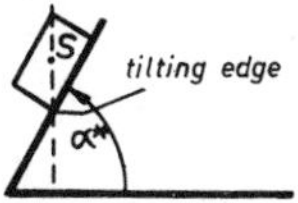

Rolling motion on an inclined plane

	unknown param.	excluding friction $f = 0$	including friction $f > 0$
I 47	$a =$	$\dfrac{g\,r^2}{r^2 + k^2}\sin\alpha$	$g r^2 \dfrac{\sin\alpha - \dfrac{f}{r}\cos\alpha}{r^2 + k^2}$
I 48	$v =$	see above I 45	
I 49	$s =$	see above I 46	
I 50	α	$0 \ldots \alpha_{max}$	α_{min}: $\tan\alpha_{min} = \dfrac{f}{r}$
I 51		$\tan\alpha = \mu_0 \dfrac{r^2 + k^2}{k^2}$	α_{max}: $\tan\alpha_{max} = \mu_0 \dfrac{r^2 + k^2 + f\,r}{k^2}$

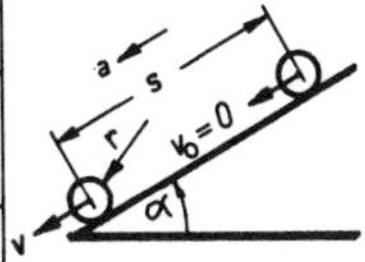

	ball	solid cylinder	pipe with low wall thickness
I 52	$k^2 = \dfrac{2}{5} r^2$	$k^2 = \dfrac{r^2}{2}$	$k^2 = \dfrac{r_1^2 + r_2^2}{2} \approx r^2$

α^* : tilting angle, where centre of gravity S vertically above tilting edge

μ : coefficient of sliding friction (see Z 7)

μ_0 : coefficient of static friction (see Z 7)

I 53 ϱ : angle of sliding fricton ($\mu = \tan\varrho$)

I 54 ϱ_0 : angle of static friction ($\mu_0 = \tan\varrho_0$)

f : lever arm of rolling resistance (see k 70 and Z 7)

k : radius of gyration

Simple Conn-Rod mechanism

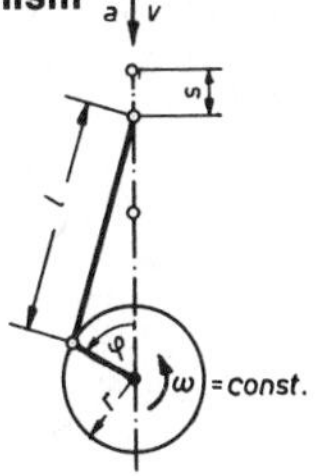

I 55 $s = r(1 - \cos\varphi) + \frac{\lambda}{2} r \sin^2\varphi$

I 56 $v = \omega r \sin\varphi(1 + \lambda \cos\varphi)$

I 57 $a = \omega^2 r(\cos\varphi + \lambda \cos 2\varphi)$

I 58 $\lambda = \frac{r}{l} = \frac{1}{4} \dots \frac{1}{6}$

I 59 $\varphi = \omega t = 2\pi n t$

(λ is called the crank ratio)

Scotch-Yoke mechanism

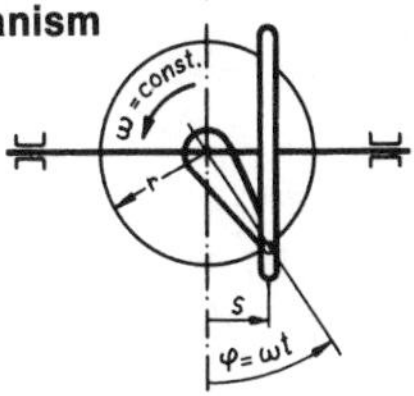

I 60 $s = r \sin(\omega t)$

I 61 $v = \omega r \cos(\omega t)$

I 62 $a = -\omega^2 r \sin(\omega t)$

I 63 $\omega = 2\pi n$

(motion is simple harmonic)

Cardan joint

For uniform drive the off-drive will be

non-uniform	uniform due to auxiliary shaft H	

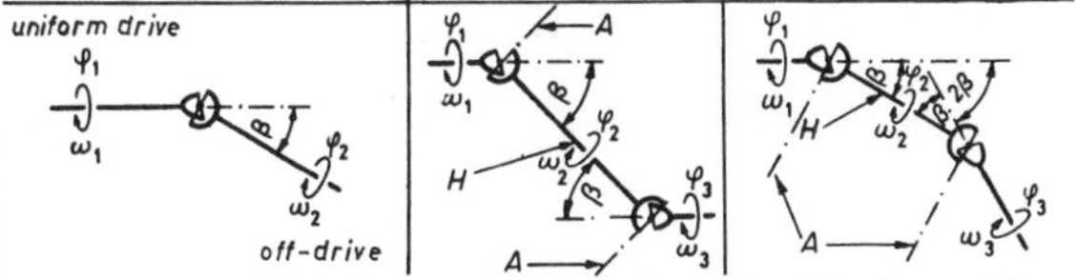

For all shafts being located in one plane the following relations apply:

	non-uniform	uniform due to auxiliary shaft H	
I 64	$\tan\varphi_2 = \tan\varphi_1 \times \cos\beta$	$\tan\varphi_3 = \tan\varphi_1$	$\tan\varphi_3 = \tan\varphi_1$
I 65	$\omega_2 = \omega_1 \frac{\cos\beta}{1 - \sin^2\beta \times \sin^2\varphi_1}$	$\omega_3 = \omega_1$	$\omega_3 = \omega_1$

I 66 $$\alpha_2 = \omega_1^2 \frac{\sin^2\beta \times \cos\beta \times \sin 2\varphi_1}{(1 - \sin^2\beta \times \sin^2\varphi_1)^2}$$

Both axes A of the auxiliary shaft joints must be parallel to each other.

The more the angle of inclination β increases, the more the max. acceleration α and the accelerating moment M_α become; therefore, in practice $\beta \leq 45^\circ$.

General

Dynamics deals with the forces acting on bodies in motion and with the terms "work, energy, and power".

The most important quantities of dynamics and their units

Mass m (is a base quantity, see preface)

Units: kg; Mg = t; g

1 kg is the mass of the international standard. A mass is measured by means of a steelyard.

Force (gravitational force) F

The force F is the product of mass m and acceleration a.

m 1 $$F = m a$$

The gravitational force W is the force acting on a mass m due to the earth's acceleration g:

m 2 $$W = m g$$

Being a gravitational force the weight W is measured by means of a spring balance.

Units: N; (kgf; lbf)

1 N is the force that, when acting on a body of a mass $m = 1$ kg for 1 s, accelerates this body to a final velocity of 1 m s^{-1} (i. e. accelerates this body at 1 m s^{-2}). 9·81 N (= 1 kgf) is the gravitational force acting on a mass of 1 kg due to the earth's attraction.

Work W

The mechanical work is the product of force F and distance s, where the constant force F acts on a body in linear motion in a direction parallel to the distance s covered ($W = F s$).

Units: N m = Joule = J = W s; (kgf m; ft lbf).

Where a force of 1 N acts over a distance of 1 m, it produces the work (energy) of 1 N m (J).

Power P

The power P is the derivative of work with respect to time. Where work (energy) increases or decreases linearily with time, power is the quotient of work and time ($P = W/t$).

Units: W (Watt); (kgf m s^{-1}; H. P.)

Where for a period of 1 s an energy of 1 J is converted at a constant rate, the corresponding power is 1 W.

$$1 \text{ W} = 1 \text{ J/s}$$

Definition of the mass moment of inertia I

The mass moment of inertia of a body about an axis has been defined as the sum of the products of mass-elements and the squares of their distance from the axis.

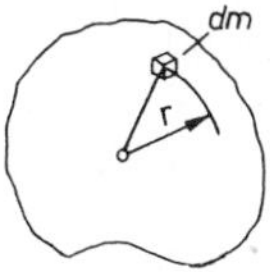

m 3 $$I = \sum r^2 \Delta m = \int r^2 \, dm \qquad \text{kg m}^2$$

Steiner's Theorem (Parallel axis theorem) (see also J 9)

Where the mass moment of inertia of a body of mass m about an axis through its centre of gravity S–S is I_{SS}, the mass moment of inertia about a parallel axis O–O at a distance l_S will be:

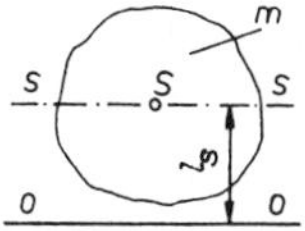

m 4 $$I_{OO} = I_{SS} + m\, l_S^2 \qquad \text{kg m}^2$$

Radius of gyration k

The radius of gyration of a body of mass m and mass moment of inertia I is the radius of an imaginary cylinder of infinitely small wall thickness having the same mass and the same mass moment of inertia as the body in question.

m 5 hence $$k = \sqrt{\frac{I}{m}} \qquad \text{m, cm, mm}$$

Flywheel effect

m 6 Flywheel effect $$W k^2 = m g k^2 = g I \qquad \text{kg cm}^3\,\text{s}^{-2}, \text{N m}^2$$

(k^2 formulae see M 3)

Equivalent mass (for rolling bodies)

m 8 $$m_{eq} = \frac{I}{k^2} \qquad \text{kg}$$

Basic formulae

	linear motion formulae	units	rotational motion formulae	units
m 9	$F_a = m\,a$	N , (kgf)	$M_a = I\,\alpha$	N m , (kgf m)
m 10	$W = F\,s$ (F = const.)	N m , (kgf m)	$W = M\,\varphi$ (M = const.)	N m , (kgf m)
m 11	$W_K = \frac{1}{2} m\,v^2$	J , (kgf m)	$W_K = \frac{1}{2} I\,\omega^2$	J , (kgf m)
m 12	$W_P = m\,g\,h$	J , (kgf m)	$\omega = 2\,\pi\,n$	s^{-1} , min^{-1}
m 13	$W_F = \frac{1}{2} F\,\Delta l$	J , (kgf m)	$W_F = \frac{1}{2} M\,\Delta\beta$	W s , (kgf m)
m 14	$P = \frac{dW}{dt} = F\,v$	W , kW	$P = \frac{dW}{dt} = M\,\omega$	W , kW

For explanation of symbols see M 4

	Mass moment of inertia about axis *a–a* (turning axis)	Mass moment of inertia about axis *b–b* passing through centre of gravity *S*	type of body
m 15	$I = m r^2$	$I = \frac{1}{2} m r^2$	circular hoop 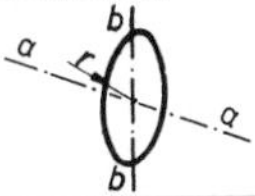
m 16	$k^2 = r^2$	$k^2 = \frac{1}{2} r^2$	
m 17	$I = \frac{1}{2} m r^2$	$I = \frac{m}{12} (3 r^2 + h^2)$	cylinder 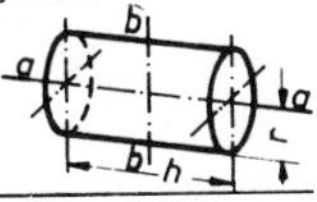
m 18	$k^2 = \frac{1}{2} r^2$	$k^2 = \frac{1}{12} (3r^2 + h^2)$	
m 19	$I = \frac{1}{2} m(R^2 + r^2)$	$I = \frac{m}{12}(3R^2+3r^2+h^2)$	hollow cylinder 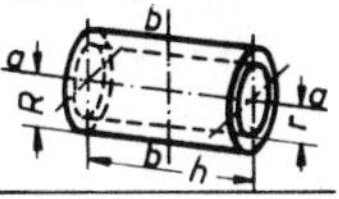
m 20	$k^2 = \frac{1}{2} (R^2 + r^2)$	$k^2 = \frac{1}{12}(3R^2+3r^2+h^2)$	
m 21	$I = \frac{3}{10} m r^2$	$I = \frac{3}{80} m(4r^2 + h^2)$	cone 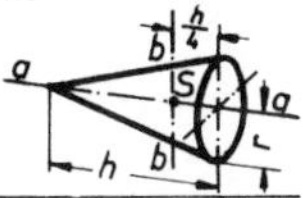
m 22	$k^2 = \frac{3}{10} r^2$	$k^2 = \frac{3}{80} (4r^2 + h^2)$	
m 23	$I = \frac{2}{5} m r^2$	$I = \frac{2}{5} m r^2$	sphere 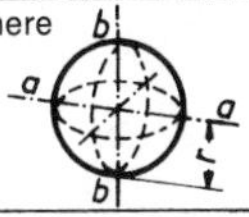
m 24	$k^2 = \frac{2}{5} r^2$	$k^2 = \frac{2}{5} r^2$	
m 25	$I = m(R^2+\frac{3}{4}r^2)$	$I = m \frac{4R^2 + 5r^2}{8}$	torus 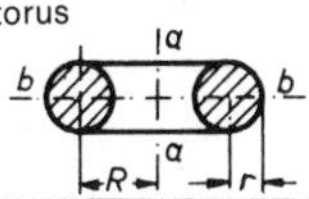
m 26	$k^2 = R^2 + \frac{3}{4}r^2$	$k^2 = \frac{1}{8}(4R^2 + 5r^2)$	
m 27	short bar: $I = \frac{m}{12}(d^2+4l^2)$ thin bar $d, c \ll l$: $I = \frac{m}{3} l^2$	$I = \frac{m}{12} (d^2 + c^2)$	short bar, thin bar 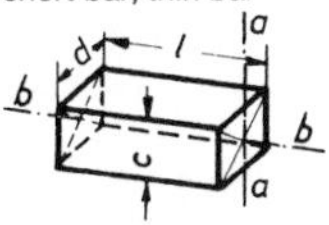
m 28	$k^2 = \frac{4}{3} l^2$	$k^2 = \frac{1}{3} (d^2 + c^2)$	

Total kinetic energy of a body

m 29 $W_K = \frac{1}{2} m v_s^2 + \frac{1}{2} I_S \omega^2$ J, [kgf m]

Kinetic energy of a rolling body – no sliding

m 30 $W_K = \frac{1}{2} (m + m_{eq}) v_s^2$ J, [kgf m]

m 31 $v_s = \omega r$ m/s, km/h

Rotational torque

m 32 $M = \frac{P}{\omega} = \frac{P}{2 \pi n}$ N m, [kgf m]

Transmission ratios

Transmission ratio

m 33 $i = \frac{d_2}{d_1} = \frac{z_2}{z_1} = \frac{n_1}{n_2} = \frac{\omega_1}{\omega_2}$

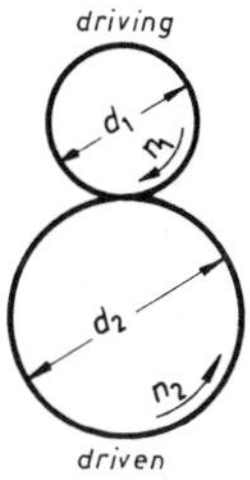

Torque ratio

m 34 $\frac{\text{moment of force}}{\text{moment of load}} = \frac{M_F}{M_L} = \frac{1}{i \eta}$

Efficiency

m 35 $\eta = \frac{\text{work produced}}{\text{work applied}} = \frac{\text{output}}{\text{input}}$

Overall efficiency for a series of transmissions

m 36 $\eta = \eta_1 \times \eta_2 \times \eta_3 \times \ldots$

---

m_{eq} : see m 8

v_s : velocity of linear motion of centre of gravity

F_a : accelerating force N, [kgf]

M_a : accelerating moment N m, [kgf m]

W_K : kinetic energy J, [kgf m]

W_P : potential energy J, [kgf m]

W_F : energy of helical spring under tension J, [kgf m]

Δl : extension of helical spring

$\Delta \hat{\beta}$: angular defection of spiral spring (in radians)

Centrifugal force

m 37 $F_z = m\,\omega^2 r = \dfrac{m\,v^2}{r}$ N, [kgf]

m 38 $ = 4\,\pi^2 m\,n^2 r$ N, [kgf]

m 39 $v = 2\,\pi\,r\,n$ m/s, km/h

m 40 $\omega = 2\,\pi\,n$ 1/s, 1/min

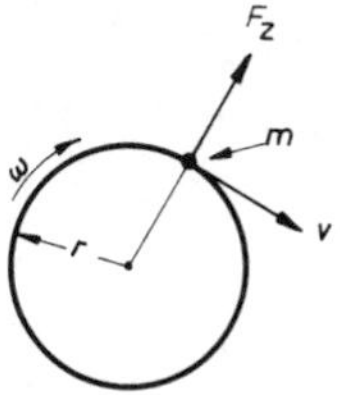

Stresses in rotation bodies (appr. formulae)

Disc

m 41 $\sigma_z = \dfrac{\omega^2 r^2 \varrho}{3} = \dfrac{v^2 \varrho}{3}$

N/m², [kgf/cm²]

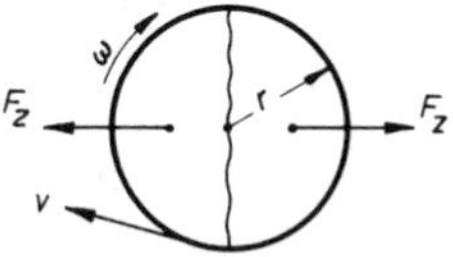

Ring

m 42 $\sigma_z = \dfrac{\omega^2 \varrho}{3}\,(r_1^{\,2} + r_1 r_2 + r_2^{\,2})$

N/m², [kgf/cm²]

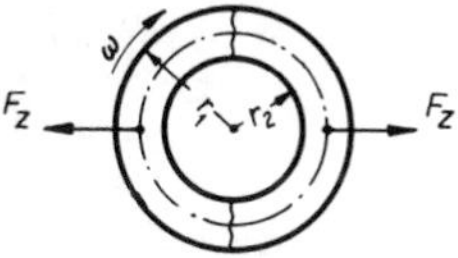

- l_s : distance from centre of gravity — m, cm, mm
- e : maximum pendulum swing — m, cm, mm
- f : instantaneous pendulum swing — m, cm, mm
- F_z : centrifugal force — N, [kgf, gf]
- I_O : mass moment of inertia about O — kg m², [kgf m s²]
- I_S : mass moment of inertia about S — kg m², [kgf m s²]
- M_1 : moment required to deflect spiral spring by 1 rad = 57·3° — N m, [kgf cm]
- σ_z : tensile stress — N/m², kgf/cm², [kgf/mm²]
- T : period of oscillation (B to B' and back) — s, min
- v_E : velocity at E — m/s, cm/s, km/h
- v_F : velocity at F — m/s, cm/s, km/h
- W_{KE}: kinetic energy at E — N m, [kgf m]

Mechanical oscillation

General

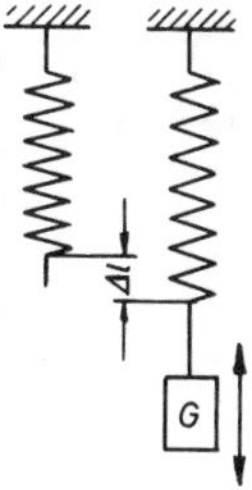

m 43 period $T = 2\pi\sqrt{\frac{m}{c}}$ s, min

m 44 stiffness $c = \frac{G}{\Delta l}$ N/m, [kgf/cm]

m 45 frequency $f = \frac{1}{T}$ (see L 1) s^{-1}, min^{-1}

m 46 angular velocity $\omega = 2\pi f = \sqrt{\frac{c}{m}}$ s^{-1}, min^{-1}

Critical speed n_c of shaft

m 47 $$n_c = \frac{1}{2\pi}\sqrt{\frac{c_q}{m}}$$

m 48 $$= 300\sqrt{\frac{10\ c_q\ \text{mm}}{9 \cdot 81\ \text{N}} \times \frac{\text{kg}}{m}} \quad \text{min}^{-1}$$

stiffness c_q of

2-bearing shaft, load symmetrical	2-bearing shaft, load asymmetrical	overhung (cantilever) shaft
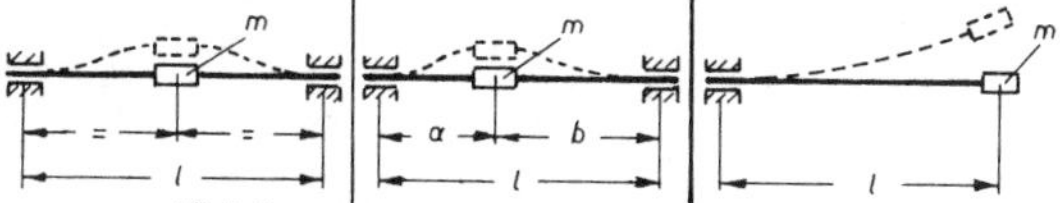		
m 49 $c_q = \frac{48\,E\,I}{l^3}$	$c_q = \frac{3\,E\,I\,l}{a^2\,b^2}$	$c_q = \frac{3\,E\,I}{l^3}$

Δl : deflection or elongation

I : second moment of area of shaft cross section

m : mass. When calculating the critical speed the mass m (e.g. of a belt disc) is assumed to be concentrated at a single point. The mass of the shaft should be allowed for by a slight increase

c_q : stiffness for transverse oscillations

Pendulum

(Explanations see L 4)

Conical pendulum

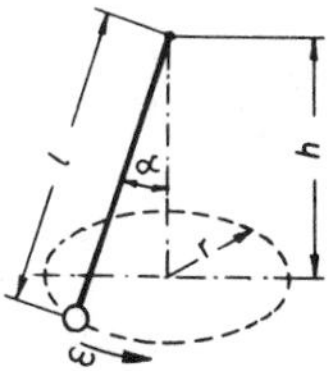

m 50 $T = 2\pi\sqrt{\dfrac{h}{g}} = 2\pi\sqrt{\dfrac{l\cos\alpha}{g}}$ s, min

m 51 $\tan\alpha = \dfrac{r\,\omega^2}{g} = \dfrac{r}{h}$

m 52 $\omega = \sqrt{\dfrac{g}{h}}$ | $h = \dfrac{g}{\omega^2}$ m, cm

Simple pendulum

The arm of a pendulum has no mass, the total mass is represented as a point.

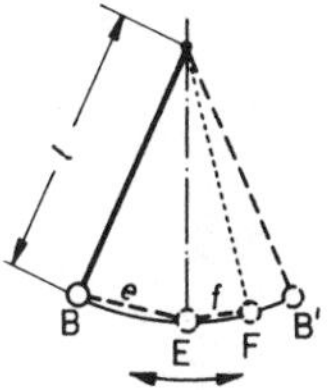

m 53 $T = 2\pi\sqrt{\dfrac{l}{g}}$ s, min

m 54 $v_E = e\sqrt{\dfrac{g}{l}}$ | $v_F = \sqrt{\dfrac{g}{l}(e^2 - f^2)}$ m/s, km/h

m 55 $W_{KE} = m\,g\,\dfrac{e^2}{2\,l}$ J, N m, [kgf cm]

Compound pendulum

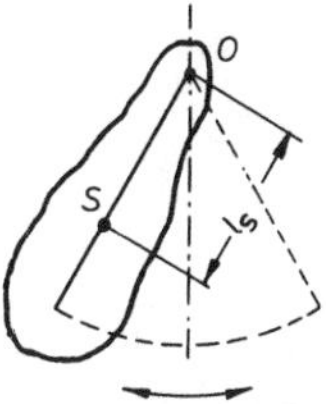

m 56 $T = 2\pi\sqrt{\dfrac{I_0}{m\,g\,l_S}} = 2\pi\sqrt{\dfrac{k_0^2}{g}}$ s, min

m 57 $I_O = I_S + m\,l_S^2$ N m s², [kgf cm s²]

m 58 $I_S = m\,g\,l_S\left(\dfrac{T^2}{4\pi^2} - \dfrac{l_S}{g}\right)$ N m s², [kgf cm s²]

If a body with centre of gravity S is suspended from O, distant l_S from S, and the period of swing determined the mass moment of inertia about O can be calculated using m 58.

Torsional pendulum

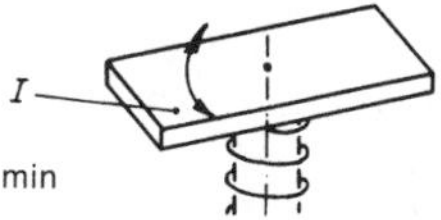

m 59 $T = 2\pi\sqrt{\dfrac{I}{M_1}}$ s, min

Impact

When two bodies of mass m_1 and m_2 and velocities v_{11} and v_{21} collide,
m 60 the total momentum $p = m\ v$ will remain constant over the whole impact period (velocities become v_{12} and v_{22}):

m 61 $$p = m_1\ v_{11} + m_2\ v_{21} = m_1\ v_{12} + m_2\ v_{22}$$

Impact-direction

direct and concentric impact	velocities parallel to normal to surfaces at point of impact	normal to surfaces at point of impact through centre of gravity of both bodies
oblique and concentric impact	any random velocities	
oblique and excentric impact		any random normal to surfaces at point of impact

Types of impact

		elastic impact⁺	plastic impact
	relative velocity	equal, before and after impact	equals zero after impact
m 62 m 63 m 64	velocity after direct and concentric impact	$v_{12} = \dfrac{v_{11}(m_1 - m_2) + 2m_2\ v_{21}}{m_1 + m_2}$ $v_{22} = \dfrac{v_{21}(m_2 - m_1) + 2m_1\ v_{11}}{m_1 + m_2}$	$v_{02} = \dfrac{m_1\ v_{11} + m_2\ v_{21}}{m_1 + m_2}$
m 65	coeff. of restitution	$\varepsilon = 1$	$\varepsilon = 0$

Coefficient of restitution ε

This indicates by what factor the relative velocities will vary before (v_{r1}) and after (v_{r2}) impact:

m 66 $$\varepsilon = \frac{v_{r2}}{v_{r1}}\ , \qquad \text{here} \qquad 0 \leqq \varepsilon \leqq 1$$

+) For an oblique, concentric, elastic impact the velocity vector v is split into a normal and a tangential component. The normal component v_n produces a direct impact (see above), the tangential component v_t has no effect on the impact.

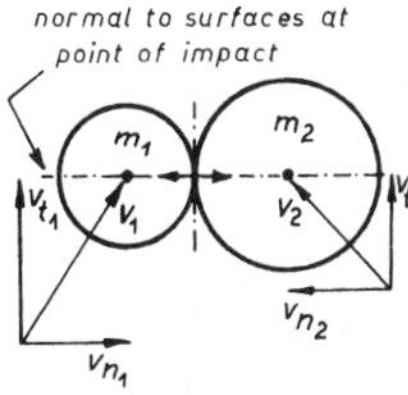

General

Hydraulics deals with the behaviour of liquids. With good approximation, liquids may be considered incompressible, which means that the influence of pressure changes on their densities is negligibly small.

Quantities

Pressure p see O 1

Density ϱ see O 1

Dynamic viscosity η EU: $\text{Pa s} = \frac{\text{kg}}{\text{m s}} = \frac{\text{N s}}{\text{m}^2} = (10\ \text{P})$

The dynamic viscosity is a material constant, which is a function of pressure and temperature:

n 1 $$\eta = f(p, t)$$

The dependence on pressure can often be neglected. Hence

n 2 $$\eta = f(t)$$ (for figures see Z 14)

Kinematic viscosity ν EU: $\text{m}^2/\text{s} = (10^4\ \text{St}) = (10^6\ \text{cSt})$

The kinematic viscosity is the quotient of dynamic viscosity η and density ϱ:

n 3 $$\nu = \frac{\eta}{\varrho}$$

Hydrostatics

Pressure distribution in a fluid

n 4 $$p_1 = p_0 + g\,\varrho\,h_1$$

n 5 $$p_2 = p_1 + g\,\varrho\,(h_2 - h_1)$$

$$= p_1 + g\,\varrho\,\Delta h$$

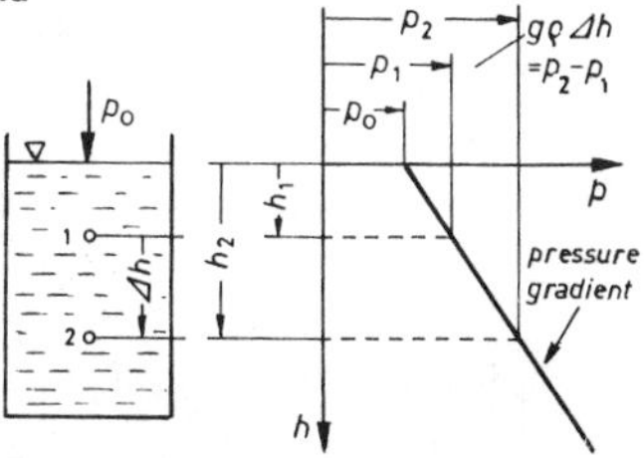

continued on N 2

Hydrostatic forces on plane surfaces

Hydrostatic force is the component acting on the surface which is caused by the weight of the fluid alone, i. e. without taking into account the atmospheric pressure p_0.

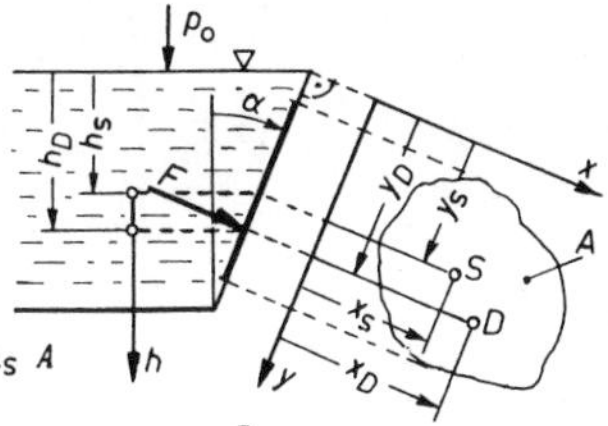

n 6 $$F = g\,\varrho\,y_s\,A\cos\alpha = g\,\varrho\,h_s\,A$$

n 7 $$y_D = \frac{I_x}{y_s\,A} = y_s + \frac{I_s}{y_s\,A}\quad ; \qquad x_D = \frac{I_{xy}}{y_s\,A}$$ m, mm

Hydrostatic forces on curved surfaces

The hydrostatic force acting on the curved surface 1, 2, is resolved into the horizontal component F_H and the vertical component F_V.

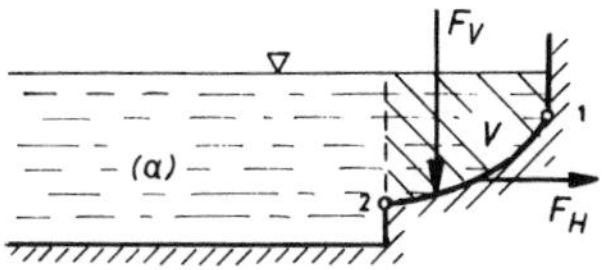

F_V is equal to the weight of the fluid in (*a*) or the equivalent weight of fluid (*b*), above the surface 1,2. The line of action runs through the centre of gravity of the volume *V*.

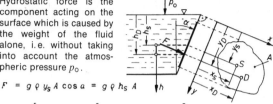

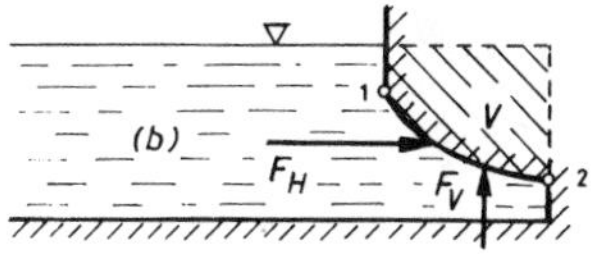

n 8 $$|F_V| = g\,\varrho\,V$$ N, kN

F_H is equal to the hydrostatic pressure force acting on the projection of the considered surface 1,2 on the plane perpendicular to F_H. Calculation is accomplished by n 6 and n 7.

S : centre of gravity of area A

D : centre of pressure = point of action of force F

I_x : second moment of area A in relation to axis x

I_s : second moment of area A in relation to an axis running parallel to axis x through the centre of gravity (see I 17 and P 10)

I_{xy} : centrifugal moment of area A in relation to axes x and y (see I 17).

Buoyancy

The buoyancy F_A is equal to the weight of the displaced fluids of densities ϱ and ϱ'.

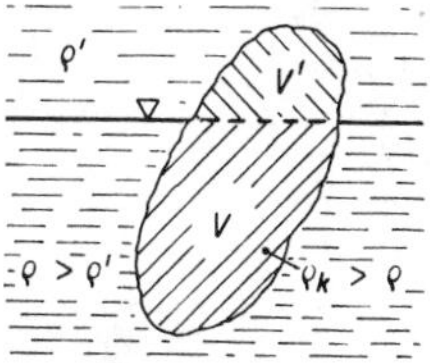

n 9 $$F_A = g\,\varrho\,V + g\,\varrho'\,V' \qquad \text{N, kN}$$

If the fluid of density ϱ' is a gas, the following formula is valid:

n 10 $$F_A \approx g\,\varrho\,V \qquad \text{N, kN}$$

With ϱ_k being the density of the body,

n 11 $\varrho > \varrho_k$ the body will float
n 12 $\varrho = \varrho_k$ " " " remain suspended } in the heavier fluid
n 13 $\varrho < \varrho_k$ " " " sink

Determination of density ϱ of solid and liquid bodies

Solid body of greater density than the fluid used	Solid body of smaller density than the fluid used	For fluids first determine F_1 and m of a deliberate body in a fluid of known density ϱ_b. This yields:
n 14 $\varrho = \varrho_F \dfrac{1}{1 - \dfrac{F}{m\,g}}$	n 15 $\varrho = \varrho_F \dfrac{1}{1 + \dfrac{F_H - F}{m\,g}}$	n 16 $\varrho = \varrho_b \dfrac{1 - \dfrac{F}{m\,g}}{1 - \dfrac{F_1}{m\,g}}$
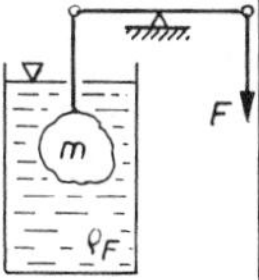	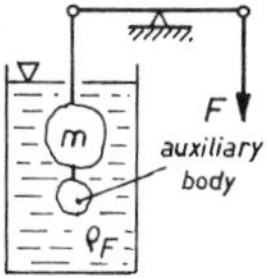	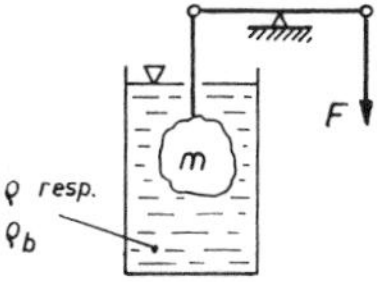

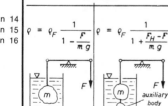

m : mass of the body remaining suspended in the fluid
F : equilibrium force necessary
F_H : equilibrium force necessary in the preliminary trial for the auxiliary body alone
ϱ_F : density of the fluid used

Hydrodynamics

(of a steady flow)

Continuity equation

Rule of conservation of mass:

n 17 $$A_1\, v_1\, \varrho_1 = A\, v\, \varrho = A_2\, v_2\, \varrho_2$$

n 18 $$= \dot{m} = \dot{V}\,\varrho \qquad \frac{\mathrm{g}}{\mathrm{s}},\ \frac{\mathrm{kg}}{\mathrm{s}}$$

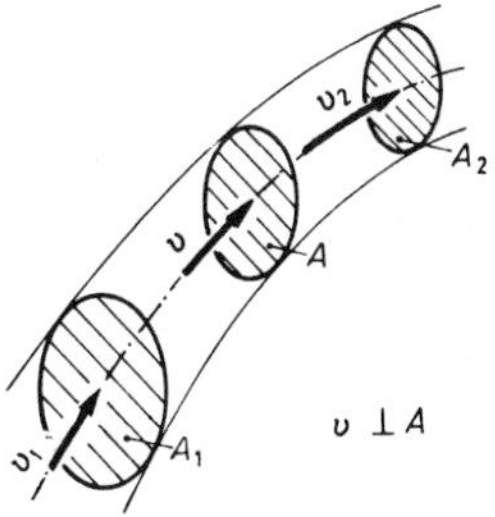

Rule of conservation of volume:

n 19 $$\dot{V} = A\, v \qquad \frac{\mathrm{m}^3}{\mathrm{s}},\ \frac{\mathrm{cm}^3}{\mathrm{s}}$$

$v \perp A$

Bernoulli's equation (Rule of conservation of energy)

No friction (ideal fluid):

n 20 $$\frac{p_1}{\varrho} + g\, z_1 + \frac{v_1^2}{2} = \frac{p}{\varrho} + g\, z + \frac{v^2}{2} = \frac{p_2}{\varrho} + g\, z_2 + \frac{v_2^2}{2} \qquad \frac{\mathrm{J}}{\mathrm{kg}}$$

$\frac{p}{\varrho}$: pressure energy per unit mass

$g\, z$: potential energy per unit mass

$\frac{v^2}{2}$: kinetic energy per unit mass

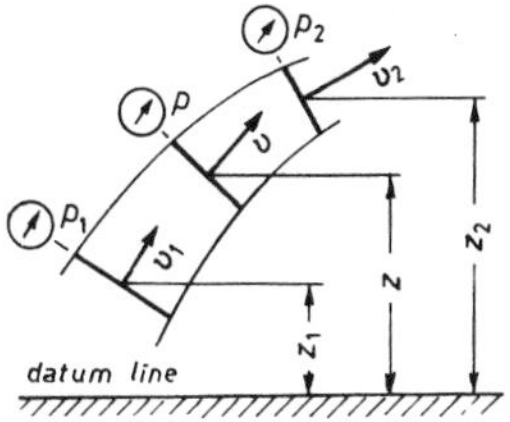

Including losses (real fluid):

n 21 $$\frac{p_1}{\varrho} + g\, z_1 + \frac{v_1^2}{2} = \frac{p_2}{\varrho} + g\, z_2 + \frac{v_2^2}{2} + w_{R\,1,2} \qquad \frac{\mathrm{J}}{\mathrm{kg}}$$

---

v : velocity

$w_{R\,1,2}$: resistance losses along path from 1 to 2 (per unit mass)

Power P of an hydraulic machine

n 22 $$P = \dot{m}\, w_{t\,1,2}$$ kW, W

technical work per unit mass:

n 23 $$w_{t\,1,2} = \frac{1}{\varrho}(p_2 - p_1) + g(z_2 - z_1) + \frac{1}{2}(v_2^2 - v_1^2) - w_{R\,1,2}$$ $\frac{\text{J}}{\text{kg}}$

n 24 for hydraulic machines: $w_{t\,1,2} < 0$

n 25 for pumps: $w_{t\,1,2} > 0$

Momentum equation

For a fluid flowing through a stationary reference volume the following vector equation is valid:

n 26 $$\Sigma \vec{F} = \dot{m}(\vec{v_2} - \vec{v_1})$$ N, kN

$\Sigma \vec{F}$ is the vector sum of the forces acting on the fluid contained in the reference volume. These can be:

volume forces (e. g. weight)
pressure forces
friction forces.

$\vec{v_2}$ is the exit velocity of the fluid leaving the reference volume

$\vec{v_1}$ is the entrance velocity of the fluid entering the reference volume.

Angular-momentum equation

In a steady state rotational flow a torque M is exerted on the fluid flowing through the reference volume, given by:

n 27 $$M = \dot{m}(v_{2,u} \times r_2 - v_{1,u} \times r_1)$$ N m

$v_{2,u}$ and $v_{1,u}$ are the circumferential components of exit velocity out of and entrance velocity into the reference volume.

r_2 and r_1 are the radii associated with v_2 and v_1.

Friction losses in pipe flow

n 28 Friction loss per unit mass $\quad w_{R\,1,2} = \Sigma(\zeta\, a \frac{v^2}{2})$, hence

n 29 Pressure loss $\quad \Delta p_V = \varrho\, w_{R\,1,2}$

Determination of coefficient of resistance ζ and coefficient of shape a:

	circular pipes	non circular pipes
n 30	$Re = \frac{v\, d\, \varrho}{\eta}$	$Re = \frac{v\, d_h\, \varrho}{\eta}$

n 31 Where $Re < 2000$, the flow is laminar.
n 32 Where $Re > 3000$, the flow is turbulent.
Where $Re = 2000 \ldots 3000$, the flow can be either laminar or turbulent.

	Flow: laminar	Flow: turbulent*)	Flow: laminar	Flow: turbulent*)
n 33	$\zeta = \frac{64}{Re}$	$\zeta = f(Re, \frac{k}{d})$	$\zeta = \varphi \frac{64}{Re}$	$\zeta = f(Re, \frac{k}{d_h})$
n 34	$a = \frac{l}{d}$ for straight pipes		$a = \frac{l}{d_h}$ for straight pipes	

n 35 $a = 1$ for fittings, unions and valves.

Determination of coefficient φ

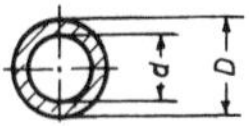

n 36 For annular cross sections:

D/d	1	3	5	7	10	30	50	70	100	∞
φ	1·50	1·47	1·44	1·42	1·40	1·32	1·29	1·27	1·25	1·00

n 37 For rectangular cross sections:

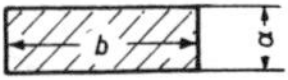

a/b	0	0·1	0·2	0·3	0·4	0·5	0·6	0·7	0·8	1·0
φ	1·50	1·34	1·20	1·10	1·02	0·97	0·94	0·92	0·90	0·89

d : internal diameter of pipe | l : length of pipe

n 38 $d_h = 4\, A/U$: hydraulic diameter

A : cross section perpendicular to fluid flow

U : wetted circumference

k/d and k/d_h : relative roughness

k : mean roughness (see Z 9)

η : dynamic viscosity (see N 1)

*) ζ is taken from diagram Z 8

Determination of coefficient

Flow of liquids from containers

Base apertures

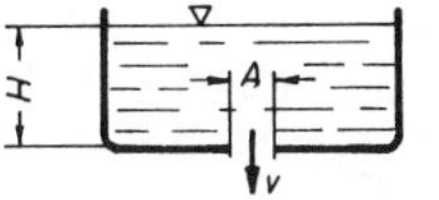

n 39 $$v = C_v \sqrt{2 g H}$$

n 40 $$Q = C_d A \sqrt{2 g H}$$

Small lateral apertures

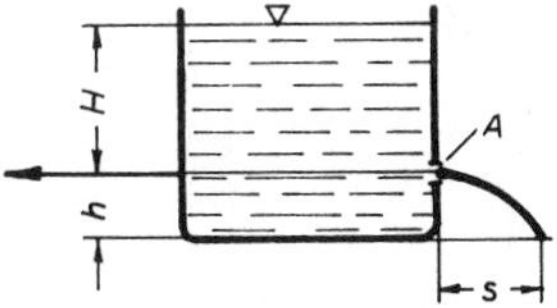

n 41 $$v = C_v \sqrt{2 g H}$$

n 42 $$s = 2 \sqrt{H h}$$

(without all friction values)

n 43 $$Q = C_d A \sqrt{2 g H}$$

n 44 $$F = \rho Q v$$

Large lateral apertures

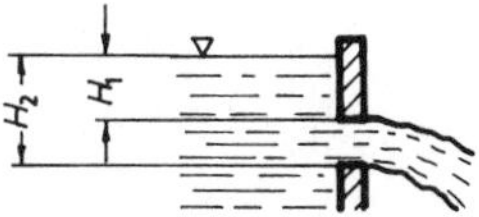

n 45 $$Q = \frac{2}{3} C_d b \sqrt{2 g} \left(H_2^{3/2} - H_1^{3/2}\right)$$

Excess pressure on surface of liquid

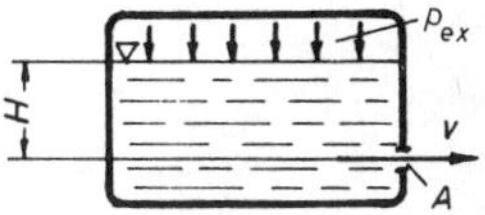

n 46 $$v = C_v \sqrt{2 \left(g H + \frac{p_{ex}}{\rho}\right)}$$

n 47 $$Q = C_d A \sqrt{2 \left(g H + \frac{p_{ex}}{\rho}\right)}$$

Excess pressure applied to an outlet point

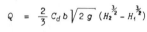

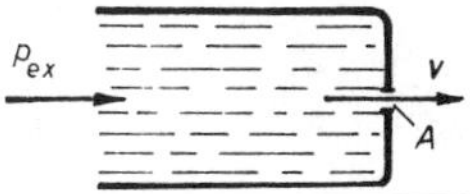

n 48 $$v = C_v \sqrt{2 \frac{p_{ex}}{\rho}}$$

n 49 $$Q = C_d A \sqrt{2 \frac{p_{ex}}{\rho}}$$

---

v : outlet velocity — m/s, km/h

p_{ex}: pressure in excess of atmosphere pressure — N/m², kgf/cm²

C_d : discharge coefficient ($C_d = C_c \times C_v$)

C_c : contraction coefficient ($C_c = 0 \cdot 62$ for sharp edge aperture)
($C_c = 0 \cdot 97$ for well rounded aperture)

C_v : velocity coefficient (for water $C_v = 0 \cdot 97$)

b : width of aperture — m, cm

F : reaction force — N, kgf

Q : volume of outlet flow — m³/s, m³/h

HEAT

Thermal variables of state

O 1

Thermal variables of state are pressure p, temperature t, and density ϱ or volume per unit mass (specific volume), respectively.

Pressure p EU: N/m^2 = Pa; bar

1 Pa = 1 N/m^2 = 10^{-5} bar = ($7 \cdot 5 \times 10^{-3}$ torr)

Pressure is the quotient of force F and area A:

o 1 $$p = \frac{F}{A}$$

The absolute pressure can be interpreted as the total result of the impacts of the molecules on the wall. The pressure measured with a pressure gauge is the pressure differential Δp in relation to the ambient pressure p_u. A state of pressure means $\Delta p > 0$, vacuum means $\Delta p < 0$. Thus, the absolute pressure p can be expressed by:

o 2 $$p = p_u + \Delta p$$

Temperature T, t (Base quantity) see explanations at front of book.

The unit of temperature T is the Kelvin K, defined by equation

o 3 $$1\ \text{K} = \frac{T_{TR}}{273 \cdot 16}$$

where T_{TR} is the temperature of pure water at the triple point. In addition to the Kelvin scale the centigrade scale is also used. The temperature of this scale has been internationally defined by:

o 4 $$t = \left(\frac{T}{\text{K}} - 273 \cdot 15\right) {}^\circ\text{C} \ ; \quad T = \left(\frac{t}{{}^\circ\text{C}} + 273 \cdot 15\right) \text{K}$$

Density ϱ EU: kg/m^3

Density is the quotient of mass m and volume V:

o 5 $$\varrho = \frac{m}{V}$$

Volume per unit mass (specific volume) v EU: m^3/kg

Specific volume is the quotient of the volume V and the mass m:

o 6 $$v = \frac{V}{m} = \frac{1}{\varrho}$$

Molecular volume V_m EU: m^3/mol

Molecular volume is the quotient of volume V and number of moles contained in the volume:

o 7 $$V_m = \frac{V}{n}$$

Amount of substance n (Base quantity) see explanations at front of book.

Heating of solid and liquid bodies

Heat (thermal energy) Q EU: J

Heat is energy exchanged between systems of different temperatures, where these systems interact with each other through diathermal walls.

Heat per unit mass q EU: J/kg

The heat per unit mass is the quotient of heat Q and mass m:

o 8 $$q = \frac{Q}{m}$$

Specific heat c EU: J/(kg K)

The specific heat c denotes the amount of heat Q to be supplied to or extracted from a substance of mass m to change its temperature by a difference Δt:

o 9 $$c = \frac{Q}{m\,\Delta t} = \frac{q}{\Delta t}$$

The specific heat is a function of temperature. For figures see Z 1 . . . Z 5.

Latent heats per unit mass l EU: J/kg – (values see Z 10)

The supply or extraction of latent heat causes a body to change its state without changing its temperature. The following latent heats exist:

<table>
<tr><td>o 10</td><td>l_f</td><td rowspan="3">latent heat of</td><td>fusion</td><td rowspan="3">is the heat necessary to convert a</td><td>solid body of the fusing temperature into a fluid</td><td rowspan="3">of the same temperature</td></tr>
<tr><td>o 11</td><td>l_d</td><td>vapourisation</td><td>fluid of the boiling temperature (dependent on pressure) into dry saturated vapour</td></tr>
<tr><td>o 12</td><td>l_s</td><td>sublimation</td><td>solid body of a temperature below its triple temperature at the sublimation temperature (dependent on pressure) directly into dry saturated vapour</td></tr>
</table>

HEAT

Heating of solid and liquid bodies

Expansion of solid bodies

A solid body changes its dimensions due to temperature changes. With α being the coefficient of linear expansion (for figures see Z 11) the following formulae are valid for:

o 13 Length: $l_2 = l_1 [1 + \alpha (t_2 - t_1)]$

o 14 $\Delta l = l_2 - l_1 \approx l_1 \alpha (t_2 - t_1)$

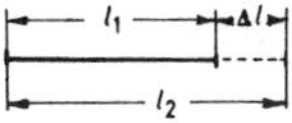

o 15 Area: $A_2 \approx A_1 [1 + 2\alpha (t_2 - t_1)]$

o 16 $\Delta A = A_2 - A_1 \approx A_1 2\alpha (t_2 - t_1)$

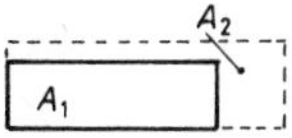

o 17 Volume: $V_2 \approx V_1 [1 + 3\alpha (t_2 - t_1)]$

o 18 $\Delta V = V_2 - V_1 \approx V_1 3\alpha (t_2 - t_1)$

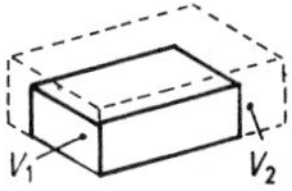

Expansion of liquid bodies

With γ being the coefficient of volume expansion (for figures see Z 11) the following formulae apply:

o 19 $V_2 = V_1 [1 + \gamma (t_2 - t_1)]$

o 20 $\Delta V = V_2 - V_1 = V_1 \gamma (t_2 - t_1)$

Bending due to heat A

Bimetallic strips are subject to bending due to heat. Bending occurs towards the side of the metal with the lower coefficient of expansion. With α_b being the "specific thermal bending" the bending due to heat can be calculating by (α_b approx. 14×10^{-6}/K, for exact values see manufacturers catalogues):

o 21 $A = \dfrac{\alpha_b L^2 \Delta t}{s}$

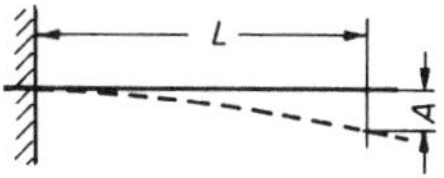

---

l_1	: length at $t = t_1$	A_1	: area at $t = t_1$		
l_2	: length at $t = t_2$	A_2	: area at $t = t_2$		
V_1	: volume at $t = t_1$	t_1	temper-	prior to	heating
V_2	: volume at $t = t_2$	t_2	ature	after	
s	: thickness	Δt	: temperature difference		

General equation of state of ideal gases

The state of ideal gas is determined by two thermal variables of state. Thus, the third variable can be calculated using the general equation of state. With R being the characteristic gas constant (different values for different gases, see Z 12) the equation reads as follows:

o 22 $$p\,v = R\,T \qquad \text{or} \qquad p\,V = m\,R\,T \qquad \text{or} \qquad p = \rho\,R\,T$$

If the gas constant is related to the mole volume, the equation reads

o 23 $$p\,V_m = R_m\,T$$

where $R_m = 8314 \cdot 3$ J/(kmol K) is the universal gas constant (valid for all ideal gases). R and R_m are related by

o 24 $$R_m = M\,R$$

where M is the molecular mass (see Z 12).

Thermal state of real (non ideal) gases and vapours

The thermal state of real gases and vapours is calculated using special equations or diagrams.

Changes of state

Changes of state are caused by interactions of the system with the surroundings. These interactions are calculated using the 1st and the 2nd law:

1st law for closed systems	1st law for open systems	2nd law for all systems
o 25 o 26 o 27 $q_{1,2} + w_{1,2} = u_2 - u_1$	$q_{1,2} + w_{t\,1,2} = h_2 - h_1 + \Delta e$	$q_{1,2} = \int_1^2 T\,ds$

In these formulae, energy input is positive (i. e. $q_{1,2}$, $w_{1,2}$, $w_{t\,1,2}$) and energy output negative.

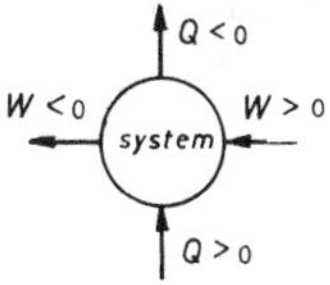

h : enthalpy per unit mass
s : entropy per unit mass
u : internal energy per unit mass
$w_{1,2}$: external work done per unit mass (see O 7)
$w_{t\,1,2}$: continuous external work done per unit mass (see O 7)
Δe : changes in kinetic or potential energies

Changes of state of ideal gases

The table on page 0 6 shows the relations for different changes of state, which have been developed from formulae o 25 to o 27. The following explanations apply:

Each change of state, may be represented by an equation

o 28 $$p\,v^n = \text{const.}$$

The various exponents n are given in column 1.

c_{pm} and c_{vm} are the mean specific heats for constant pressure and constant volume, respectively, in the temperature range between t_1 and t_2. There, the following relations apply (values for c_{pm} see Z 13):

o 29 $$c_{pm} = c_{pm}\Big|_{t_1}^{t_2} = \frac{c_{pm}\Big|_0^{t_2} t_2 - c_{pm}\Big|_0^{t_1} t_1}{t_2 - t_1}$$

o 30 $$c_{vm} = c_{vm}\Big|_{t_1}^{t_2} = c_{pm}\Big|_{t_1}^{t_2} - R$$

o 31 $$\gamma = \gamma_m = \gamma_m\Big|_{t_1}^{t_2} = c_{pm}\Big|_{t_1}^{t_2} \Big/ c_{vm}\Big|_{t_1}^{t_2}$$

The change of entropy occurring during the change of state is given by:

o 32 $$s_2 - s_1 = c_{pm}\ln\left(\frac{T_2}{T_1}\right) - R\ln\left(\frac{p_2}{p_1}\right) = c_{vm}\ln\left(\frac{T_2}{T_1}\right) + R\ln\left(\frac{v_2}{v_1}\right)$$

Changes of state of real gases and vapours

The table below shows the relations for different changes of state, which have been developed from formulae o 25 to o 27. The thermal variables of state, p, v, T as well as the properties, u, h, s are generally taken from appropriate diagrams.

	change of state, const. quantity	external work $w_{1,2} = -\int_1^2 p\,dv$	continuous external work $w_{t1,2} = \int_1^2 v\,dp$	heat per unit mass $q_{1,2}$
o 33	isochoric, v = const.	0	$v(p_2 - p_1)$	$u_2 - u_1 = (h_2 - h_1) - v(p_2 - p_1)$
o 34	isobaric, p = const.	$p(v_1 - v_2)$	0	$h_2 - h_1$
o 35	isothermal, T = const.	$(u_2 - u_1) - T(s_2 - s_1) = (h_2 - h_1) - T(s_2 - s_1) - (p_2 v_2 - p_1 v_1)$	$(h_2 - h_1) - T(s_2 - s_1)$	$T(s_2 - s_1)$
o 36	isentropic, s = const.	$u_2 - u_1 = (h_2 - h_1) - (p_2 v_2 - p_1 v_1)$	$h_2 - h_1$	0

process details, exponent	relation between state 1 and state 2	closed system, reversible $w_{1,2} = -\int_1^2 p\, dv$	open system, reversible $w_{t1,2} = \int_1^2 v\, dp$	heat transfer per unit mass $q_{1,2}$	p–v–diagram	T–s–diagram
isochoric v = const. $n = \infty$ (o 37)	$\frac{p_2}{p_1} = \frac{T_2}{T_1}$	0	$v(p_2 - p_1)$ $= R(T_2 - T_1)$	$c_{vm}(T_2 - T_1)$	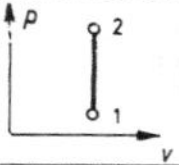	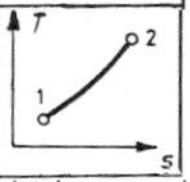
isobaric p = const. $n = 0$ (o 38)	$\frac{v_2}{v_1} = \frac{T_2}{T_1}$	$p(v_1 - v_2)$ $= R(T_1 - T_2)$	0	$c_{pm}(T_2 - T_1)$	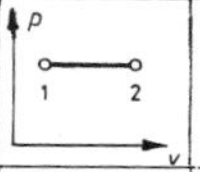	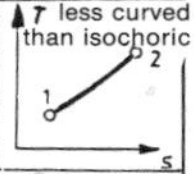
isothermal T = const. $n = 1$ (o 39)	$\frac{p_2}{p_1} = \frac{v_1}{v_2}$	$R\,T \ln\left(\frac{v_1}{v_2}\right)$ $= R\,T \ln\left(\frac{p_2}{p_1}\right)$	$w_{1,2}$	$-w_{1,2}$	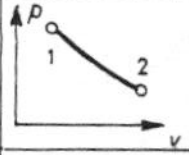	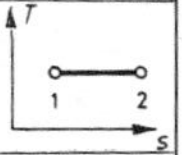
isentropic s = const. $n = \gamma$ (o 40)	$\frac{p_2}{p_1} = \left(\frac{v_1}{v_2}\right)^{\gamma}$ $\frac{p_2}{p_1} = \left(\frac{T_2}{T_1}\right)^{\frac{\gamma}{\gamma-1}}$ $\frac{v_2}{v_1} = \left(\frac{T_1}{T_2}\right)^{\frac{1}{\gamma-1}}$	$u_2 - u_1 = c_{vm}(T_2 - T_1)$ $= c_{vm}\, T_1 \left[\left(\frac{p_2}{p_1}\right)^{\frac{\gamma-1}{\gamma}} - 1\right]$ $= \frac{1}{\gamma-1} R\, T_1 \left[\left(\frac{p_2}{p_1}\right)^{\frac{\gamma-1}{\gamma}} - 1\right]$	$h_2 - h_1 = c_{pm}(T_2 - T_1)$ $= c_{pm}\, T_1 \left[\left(\frac{p_2}{p_1}\right)^{\frac{\gamma-1}{\gamma}} - 1\right]$ $= \frac{\gamma}{\gamma-1} R\, T_1 \left[\left(\frac{p_2}{p_1}\right)^{\frac{\gamma-1}{\gamma}} - 1\right]$	0	steeper than isothermal 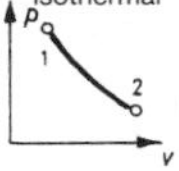	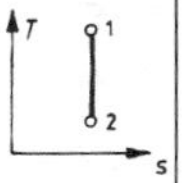
polytropic process n = const. (o 41)	$\frac{p_2}{p_1} = \left(\frac{v_1}{v_2}\right)^{n}$ $\frac{p_2}{p_1} = \left(\frac{T_2}{T_1}\right)^{\frac{n}{n-1}}$ $\frac{v_2}{v_1} = \left(\frac{T_1}{T_2}\right)^{\frac{1}{n-1}}$	$\frac{1}{n-1} R\,(T_2 - T_1)$ $= \frac{1}{n-1} R\, T_1 \left[\left(\frac{p_2}{p_1}\right)^{\frac{n-1}{n}} - 1\right]$	$\frac{n}{n-1} R\,(T_2 - T_1)$ $= \frac{n}{n-1} R\, T_1 \left[\left(\frac{p_2}{p_1}\right)^{\frac{n-1}{n}} - 1\right]$	$c_{vm} \frac{n-\gamma}{n-1}(T_2 - T_1)$	drawn to fit existing process	

p–v diagram

For reversible processes the area between the curve of the variation of state and the v-axis represents the external work per unit mass, the area between the curve and the p-axis represents the continuous external work per unit mass.

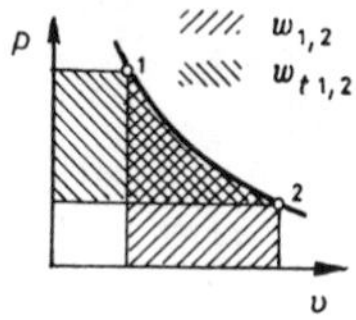

T–s diagram

For reversible processes the area between the curve and the s-axis represents the heat transfer per unit mass.

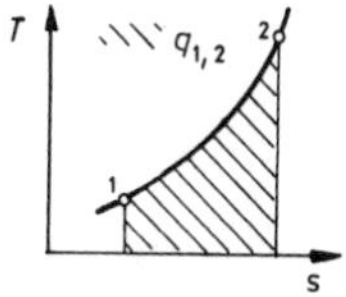

Total transfer of heat

The heat added to or removed from a closed system during a single variation of state is:

o 42 $$Q_{1,2} = m\,q_{1,2} \qquad \text{J}$$

The heat flow continuously added to or removed from an open system is:

o 43 $$\Phi_{1,2} = \dot{Q}_{1,2} = \dot{m}\,q_{1,2} \qquad \text{W}$$

where $\dot{m}$ is the mass flow (EU: kg/s).

Total transfer of work

The external work added to or done by a closed system during a single variation of state is:

o 44 $$W_{1,2} = m\,w_{1,2} \qquad \text{J}$$

The external power continuously added to or done by an open system is given by:

o 45 $$P_{1,2} = \dot{m}\,w_{t\,1,2} \qquad \text{W}$$

Mass m of a mixture of components m_1, m_2, ...

o 46 $$m = m_1 + m_2 + \ldots + m_n = \sum_{i=1}^{i=n} m_i$$

Mass fractions ξ_i of a mixture

o 47 $$\xi_i = \frac{m_i}{m} \quad \text{and} \quad \sum_{i=1}^{i=n} \xi_i = 1$$

Number of moles n of a mixture of components n_1, n_2, ...

o 48 $$n = n_1 + n_2 + \ldots + n_n = \sum_{i=1}^{i=n} n_i$$

Mole fractions ψ_i of a mixture

o 49 $$\psi_i = \frac{n_i}{n} \quad \text{and} \quad \sum_{i=1}^{i=n} \psi_i = 1$$

Equivalent molecular mass M of a mixture

For the molecular mass the following formulae apply:

o 50 $$M_i = \frac{m_i}{n} \quad \text{and} \quad M = \frac{m}{n}$$

where the equivalent molecular mass M of the mixture can be calculated as follows:

o 51 $$M = \sum_{i=1}^{i=n} (M_i \times \psi_i) \quad \text{and} \quad \frac{1}{M} = \sum_{i=1}^{i=n} \left(\frac{\xi_i}{M_i} \right)$$

Conversion between mass- and mole-fractions

o 52 $$\xi_i = \frac{M_i}{M} \psi_i$$

Pressure p of the mixture and partial pressures p_i of the components

o 53 $$p = \sum_{i=1}^{i=n} p_i \quad \text{where} \quad p_i = \psi_i \times p$$

continued on page O 9

Continuation of page O 8

Volume fractions r_i of a mixture

o 54 $$r = \frac{V_i}{V} = \psi_i \quad \text{and} \quad \sum_{i=1}^{i=n} r_i = 1$$

Here, by partial volume V_i we mean the volume the component would occupy at the temperature T and the total pressure p of the mixture. For ideal gases the following formulae apply:

o 55 $$V_i = \frac{m_i R_i T}{p} = \frac{n_i R_m T}{p} \quad \text{and} \quad \sum_{i=1}^{i=n} V_i = V$$

Internal energy u and enthalpy h of a mixture

o 56 $$u = \sum_{i=1}^{i=n} (\xi_i \times u_i) \quad ; \qquad h = \sum_{i=1}^{i=n} (\xi_i \times h_i)$$

Using these formulae, the temperature of the mixture can be determined, for real gases and vapours by using diagrams, and for ideal gases as follows:

o 57 internal energy	$t = \dfrac{c_{v_{m1}} \times t_1 m_1 + c_{v_{m2}} \times t_2 m_2 + \ldots + c_{v_{mn}} \times t_n m_n}{c_{v_m} \times m}$
o 58 enthalpy	$t = \dfrac{c_{p_{m1}} \times t_1 m_1 + c_{p_m} \times t_2 m_2 + \ldots + c_{p_{mn}} \times t_n m_n}{c_{p_m} \times m}$

where the specific heats of the mixture are determined as follows:

o 59 $$c_{v_m} = c_{p_m} - R$$

o 60 $$c_{p_m} = \sum_{i=1}^{i=n} (\xi_i \times c_{p_{m\,i}})$$

Due to the temperature difference between two points heat flows from the point of higher temperature towards the point of lower temperature. The following kinds of heat transmission must be distinguished:

Conduction

o 61 in a plane wall: $$\Phi = \dot{Q} = \lambda A \frac{t_{w_1} - t_{w_2}}{s}$$

o 62 in the wall of a pipe: $$\Phi = \dot{Q} = \lambda A_m \frac{t_{w_1} - t_{w_2}}{s}$$

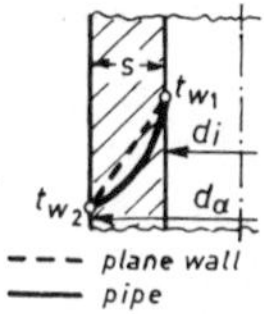

The mean logarithmic area is

o 63 $$A_m = \pi\, d_m L\ ; \quad \text{where} \quad d_m = \frac{d_a - d_i}{\ln\left(\dfrac{d_a}{d_i}\right)}$$

L : length of the pipe

Convection

By heat convection we mean the heat transfer in a fluid. Due to their flow the molecules as carriers of the mass are also the carriers of the heat.

Where the flow originates by itself, the convection is called natural convection. The convection taking place in a flow is called forced convection.

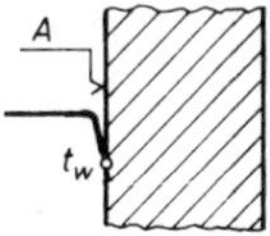

o 64 $$\Phi = \dot{Q} = \alpha A (t - t_w)$$

Radiation

This kind of heat transmission does not require mass as a carrier (e.g. the radiation of the sun through space). For calculations formula o 64 is used.

Heat transfer

By heat transfer we mean the combined result of the different processes contributing to the heat transmission:

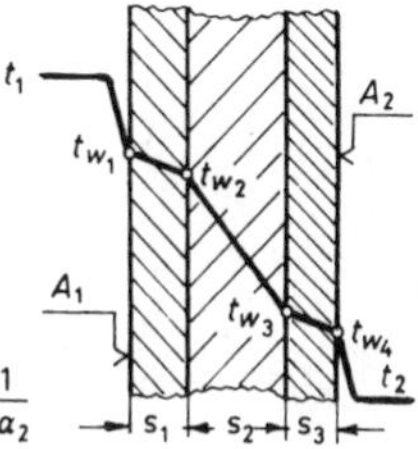

o 65 $$\Phi = \dot{Q} = k A (t_1 - t_2)$$

The heat transfer coefficient k is given by (for approx. values see Z 11):

o 66 plane wall: $$\frac{1}{k} = \frac{1}{\alpha_1} + \sum_{i=1}^{i=n} \left(\frac{s}{\lambda}\right)_i + \frac{1}{\alpha_2}$$

o 67 pipe: $$\frac{1}{k A} = \frac{1}{\alpha_1 A_1} + \sum_{i=1}^{i=n} \left(\frac{s}{\lambda A_m}\right) + \frac{1}{\alpha_2 A_2}$$

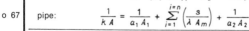

λ : thermal conductivity (for values see Z 1 ... Z 5)

α : heat transfer coefficient (for calculation see O 12)

Heat exchanger

A heat exchanger transmits heat from one fluid to another. The heat flow may be calculated by:

o 68 $$\Phi = \dot{Q} = k\,A\,\Delta t_m$$

Here, Δt_m is the logarithmic mean temperature difference. The following formula applies for both parallel-flow and counter-flow heat exchangers:

o 69 $$\Delta t_m = (\Delta t_{great} - \Delta t_{small}) \Big/ \ln \frac{\Delta t_{great}}{\Delta t_{small}}$$

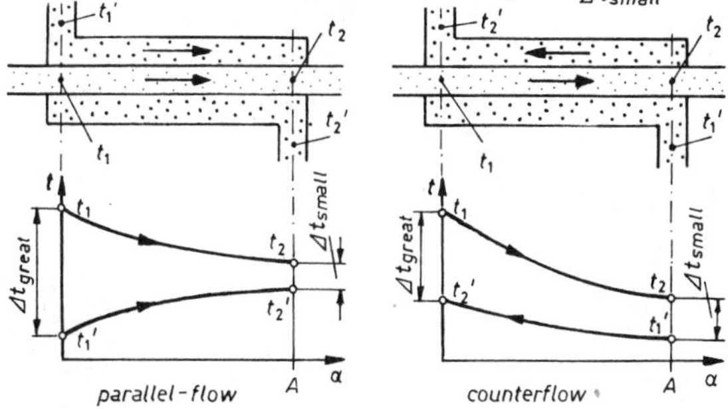

In counterflow operation Δt_{great} and Δt_{small}, can occur on the opposite ends of the exchanger, to that shown in the figure.

Symbols used on page O 12

A_1 : surface of the smaller body
A_2 : surface of the bigger body
d : inside diameter of pipe
D : outer diameter of pipe
Gr : Grashof'number
H : height of plate
L : length of plate
v : velocity
C_1 and C_2 : radiation constants of the surfaces exchanging radiation (for values see Z 12)
$C_s = 5{\cdot}67 \times 10^{-8}$ W/(m² K⁴): radiation constant of the black body
o 70 Pr : Prandtl-Number; $Pr = (\eta\, c_p)/\lambda$ (for values see Z 14)
$\Delta t = |\,t_w - t_\infty\,|$ absolute temperature difference between wall and fluid in the thermally not affected region
t_∞: ambient temperature
ν : kinematic viscosity ($\nu = \eta/\varrho$)
η : dynamic viscosity (for values see Z 14)
η_{Fl} : dynamic viscosity at mean temperature of fluid
η_W : dynamic viscosity at wall temperature
λ : thermal conductivity of fluid (for values see Z 5, Z 6)
γ : volume expansion coefficient (see Z 11 and o 77)
β^* : temperature factor

Calculation of heat transfer coefficient α [1)]

For free convection (according to Grigull)

o 71	on a vertical plate	$\alpha = \frac{Nu\,\lambda}{H}$	$Nu = 0{\cdot}55\sqrt[4]{Gr\,Pr}$, for $1700 < Gr\,Pr < 10^8$
o 72			$Nu = 0{\cdot}13\sqrt[3]{Gr\,Pr}$, for $Gr\,Pr > 10^8$
o 73			$Gr = \frac{g\,\gamma\,\Delta t\,H^3}{\nu^2} = \frac{g\,\gamma\,\Delta t\,\varrho^2\,H^3}{\eta^2}$
o 74	on a horizontal plate	$\alpha = \frac{Nu\,\lambda}{D}$	$Nu = 0{\cdot}41\sqrt[4]{Gr\,Pr}$, for $Gr\,Pr > 10^5$
o 75			$Gr = \frac{g\,\gamma\,\Delta t\,D^3}{\nu^2} = \frac{g\,\gamma\,\Delta t\,\varrho^2\,D^3}{\eta^2}$

o 76 Fluid properties must be related to reference temperature $t_B = \frac{t_W + t_\infty}{2}$

o 77 The expansion coefficient of gases is: $\beta_{gas} = 1/T_\infty$

For forced convection inside pipes (according to Hausen)

o 79 $$\alpha = Nu\,\lambda/d$$

o 80	flow	laminar $Re < 2000$	$Nu = \left[3{\cdot}65 + \frac{0{\cdot}0668\left(Re\,Pr\,\frac{d}{L}\right)}{1 + 0{\cdot}045\left(Re\,Pr\,\frac{d}{L}\right)^{2/3}}\right]\left(\frac{\eta_{Fl}}{\eta_W}\right)^{0,14}$
o 81			if $10^4 > Re\,Pr\frac{d}{L} > 10^{-1}$, where $Re = \frac{v\,d\,\varrho}{\eta}$
		turbulent $Re > 3000$	$Nu = 0{\cdot}116(Re^{2/3} - 125)Pr^{1/3}\left[1 + \left(\frac{d}{L}\right)^{2/3}\right]\left(\frac{\eta_{Fl}}{\eta_W}\right)^{0,14}$
			if $2320 < Re < 10^6$; $0{\cdot}6 < Pr < 500$; $1 > L/d < \infty$

With the exception of η_W all material values are related to the mean temperature of the fluid.

For gases factor $(\eta_{Fl}/\eta_W)^{0.14}$ must be omitted.

For radiation (heat transfer coefficient: α_{Str})

o 82 $$\alpha_{Str} = \beta^*\,C_{1,2}$$

o 83	between	parallel	surfaces	$\beta^* = \frac{T_1^4 - T_2^4}{T_1 - T_2}$	$C_{1,2} = \frac{1}{\frac{1}{C_1} + \frac{1}{C_2} - \frac{1}{C_s}}$
o 84					
o 85		enveloping			$C_{1,2} \approx \frac{1}{\frac{1}{C_1} + \frac{A_1}{A_2}\left(\frac{1}{C_2} - \frac{1}{C_s}\right)}$

1) α in J/(m² s K) or W/(m² K)

For explanation of symbols see O 11

STRENGTH

General terms

Stress

Stress is the ratio of applied force F and cross section A.

Tensile and compressive stresses occur at right angles to the cross section.

p 1 $$\sigma \text{ or } f = \frac{F}{A} \qquad \text{N/mm}^2$$

In calculations $\dfrac{\text{tensile}}{\text{compressive}}$ stresses are usually $\dfrac{\text{positive}}{\text{negative}}$

Shear stresses act along to the cross section.

p 2 $$\tau \text{ or } q = \frac{F}{A} \qquad \text{N/mm}^2$$

Stress-strain diagrams (tensile test)

Materials with

yield point (e. g. mild steel)	plastic yield (e. g. aluminium alloy)

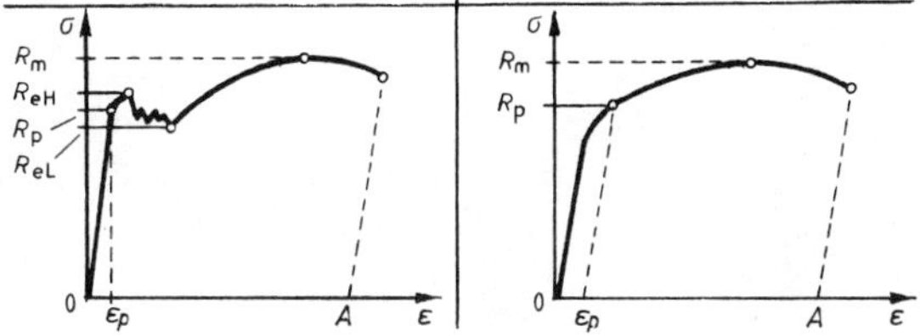

Notation: The standard symbols are from BS 18 and DIN 50145.

p 3 $$R_m = \frac{F}{S_o}\,;\quad \left[\sigma_B = \frac{F}{A_o}\right] \quad : \text{ tensile stress, where}$$

F : tensile force

S_o ; $[A_o]$: original cross section (of unloaded specimen)

p 4 $$\varepsilon = \frac{\Delta L}{L_o} \times 100\%\,;\quad \left[\varepsilon = \frac{\Delta l}{l_o} \times 100\%\right] \quad : \text{ strain, where}$$

L_o ; $[l_o]$: original length (of unloaded specimen)

ΔL ; $[\Delta l]$: change in length of loaded specimen

continued on P 2

STRENGTH

General terms

continued from P 1 (stress-strain diagram)

R_p ; $[\sigma_p]$: Proof stress or yield strength (offset)

The limit of proportionality is sometimes known as the elastic limit.

$\varepsilon_p \simeq 0 \cdot 01\% \Longrightarrow R_{p\,0\cdot01}$; $[\sigma_p]$

Yield point (ferrous metals)

R_{eH} ; $[\sigma_{So}]$: upper yield stress or upper yield point

R_{eL} ; $[\sigma_{Su}]$: lower yield stress or lower yield point.

Proof stress (non-ferrous metals)

$\varepsilon_p \simeq 0 \cdot 2\% \Longrightarrow R_{p\,0\cdot2}$; $[\sigma_{p\,0\cdot2}]$

p 5 $R_m = \dfrac{F_m}{S_o}$; $\left[\sigma_B = \dfrac{F_{max}}{A_o}\right]$: tensile strength

p 6 $A = \dfrac{\Delta L}{L_o} \times 100\%$; $\left[\delta = \dfrac{\Delta l_{max}}{l_o} \times 100\%\right]$: percentage elongation after fracture.

For specimens with circular cross sections, percentage elongations may be quoted, based on gauge lengths, e.g. A_5; $[\delta_5]$ is based on a gauge length of $5 \times \sqrt{\dfrac{4\,S_o}{\pi}}$ mm.

Permissible stress (allowable stress)

Must be below the elastic limit R_p, thus

the permissible stress is: $p_t = \dfrac{R_m}{\nu}$

R_m : yield strength of material

ν : safety factor, always greater than 1.

Ultimate safety factor against fracture)	Proof safety factor (against yield or 0·2 proof)
$\nu = 2 \ldots 3 \ldots 4$	$\nu = 1{\cdot}2 \ldots 1{\cdot}5 \ldots 2$

Loads

type	nature of stress	load diagram
I	dead	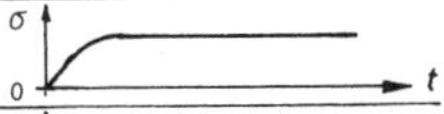
II	undulating	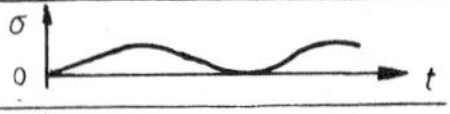
III	alternating	

Modulus of elasticity E: The relationship between σ and ε (Hooke's law) is applicable to the elastic range, i.e. below the elastic limit (see Z 16/17 for values of E). E is known as "Young's modulus".

p 7 $\sigma = E \times \varepsilon = E \times \Delta l / l_0 \; ; \qquad E = \sigma/\varepsilon = \sigma \times l_0/\Delta l$

Tensile and compressive stresses σ_t **and** σ_c

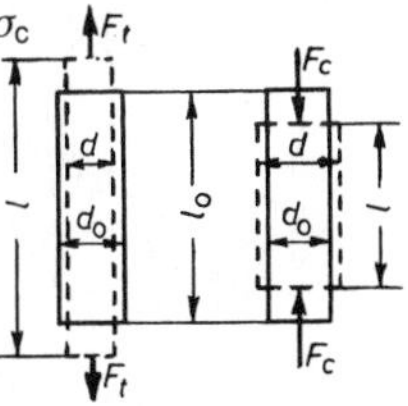

p 8 $\sigma_t = \dfrac{F_t}{A} \leqslant p_t \; ; \qquad \sigma_c = \dfrac{F_c}{A} \leqslant p_c$

Strain ε under tension

p 9 $\varepsilon = \dfrac{\Delta l}{l_0} = \dfrac{l - l_0}{l_0} = \dfrac{\sigma_t}{E} = \dfrac{F_t}{E \times A}$

Compressive strain ε_c **under compression**

p10 $\varepsilon_c = \dfrac{\Delta l}{l_0} = \dfrac{l_0 - l}{l_0} = \dfrac{\sigma_c}{E} = \dfrac{F_c}{E \times A}$ | $E \times A$ = tensile or compressive stiffness

Transverse contraction under tension (Poisson's ratio)

For circular cross section

p11 $\mu = \dfrac{\varepsilon_{cross}}{\varepsilon_{along}} \; ; \qquad$ where $\varepsilon_{along} = \dfrac{l - l_0}{l_0}$ and $\varepsilon_{cross} = \dfrac{d_0 - d}{d_0}$

For most metals Poisson's ratio can be assumed to be $\mu = 0{\cdot}3$

Thermal stresses: Tensile or compressive stress is caused by restricting thermal expansion (see also o 13/14):

p12 $\sigma_{th} = E \times \varepsilon_{th} = E \times \alpha \times \Delta t \qquad (\varepsilon_{th} = \alpha \times \Delta t)$

Δt is the temperature difference between the unstressed original state and the state considered.

$\Delta t > 0$ tensile stress, positive
$\Delta t < 0$ compressive stress, negative.

For prestressed members subjected to thermal stress the total strain comprises:

p13 $\varepsilon_{tot} = \varepsilon_{el} + \varepsilon_{th} = F/(E \times A) + \alpha \, \Delta t \; ; \qquad \varepsilon_{el} = F/(E \times A)$

Tensile and compressive stresses in thin-wall cylinders

(boiler formula):

Hoop stress $\sigma = p\,d/(2s)$

p14 Tensile stress $\sigma = p_i\,d_i/(2s)$
p15 Compressive stress $\sigma = -p_a\,d_a/(2s)$ } valid for $\dfrac{d_a}{d_i} \leqslant 1{\cdot}2$

p_i and p_a: internal and external pressures
d_i and d_a: inside and outside diameters
$s = 0{\cdot}5(d_a - d_i)$: wall thickness

Tensile stresses in rotating bodies: see M 5.

Tensile stress in a shrunk-on ring (approximate formulae)

Shrunk-on ring on a rotating shaft:

The shrinkage force F_H of the ring must be at least twice the centripetal force F_C.

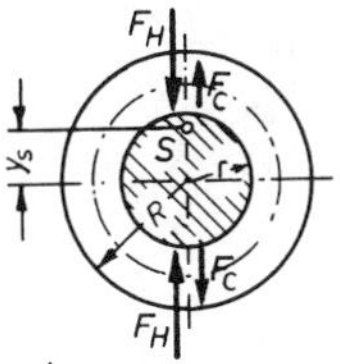

p 16 $$F_H \geqslant 2\,F_C$$

p 17 $$F_C = m\,y_s\,\omega^2 = 4\,\pi^2\,m\,y_s\,n^2$$

p 18 $$y_s = \frac{4}{3\,\pi} \times \frac{R^3 - r^3}{R^2 - r^2}$$

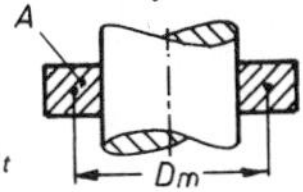

p 19 Cross section $$A = \frac{F_H}{2\,p_t}$$

p 20 Shrinkage allowance $$\lambda = \frac{1}{E}\,D_m \times p_t$$

(λ = outside diameter of shaft – inside diameter of ring)

Shrunk-on ring for clamping

Split, rotating clamped parts.

F_C comprises:

Centripetal force F_{CR} for ring

Centripetal force F_{CM} for clamped parts,

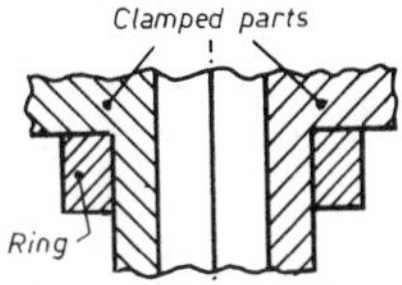

p 21 or $$F_H \geqslant 2(F_{CR} + F_{CM})\,;$$ then as p 19 and p 20

Energy of deformation U (Strain energy)

The energy stored in a deformed component is:

p 22 $$U = w\,V\;;\text{ where}$$

p 23 $$w = \frac{1}{2}\,\sigma\,\varepsilon = \frac{1}{2}\,E\,\varepsilon^2 = \frac{\sigma^2}{2\,E}\quad;\; V:\text{ volume of component}$$

Limit cross section for similar types of stress

Where a tension (or compression) force is applied at a point within the dotted core area, only tension (or compression) forces will occur over the whole cross section. If applied at any other point, bending stress, i.e. simultaneous tension and compression stress will occur.

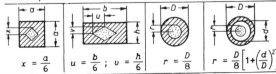

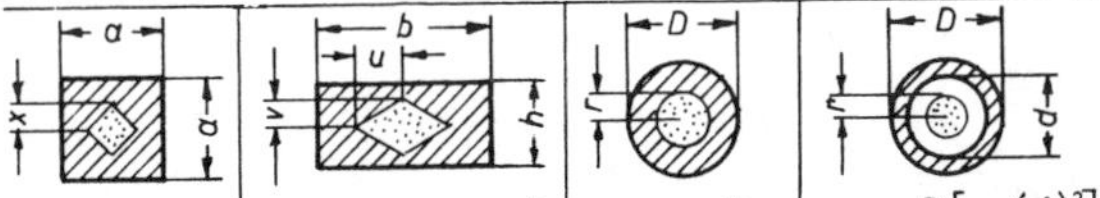

p 24 $x = \dfrac{a}{6}$ | $u = \dfrac{b}{6}\;;\; v = \dfrac{h}{6}$ | $r = \dfrac{D}{8}$ | $r = \dfrac{D}{8}\left[1 + \left(\dfrac{d}{D}\right)^2\right]$

S : centre of mass of half ring (see K 7)

D_m : mean diameter $(D_m = R + r)$

Explanation

All external loads on a beam (including support reactions and its own weight) produce internal forces and moments which stress the material. By taking a section through the beam at a point x it is possible to show the internal loads: Vertical shear forces V and bending moments M.

End loads P **and torsion** T are considered separately.

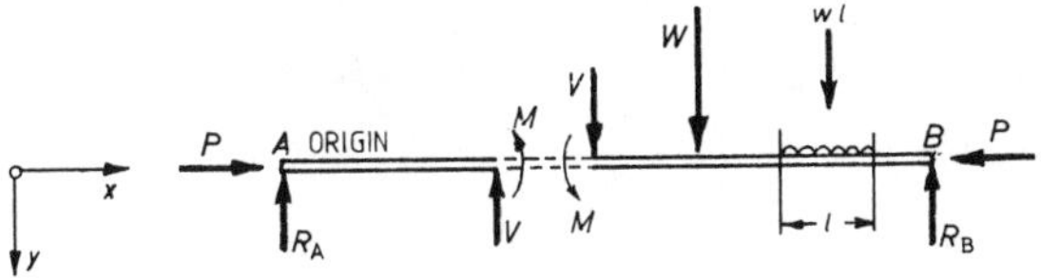

Referring to the x–y **plane** (z axis is at right angles):

Forces in direction of	x-axis	produce	end loads	P
	y-axis		shear forces	V
Moments about the	z-axis		bending moments	M
	x-axis		torsion	T

Always consider the left-hand side of the section.

In each part of the beam there must be equilibrium between all external and internal forces and moments:

Considered separately:

p 25/26 $$V + \sum_{i=1}^{n} V_i = 0 \qquad P + \sum_{i=1}^{n} P_i = 0$$

p 27/28 $$M + \sum_{i=1}^{n} M_i = 0 \qquad T + \sum_{i=1}^{n} T_i = 0$$

Method of calculation

1. Calculate the reactions.
2. Section the beam at the following places:
   2.1 Points of action of point loads W and beginning and end of distributed loads w.
   2.2 Points where the beam axis changes direction or the cross section changes.
   2.3 Any other convenient places.

continued on P 6

continued from P 5

3. Find the forces and moments on the left hand side of the section as in p 25 ... p 28
4. Plot shear force and bending moment diagrams.

Relations between w, V **and** M at any point x

p 29/30

$$\frac{dM}{dx} = V \qquad \frac{dV}{dx} = -w$$

Rules:

M is a maximum when	$V = 0$
In sections with no loads	$V =$ constant.

Example: Simply supported beam with end load. (Fixed at A) The reactions are:

$R_A = 2{\cdot}5$kN; $P_A = 3$ kN; $R_B = 1{\cdot}5$ kN

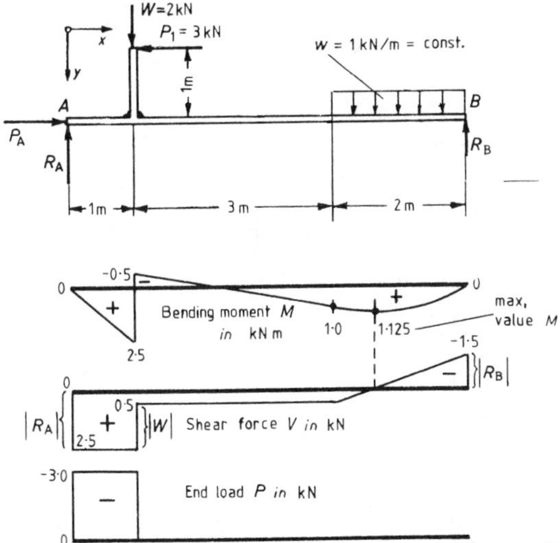

Calculations see P 7

continued from P 6

Range	from equation p ...	
$0 < x < 1$ m	27	$M - R_A \times x = 0$; $M = R_A \times x = 2{\cdot}5\,\text{kN} \times x$ *)
	26	$V - R_A = 0$; $V = R_A = 2{\cdot}5\,\text{kN} = \text{const.}$
	29	or $V = \frac{d}{dx}(2{\cdot}5\,\text{kN} \times x) = 2{\cdot}5\,\text{kN} = \text{const.}$
	25	$P + P_A = 0$; $P = -P_A = -3\,\text{kN}$
$1\,\text{m} < x < 4$ m	27	$M - R_A \times x + W(x - 1\text{m}) + P_1 \times 1\text{m} = 0$; $M = 0{\cdot}5\,\text{kN}\,x - 1\,\text{kN m}$ *)
	26	$V - R_A + W = 0$; $V = R_A - W = 2{\cdot}5\,\text{kN} - 2\,\text{kN} = 0{\cdot}5\,\text{kN} = \text{const.}$
	29	or $V = \frac{d}{dx}(0{\cdot}5\,\text{kN} \times x - 1\,\text{kN m}) = 0{\cdot}5\,\text{kN} = \text{const.}$
	25	$P + P_A - P_1 = 0$; $P = P_1 - P_A = 3\,\text{kN} - 3\,\text{kN} = 0$
$4\,\text{m} < x < 6$ m	27	$M - R_A \times x + W(x - 1\,\text{m}) + P_1 \times 1\,\text{m} + w\frac{(x - 4\,\text{m})^2}{2} = 0$ $M = -9\,\text{kNm} + 4{\cdot}5\,\text{kN} \times x - 0{\cdot}5\,\frac{\text{kN}}{\text{m}} \times x^2$ **)
	26	$V - R_A + W + w(x-4\text{m}) = 0$; $V = R_A - W - w(x-4\text{m}) = 2{\cdot}5\text{kN} - 2\text{kN} - 1\frac{\text{kN}}{\text{m}}(x-4\text{m}) = 4{\cdot}5\text{kN} - 1\frac{\text{kN}}{\text{m}}x$ *)
	29	or $V = \frac{d}{dx}(-9\,\text{kN m} + 4{\cdot}5\,\text{kN}\,x - 0{\cdot}5\,\frac{\text{kN}}{\text{m}}x^2) = 4{\cdot}5\,\text{kN} - 1\,\frac{\text{kN}}{\text{m}}\,x$
	25	$P + P_A - P_1 = 0$; $P = P_1 - P_A = 3\,\text{kN} - 3\,\text{kN} = 0$

*) Straight line

**) Parabola

continued on P 8

continued from P 7

Example:

Curved cantilever beam
(r = const.)

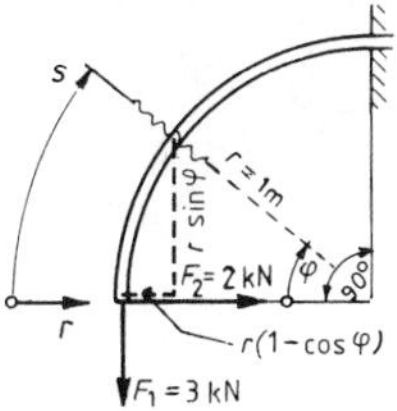

The limits are: $0 \leqslant \varphi \leqslant 90^\circ$

or: $0 \leqslant s \leqslant r\dfrac{\pi}{2}$

Bending moment:

p 31 $M + F_1\, r(1-\cos\varphi) + F_2\, r \sin\varphi = 0$

p 32 $M = -F_1\, r + F_1\, r \cos\varphi - F_2\, r \sin\varphi$

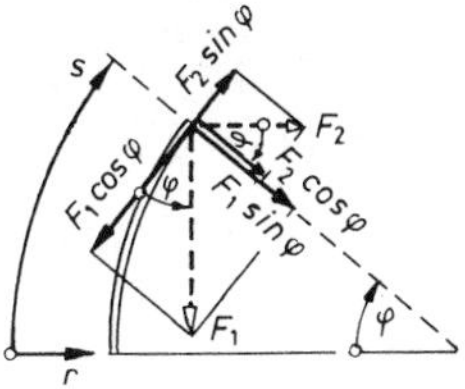

At the section φ, F_1 and F_2 are resolved into tangential and radial components.

Shear force (radial):

p 33 $F_q + F_1 \sin\varphi + F_2 \cos\varphi = 0$

p 34 $F_q = -F_1 \sin\varphi - F_2 \cos\varphi$; or from p 30:

p 35 $F_q = \dfrac{dM}{ds} = \dfrac{1}{r} \times \dfrac{dM}{d\varphi}$ (because $s = r\,\varphi$; $ds = r\, d\varphi$)

p 36 $= \dfrac{1}{r} \times \dfrac{d(-F_1\, r + F_1\, r\cos\varphi - F_2\, r \sin\varphi)}{d\varphi} = -F_1 \sin\varphi - F_2 \cos\varphi$

Normal force (tangential):

p 37 $F_n - F_1 \times \cos\varphi + F_2 \times \sin\varphi = 0$

p 38 $F_n = F_1 \times \cos\varphi - F_2 \times \sin\varphi$

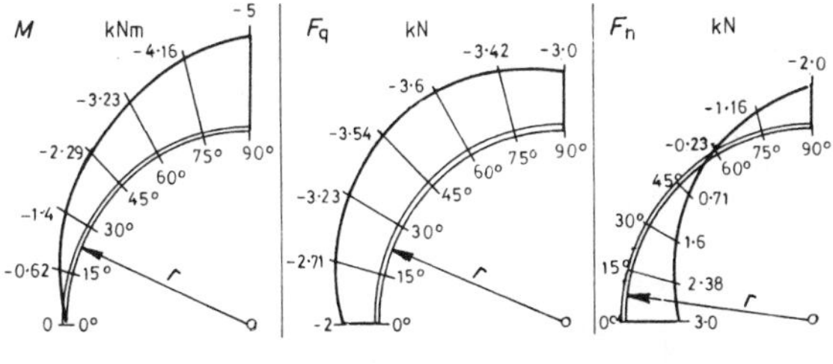

For a graphical method of determining bending moment see K 4.

STRENGTH
Bending

Maximum bending stress

p 39 $$\sigma_{bt\,max} = \frac{M\, y_{max}}{I_{xx}}$$

p 40 $$= \frac{M}{Z_{min}} \leqslant p_b$$

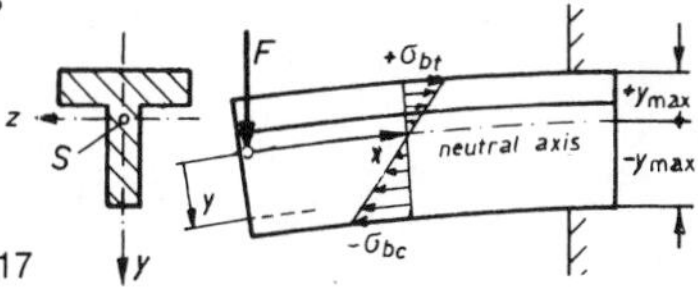

Values for p_b see Z 16/17

$+y_{max}$ (tension)	distance from surface fibre to the x-axis passing through the centroid S or neutral axis
$-y_{max}$ (compression)	

I_{xx} : Second moment of area about axis S_z or about plane of neutral axis.

Bending stress at distance y from the neutral axis

p 41 $$\sigma_b = \frac{M}{I}\, y$$

Section modulus Z_{min}

p 42 $$Z_{min} = \frac{I}{y_{max}}$$

Second moments of area

Axial second moment of area see J 10 and table P 10
Polar second moment of area see J 10
Product moment see J 10

Principal second moments of area and principal axes

The principal second moments of area $I_1 = I_{max}$ and $I_2 = I_{min}$ are applicable to asymmetric sections, when the principal axes are rotated through the angle φ_o.

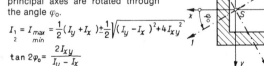

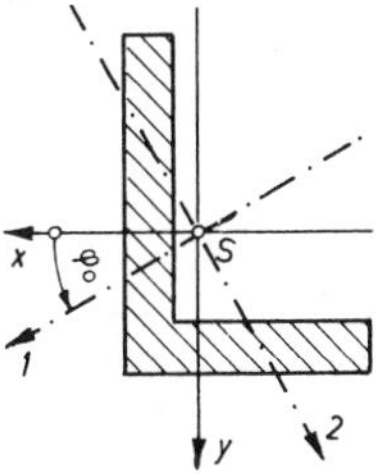

p 43 $$I_{1,2} = I_{max,\,min} = \frac{1}{2}(I_y + I_x) \pm \frac{1}{2}\sqrt{(I_y - I_x)^2 + 4 I_{xy}^2}$$

p 44 $$\tan 2\varphi_o = \frac{2 I_{xy}}{I_y - I_x}$$

For calculations of I_{xy} see J 10/11.

The principal axes are always perpendicular to each other.

The axis of symmetry of a symmetrical section is one principal axis e.g. $I_1 = I_x$.

Values of I and Z for some common sections

(see p 41 and p 42)

For position of centroid S (or neutral axis) see K 7

	I_x and I_y	Z_x and Z_y	Cross section A
p 45 p 46	$I_x = \frac{b\,d^3}{12}$ $I_y = \frac{d\,b^3}{12}$	$Z_x = \frac{b\,h^2}{6}$ $Z_y = \frac{h\,b^2}{6}$	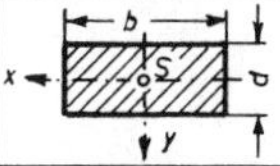
p 47	$I_x = I_y = \frac{\pi\,D^4}{64}$	$Z_x = Z_y = \frac{\pi\,D^3}{32} \simeq \frac{D^3}{10}$	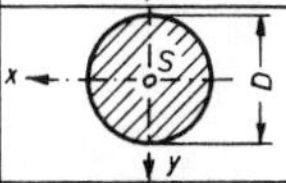
p 48	$I_x = I_y = \frac{\pi}{64}(D^4 - d^4)$	$Z_x = Z_y$ $= \frac{\pi}{32} \times \frac{D^4 - d^4}{D} \simeq \frac{D^4 - d^4}{10D}$	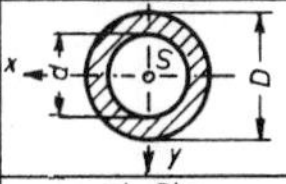
p 49 p 50 p 51 p 52	$I_x = I_y = 0{\cdot}06014\,s^4$ $= 0{\cdot}5412\,R^4$	$Z_x = 0{\cdot}1203\,s^3$ $= 0{\cdot}6250\,R^3$ $Z_y = 0{\cdot}1042\,s^3$ $= 0{\cdot}5413\,R^3$	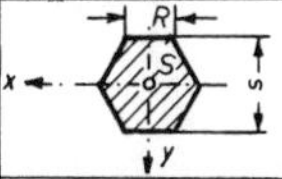
p 53 p 54	$I_x = \frac{\pi\,a\,b^3}{4}$ $I_y = \frac{\pi\,a^3\,b}{4}$	$Z_x = \frac{\pi\,a\,b^2}{4}$ $Z_y = \frac{\pi\,a^2\,b}{4}$	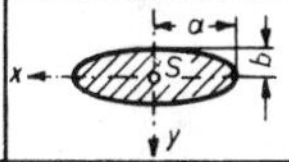
p 55 p 56	$I_x = \frac{b\,h^3}{36}$ $I_y = \frac{b^3\,h}{48}$	$Z_x = \frac{b\,h^2}{24}$ $Z_y = \frac{b^2\,h}{24}$	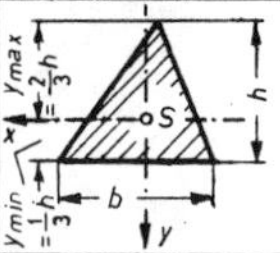
p 57 p 58 p 59	$I_x = \frac{h^3}{36} \times \frac{(a+b)^2 + 2ab}{a+b}$ $y_{max} = \frac{h}{3} \times \frac{2a+b}{a+b}$ $y_{min} = \frac{h}{3} \times \frac{a+2b}{a+b}$	$Z_x = \frac{h^2}{12} \times \frac{(a+b)^2 + 2ab}{2a+b}$	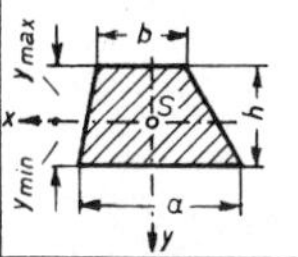

Steiner's theorem

(Parallel axis theorem for second moments of area).

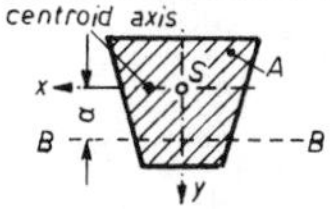

p 60 $I_{B\text{-}B} = I_x + A\,a^2$

Beams of uniform cross section

Equation of the elastic curve

The following apply to each section of the beam (see P 5, Method of calculation, Item 2):

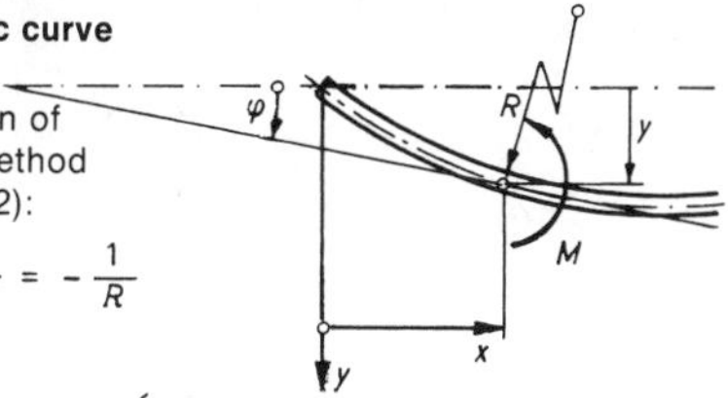

p 61 $$\frac{d^2y}{dx^2} = y'' = -\frac{M}{E\,I} = -\frac{1}{R}$$

p 62 $$E \times I \times y'' = -M$$

p 63 $$E \times I\,\frac{dy}{dx} = E \times I \times y' = -\int M\,dx + C_1$$

p 64 $$E \times I \times y = -\iint M\,dx \times dx + C_1 \times x + C_2$$

R : radius of curvature of the elastic curve at point x.

p 65 $y' = \tan\varphi$: inclination of the tangent to the elastic curve at point x.

y = deflection of beam at point x.

C_1 and C_2 are constants of integration and are determined from known factors.

e.g. $y = 0$ at the support.

$y_i = y_{i+1}$ at junction between sections i and $(i+1)$.

$y' = 0$ at the support of a cantilever beam and at the centre of a beam with symmetrical loading.

$y'_i = y'_{i+1}$ at the junction between sections i and $(i+1)$.

Strain energy due to bending U:

For a beam of length l:

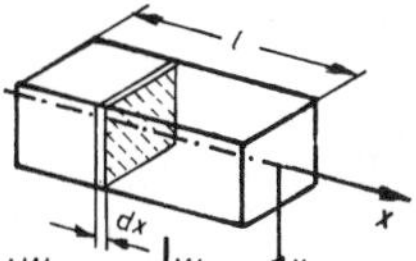

p 66 $$U = \frac{1}{2}\int_0^l \frac{M^2}{E\,I}\,dx$$

A beam with discontinuous loads may be divided into n length:

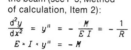

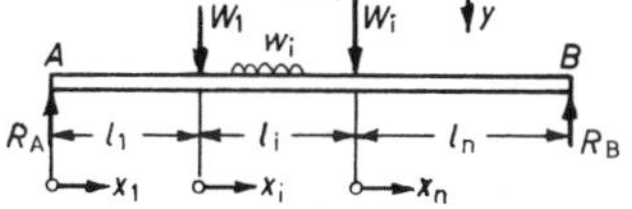

p 67 $$U_{tot} = \frac{1}{2\,E}\left(\int_{x_1=0}^{x_1=l_1} \frac{M^2}{I_1}\,dx_1 + \ldots + \int_{x_n=0}^{x_n=l_n} \frac{M^2}{I_n}\,dx_n\right)$$

p 71 p 70 p 69 p 68

Loading	Support reactions	M_{max} at point (...)	Deflection, y / Slope, $y' = \tan\varphi$	Deflection at C, y_c / Max. y_m
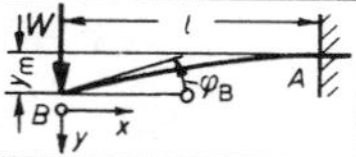	$\uparrow R_A = W$ $M_A = W\,l$	$W\,l$ (A)	$y = \frac{W l^3}{6EI}\left(2 - 3\frac{x}{l} + \frac{x^3}{l^3}\right)$ $\tan\varphi_B = -\frac{W l^2}{2EI}$	$y_m = \frac{W l^3}{3EI}$
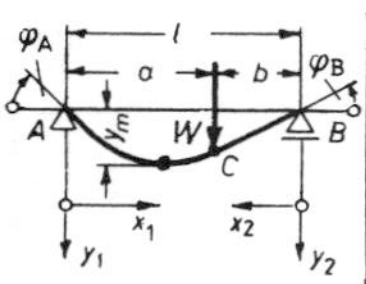 Note $a > b$	$\uparrow R_A = W\frac{b}{l}$ $\uparrow R_B = W\frac{a}{l}$	$W\frac{a\,b}{l}$ (C)	$y_1 = \frac{W l^3}{6EI}\frac{a}{l}\frac{b^2}{l^2}\frac{x_1}{l}\left(1 + \frac{l}{b} - \frac{x_1^2}{a\,b}\right)$ $y_2 = \frac{W l^3}{6EI}\frac{b}{l}\frac{a^2}{l^2}\frac{x_2}{l}\left(1 + \frac{l}{a} - \frac{x_2^2}{a\,b}\right)$ $\tan\varphi_A = \frac{y_C}{2a}\left(1 + \frac{l}{b}\right)$ $\tan\varphi_B = \frac{y_C}{2b}\left(1 + \frac{l}{a}\right)$	$y_c = \frac{W l^3}{3EI}\frac{a^2}{l^2}\frac{b^2}{l^2}$ $y_m = y_c\frac{l+b}{3b}\sqrt{\frac{l+b}{3a}}$ y_m at point $x_1 = a\sqrt{\frac{l+b}{3a}}$
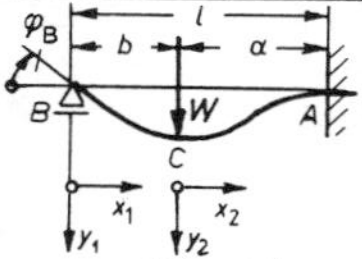 Note $a > b$	$\uparrow R_A = W - R_B$ $\uparrow R_B = W\frac{a^2}{l^2}\left(\frac{a}{l} + \frac{3b}{2l}\right)$ $M_A = W\frac{l+b}{2}\frac{a}{l}\frac{b}{l}$	M_A (A) when $b = 0{\cdot}414\,l$: $0{\cdot}171\,W l$ (C) $-0{\cdot}171\,W l$ (A)	$y_1 = \frac{R_B l^3}{6EI}\left(3\frac{x_1}{l} - \frac{x_1^3}{l^3}\right) - \frac{W a^2 x_1}{2EI}$ $y_2 = \frac{W a^3}{6EI}\left(2 - 2\frac{x_2}{a} + \frac{x_2^3}{a^3}\right) - \frac{R_B l^3}{6EI}\left(2 - 3\frac{x_1}{l} + \frac{x_1^3}{l^3}\right)$ $\tan\varphi_B = W a^2 b/(4EI\,l)$	$y_c = \frac{R_B l^3}{6EI}\left(3\frac{b}{l} - \frac{b^3}{l^3}\right) - \frac{W a^2 b}{2EI}$
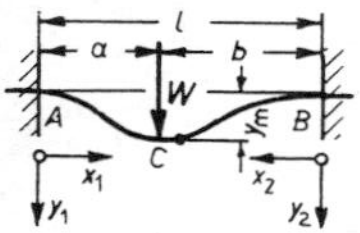 Note $a < b$	$\uparrow R_A = W\frac{b^2}{l^2}\left(3 - 2\frac{b}{l}\right)$ $\uparrow R_B = W\frac{a^2}{l^2}\left(3 - 2\frac{a}{l}\right)$ $M_A = W\,a\,b^2/l^2$ $M_B = W\,b\,a^2/l^2$	$-Wa\frac{b^2}{l^2}$ (A) $-Wb\frac{a^2}{l^2}$ (B) $2Wl\frac{a^2\,b^2}{l^2\,l^2}$ (C)	$y_1 = \frac{W}{6EI}\frac{b^2}{l^2}\left(3ax_1^2 - 3x_1^3 + 2\frac{b}{l}x_1^3\right)$ $y_2 = \frac{W}{6EI}\frac{a^2}{l^2}\left(3bx_2^2 - 3x_2^3 + 2\frac{a}{l}x_2^3\right)$ $\tan\varphi_A = \tan\varphi_B = 0$	$y_m = \frac{2Fl^3}{3EI}\times\frac{b^3}{l^3}\times\frac{a^2}{l^2}\left(\frac{l}{2b+l}\right)^2$ y_m at point $z_2 = \frac{2l\times b}{2b+l}$ $y_c = \frac{F}{3EI}\times\frac{a^3\,b^3}{l^3}$

STRENGTH

Deflection of beams in bending

P 13

p 75 p 74 p 73 p 72

Loading	Support reactions	M_{max} at point (...)	Deflection, y / Slope, $y' = \tan\varphi$	Deflection at C, y_C / Max. y_m
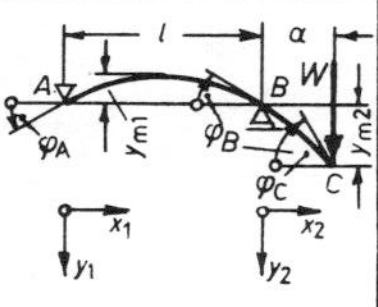	$R_A = W\frac{a}{l}$ $R_B = W\left(1+\frac{a}{l}\right)$	$-Wa$ (B)	$y_1 = \frac{W a l^2}{6EI}\left(\frac{x^3}{l^3} - \frac{x}{l}\right)$ $y_2 = \frac{W}{6EI}(2al\,x_2 + 3a\,x_2^2 - x_2^3)$ $\tan\varphi_A = -\frac{Wal}{6EI}$; $\tan\varphi_B = \frac{Wal}{3EI}$ $\tan\varphi_C = \frac{Wa}{6EI}(2l+3a)$	$y_{m_1} = -\frac{W a l^2}{9\sqrt{3}\,EI}$ y_{m_1} at point $x_1 = 0{\cdot}577\,l$ $y_{m_2} = y_C = \frac{W a^2}{3EI}(l+a)$
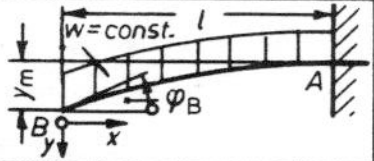	$R_A = wl$ $M_A = \frac{1}{2}wl^2$	$\frac{1}{2}wl^2$ (A)	$y = \frac{wl^4}{24EI}\left(\frac{x^4}{l^4} - 4\frac{x}{l} + 3\right)$ $\tan\varphi_B = -\frac{wl^3}{6EI}$	$y_m = \frac{wl^4}{8EI}$
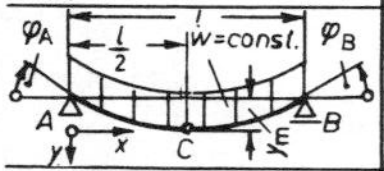	$R_A = R_B = \frac{wl}{2}$	$\frac{wl^2}{8}$ (C)	$y = \frac{wl^4}{24EI}\,\frac{x}{l}\left(1 - 2\frac{x^2}{l^2} + \frac{x^3}{l^3}\right)$ $\tan\varphi_A = \frac{wl^3}{24EI} = -\tan\varphi_B$	$y_m = \frac{5wl^4}{384EI}$
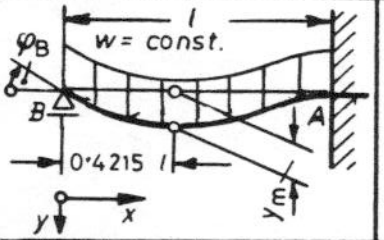	$R_B = \frac{3}{8}wl$ $R_A = \frac{5}{8}wl$ $M_A = \frac{1}{8}wl^2$	M_A (A)	$y = \frac{wl^4}{48EI}\left(\frac{x}{l} - 3\frac{x^3}{l^3} + 2\frac{x^4}{l^4}\right)$ $\tan\varphi_B = \frac{wl^3}{48EI}$	$y_m = \frac{wl^4}{185EI}$ y_m at point $x = 0{\cdot}4215\,l$
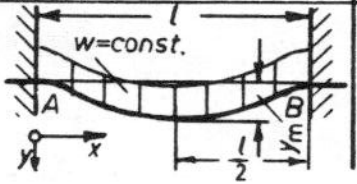	$R_A = w \times l/2$ $R_B = w \times l/2$ $M_A = w \times l^2/12$ $M_B = w \times l^2/12$	$-\frac{wl^2}{12}$ (A, B) $\frac{wl^2}{24}$ (C)	$y = \frac{wl^4}{24EI}\left(\frac{x^2}{l^2} - 2\frac{x^3}{l^3} + \frac{x^4}{l^4}\right)$ $\tan\varphi_A = \tan\varphi_B = 0$	$y_m = \frac{wl^4}{384EI}$

Formulae for y and y_m do not allow for shear deflection.

Mohr's analogy

Graphical method

1. Determine the bending moment curve by constructing a link polygon (see also K 4).

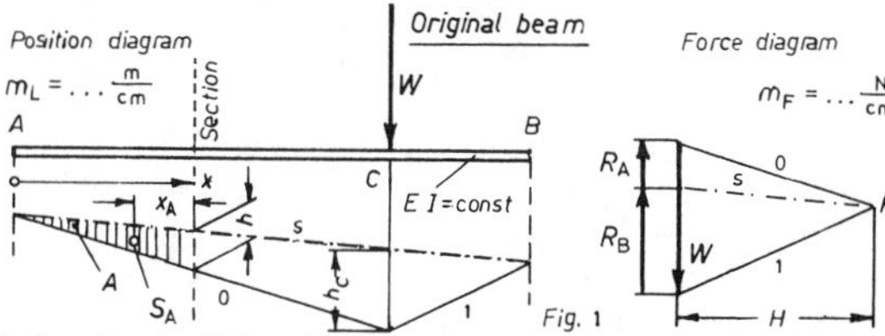

Fig. 1

2. Construct a link polygon as the equivalent distributed load w^* on the "equivalent beam". Another link polygon will give the tangents to the elastic curve.

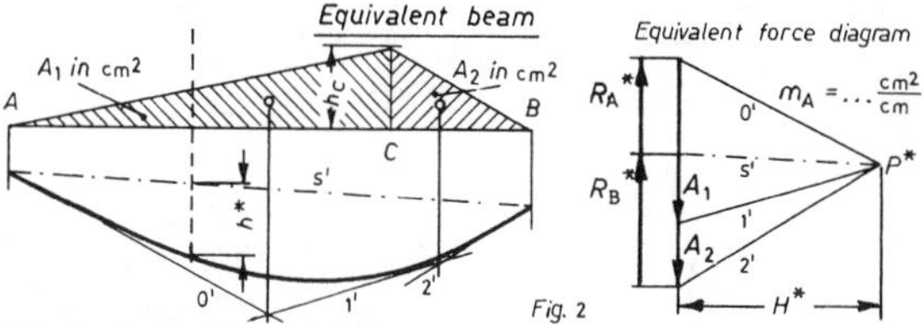

Fig. 2

3. Deflection of the original beam at point x:

p 76 $$y = h^* \frac{H\,H^*}{E\,I}\, m_F\, m_A\, m_L^3.$$

Slope at support A and B:

p 77 $$\tan\varphi_A = R_A^* \frac{H}{E\,I}\, m_F\, m_A\, m_L^2 \text{ resp. } \tan\varphi_B = R_B^* \frac{H}{E\,I}\, m_F\, m_A\, m_L^2$$

Mathematical method

1. Calculate the equivalent support reaction R_A^* of the "equivalent
p 78 beam" carrying the equivalent distributed load $w^* = A_1 + \ldots + A_n$ (see fig. 2).
2. Calculate the equivalent bending moment M^* and the equivalent shear force V^* at the point x:

p 79 $$M^* = R_A^* z - A\, z_A\,; \qquad V^* = R_A^* - A \qquad \text{(see fig. 1 + 2)}$$

x_A : Distance between center of gravity of equivalentdistributed load A and the section x.

p 80 3. Deflection $y = M^*/E\,I$; Slope $y = V^*/E\,I$

continued on P 15

continued from P 14 (Mohr's analogy)

Choice of equivalent beam

The supports of the equivalent beam must be such that its maximum equivalent bending moment M^*_{max} coincides with the point of maximum deflection in the original beam.

	Original beam	Equivalent beam
Simple beam		
Cantilever beam		

Beam of varying cross section

Fig. 1
Original beam e.g. shaft

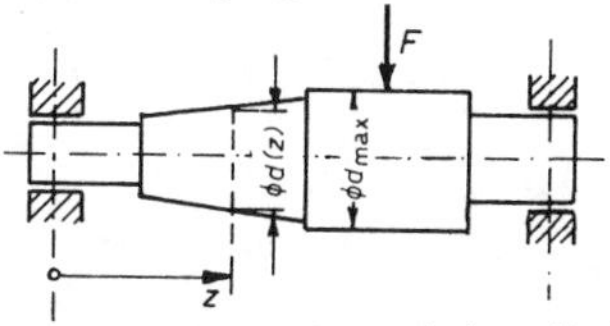

Plot the bending moment curve as the equivalent distributed load $w^*(z)$ on a uniform equivalent beam of cross section equal to the maximum second moment of area $I_{x\,max}$ of the original beam. (See P 14, item 1).

Plot $w^*(z)$ according to the ratio $\dfrac{I_{x\,max}}{I_x(z)}$:

p 81

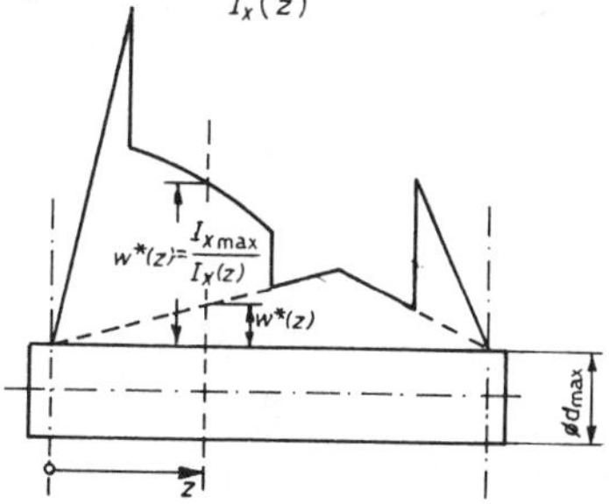

Fig. 2
Equivalent beam of the shaft in Fig. 1

Then calculate according to P 14 (items 2 and 3) or p 78 ... p 80.

	maximum section dimension	typical dimension $x =$ resp. $y =$	maximum deflection $f =$	type of beam
p 82	$h = \sqrt{\dfrac{6\,W\,l}{b\,p_{bt}}}$	$\sqrt{\dfrac{6\,W\,z}{b\,p_{bt}}}$	$\dfrac{8\,W}{b\,E}\left(\dfrac{l}{h}\right)^3$	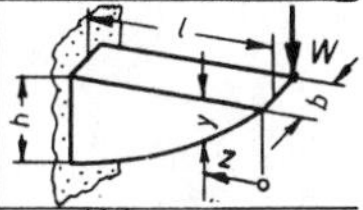 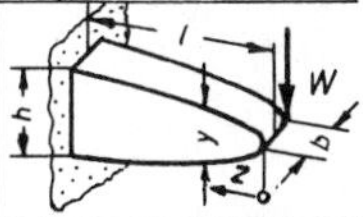
p 83	$b = \dfrac{6\,W\,l}{h^2\,p_{bt}}$	$\dfrac{6\,W\,z}{h^2\,p_{bt}}$	$\dfrac{6\,W}{b\,E}\left(\dfrac{l}{h}\right)^3$	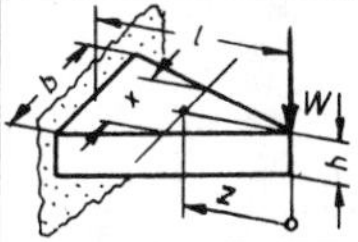
p 84	$h = \sqrt{\dfrac{3\,w\,l^2}{b\,p_{bt}}}$	$z\sqrt{\dfrac{3\,w\,l}{b\,l\,p_{bt}}}$	$\dfrac{3\,w\,l}{b\,E}\left(\dfrac{l}{h}\right)^3$	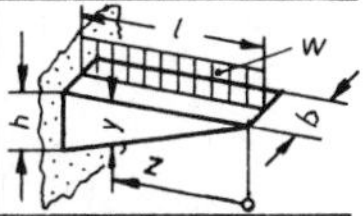
p 85	$b = \dfrac{3\,w\,l^2}{h^2\,p_{bt}}$	$\dfrac{3\,w\,l\,z^2}{h^2\,l\,p_{bt}}$		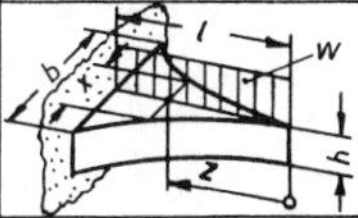
p 86	$h = \sqrt{\dfrac{3\,w\,l^2}{4\,b\,p_{bt}}}$	$\sqrt{\dfrac{3\,w\,l^2}{4\,b\,p_{bt}}\left(1-\dfrac{4z^2}{l^2}\right)}$	$\dfrac{w\,l^4}{64\,E\,I}$	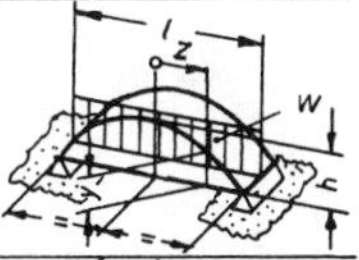
p 87	$d = \sqrt[3]{\dfrac{32\,W\,l}{\pi\,p_{bt}}}$	$\sqrt[3]{\dfrac{32\,W\,z}{\pi\,p_{bt}}}$	$\dfrac{192}{5} \times \dfrac{W\,l^3}{E\,\pi\,d^4}$	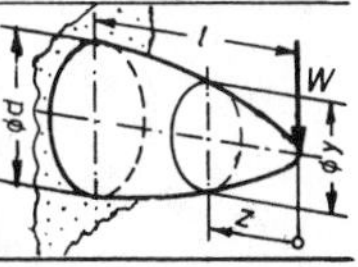

W : point load — kN

w : uniformly distributed load — kN/m

p_{bt} : permissible bending stress — N/mm^2 (see Z 17)

STRENGTH

Statically indeterminate beams

Convert a statically indeterminate beam (fig. 1) into a statically determinate one (fig. 2) by replacing one support by its support reaction (R_C in fig. 2).

Divide into two separate beams or subsystems. Determine the deflections at the point of the statically indeterminate support (see P 11 to P 15) from each subsystem, in terms of R_C.

Since no deflection can occur at support C:

p 88

$$|y_{C1}| = |y_{C2}|$$

Fig. 1

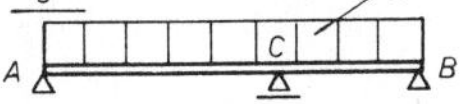

Fig. 2

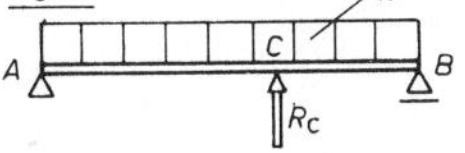

1st subsystem

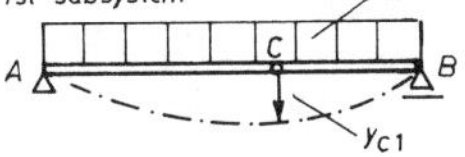

2nd subsystem

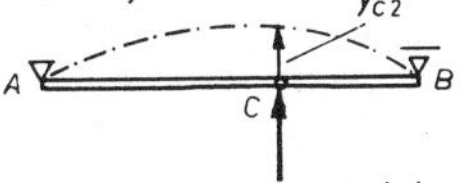

Hence, calculate the support force at C, R_C and then the remaining support reactions.

Method of solution for simple statically indeterminate beams

Statically indeterminate beam	Statically determinate beam	1st subsystem	2nd subsystem

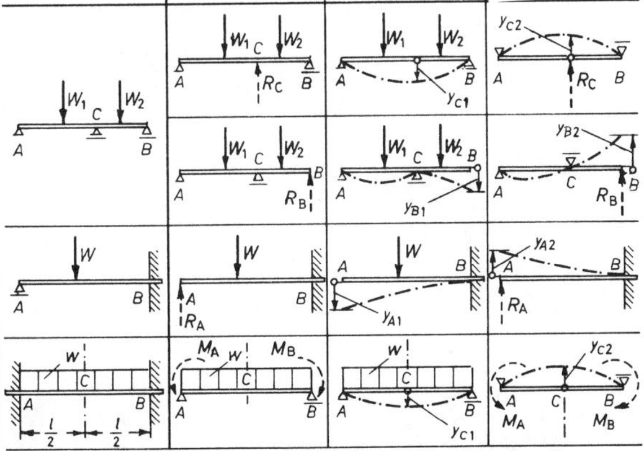

: statically indeterminate support reactions and moments

Hooke's law for shear stress

p 89 q or $\tau = G\gamma$

G : shear modulus
γ : shear strain

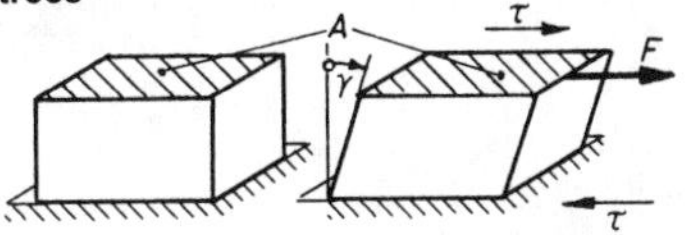

Relation between shear modulus and modulus of elasticity or Young's modulus

p 90 $G = \dfrac{E}{2(1+\mu)} \doteq 0{\cdot}385\,E$; correspond to P 3 with $\mu = 0{\cdot}3$

Mean shear stress q_f or τ_f

p 91 q_f or $\tau_f = \dfrac{F}{A} \leqslant p_q$

Permissible shear strength p_q (for values see Z 16)

type of load (see P 2)		$p_q \doteq$
	dead	$p_t/1{\cdot}5$
	undulating	$p_t/2{\cdot}2$
	alternating	$p_t/3{\cdot}0$

Ultimate shear stress q_s

p 92 $q_s = \dfrac{Q_{max}}{A}$

for ductile metals: $q_s = 0{\cdot}8\,R_m$
for cast iron: $q_s = R_m$

Shearing force Q

applied by	
guillotine shears	cutting tools (punch ...)
p 93 $Q \doteq 1{\cdot}2\,q_s\,l\,s$	$Q \doteq 1{\cdot}2\,q_s\,l_p\,s$

l : length of cut; l_p : perimeter of cut

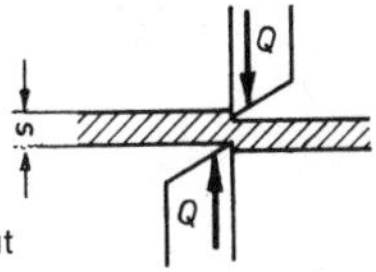

Theorem of related shear stresses

The shear stresses on two perpendicular faces of an element are equal in magnitude, perpendicular to their common edge and act either towards it or away from it.

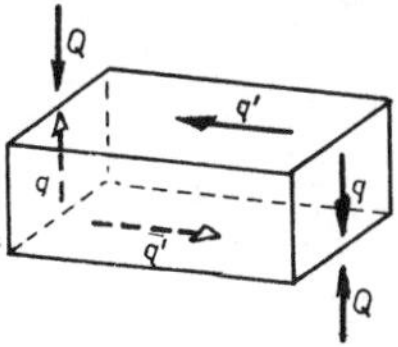

p 94 $q = q'$

q : transverse shear stress (transverse to beam axis) resulting from shear forces Q
q': axial shear stress (parallel to beam axis) "complementary shear"

Axial shear stress due to shear forces

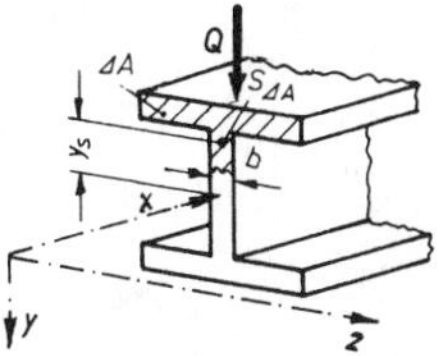

p 95 $$q = \frac{Q\,M}{b\,I} = q'$$

p 96 $$M = \Delta A\ y_s$$

$q = 0$ for $\sigma_b = \sigma_{b\,max}$.
q_{max} occurs when $\sigma_b = 0$.
i.e. on neutral axis.

Max. shear stress for different cross sections

$$q_{max} = q'_{max} = k\,\frac{Q}{A}$$

		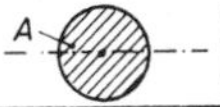	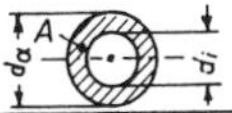
k	$\frac{3}{2}$	$\frac{4}{3}$	$\frac{4}{3} \times \frac{d_a^2 + d_a\,d_i + d_i^2}{d_a^2 + d_i^2}$; for thin-wall tubes: 2 ($d_a \approx d_i$)

p 97

Strain energy u due to shear

p 98 $$u = \frac{1}{2}\,q\,\gamma = \frac{q^2}{2\,G}$$

Shear deflection of a beam

p 99 $$y = \varkappa\,\frac{M}{G\,A} + C = \varkappa\,\frac{2{\cdot}6\,M}{E\,A} + C$$

Determine the constant C from known factors, e.g. $y = 0$ at the supports.

p100 The factor $\varkappa = A\int_{(A)}\left[\frac{M}{b\,I}\right]^2 dA$ allows for the form of cross section. For example:

			I 80	I 240	I 500
$\varkappa$ =	1·2	1·1	2·4	2·1	2·0

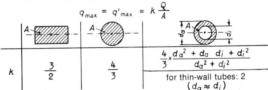

Q : shear force at point x of the beam
M : bending moment at the section A
I : second moment of area of the total cross section A about the z axis
b : width of section at point y
$S_{\Delta A}$: centre of area of section ΔA

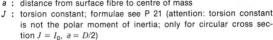

General

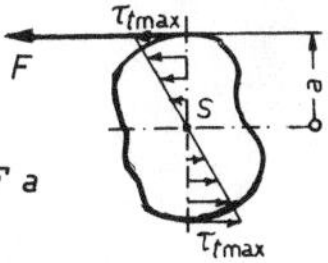

p 101 Shear stress due torsion $\tau_t = \dfrac{T\,a}{J} \leqslant p_{qt}$

p 102 Torque $T = \dfrac{P}{\omega} = \dfrac{P}{2\,\pi\,n} = F\,a$

P : power
a : distance from surface fibre to centre of mass
J : torsion constant; formulae see P 21 (attention: torsion constant is not the polar moment of inertia; only for circular cross section $J = I_p$, $a = D/2$)

Bars of circular cross section

Angle of twist φ (see e 5)

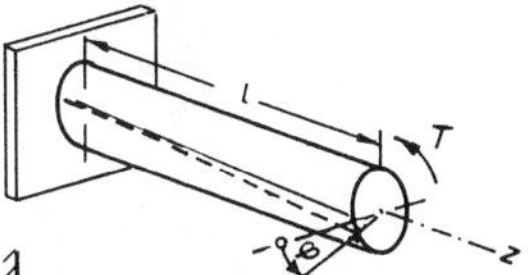

p 104 $\varphi = \dfrac{T\,l}{I_p\,G} = \dfrac{180^\circ}{\pi} \times \dfrac{T\,l}{I_p\,G}$

Stepped shafts:

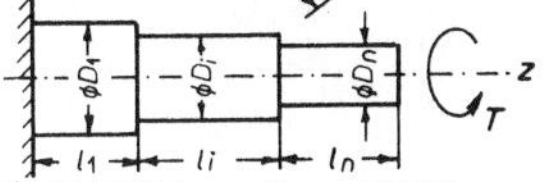

p 105 $\varphi = \dfrac{T}{G} \times \sum_{i=1}^{n} \dfrac{l_i}{I_{pi}}$

p 106 $\quad = \dfrac{180^\circ}{\pi} \times \dfrac{T}{G} \times \sum_{i=1}^{n} \dfrac{l_i}{I_{pi}}$

	polar moment of inertia I_p	max. shear stress τ_t	cross section
p 107	$\dfrac{\pi\,D^4}{32}$	$\simeq 5{\cdot}1 \dfrac{T}{D^3}$	
p 108	$\dfrac{\pi}{32}(D^4 - d^4)$	$\simeq 5{\cdot}1\,T\,\dfrac{D}{D^4 - d^4}$	

Bars of non-circular, solid or thin-wall hollow section

p 109 **Angle of twist** $\varphi = \dfrac{T\,l}{I_t\,G} = \dfrac{180^\circ}{\pi} \times \dfrac{T\,l}{I_t\,G}$

	torsion constant J	position and magnitude of τ_t	cross section
p 110	$c_1\,h\,b^3$	in 1: $\tau_{t1} = \tau_{t\,max} = \dfrac{c_2\,T}{c_1\,h\,b^2}$ in 2: $\tau_{t2} = c_3\,\tau_{t\,max}$ in 3: $\tau_{t3} = 0$	$h/b \geq 1$ 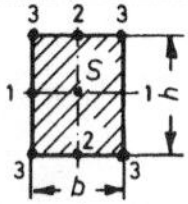

$h/b =$	1	1·5	2	3	4	6	8	10	∞
c_1	0·141	0·196	0·229	0·263	0·281	0·298	0·307	0·312	0·333
c_2	0·675	0·852	0·928	0·977	0·990	0·997	0·999	1·000	1·000
c_3	1·000	0·858	0·796	0·753	0·745	0·743	0·743	0·743	0·743

continued on P 21

STRENGTH

Torsion

P 21

	torsion constant $J =$	position and magnitude of τ_t	cross section
p 111 p 112	$\frac{a^4}{46.19} \simeq \frac{h^4}{26}$	at 1: $\tau_{t1} = \tau_{t\,max} = \frac{20\,T}{a^3} \approx \frac{13\,T}{h^3}$ at 2: $\tau_{t2} = 0$	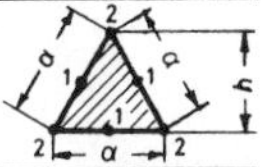
p 113 p 114 p 115	$0.1154\,s^4$ $= 0.0649\,d^4$	at 1: $\tau_{t1} = \tau_{t\,max}$ $= 5.297\,\frac{T}{s^3}$ $= 8.157\,\frac{T}{d^3}$	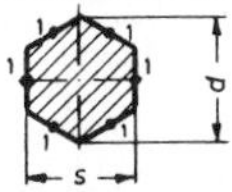
p 116 p 117	$\frac{\pi}{16} \times \frac{D^3\,d^3}{D^2 + d^2}$	at 1: $\tau_{t1} = \tau_{t\,max}$ $\simeq 5.1\,\frac{T}{D\,d^2}$ at 2: $\tau_{t2} = \tau_{t\,max}\,\frac{d}{D}$	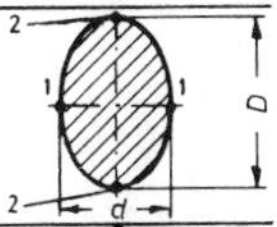
p 118 p 119 p 120	$\frac{\pi}{16} \times \frac{n^3(d^4 - d_i^4)}{n^2 + 1}$ $D/d = D_i/d_i = n \geqslant 1$	at 1: $\tau_{t1} = \tau_{t\,max}$ $\simeq 5.1\,\frac{T\,d}{n(d^4 - d_i^4)}$ at 2: $\tau_{t2} = \tau_{t\,max}\,\frac{d}{D}$	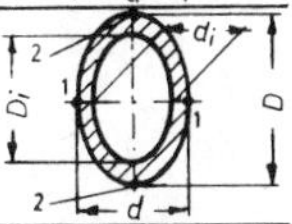
p 121 p 122 p 123 p 124	with varying wall thickness: $\frac{4\,A_m^2}{\sum_{i=1}^{n} \frac{s_i}{t_i}}$ with thin, uniform wall: $\frac{4\,A_m^2\,t}{s_m}$	at 1: $\tau_{t1} = \tau_{t\,max} = \frac{T}{2\,A_m\,t_{min}}$ at i: $\tau_{t2} = \frac{T}{2\,A_m\,t_i}$ $\tau_t = \frac{T}{2\,A_m\,t}$	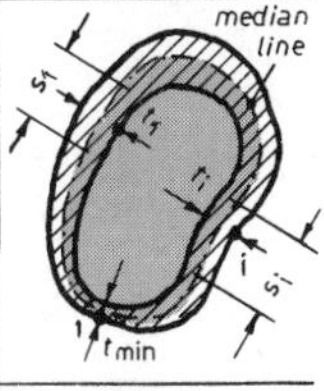
p 125	$\frac{\eta}{3} \sum_{i=1}^{n} b_i^3\,h_i$ Föppl's factor: (see below)	$\tau_{t\,max} = \frac{T\,b_{max}}{I_t}$ midway along the long side h of the rectangular section of max. thickness b_{max} (e.g. position 1 in the sketch).	profiles built up of rectangular cross sections 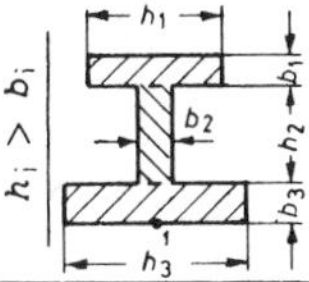

Föppl's factor:

	I $n=3$	L $n=2$	⊥ $n=2$
η	$\simeq 1.3$	$\simeq 1.0$	1.12
	[$n=3$		+ $n=2$
η	$1 < \eta < 1.3$		1.17

A_m : area enclosed by the median line
s : length of the median line
t (t_{min}) : wall thickness, (min. wall thickness)
s_i : part length of median line when wall thickness t_i = const.

Euler's formula

Applies for elastic instability of struts. Minimum load P_e at which buckling occurs:

p 126 $$P_e = \pi^2 \frac{E\, I_{min}}{l_k^2}$$

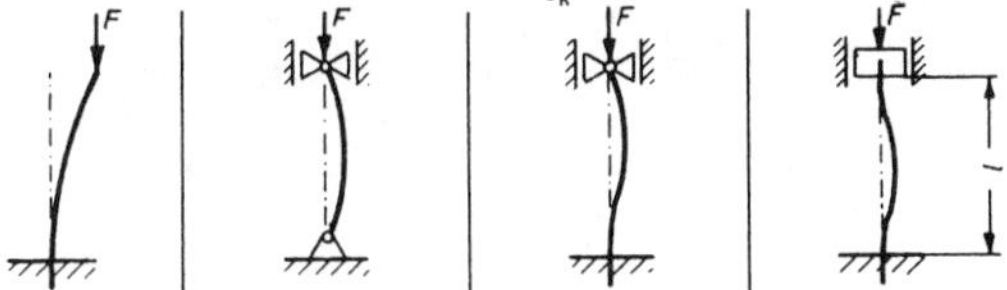

p 127 $l_k = 2\,l$ | $l_k = l$ | $l_k = 0{\cdot}707\,l$ | $l_k = 0{\cdot}5\,l$

p 128 Permissible working load $F = P_e / \nu_k$

p 129 Slenderness ratio $\lambda = \frac{l}{k} = l\sqrt{\frac{A}{I_{min}}}$

Limit 1 based on $R_{p0{\cdot}01}$	Limit 2 based on $R_{p0{\cdot}2}$
p 130 $\lambda_{lim\,0{\cdot}01} = \pi\sqrt{\frac{E}{R_{p\,0{\cdot}01}}}$	$\lambda_{lim\,0{\cdot}2} = \pi\sqrt{\frac{E}{R_{p\,0{\cdot}2}}}$

Tetmajer formula

Valid in the range $R_{p\,0{\cdot}01} \leqslant \frac{P_e}{A} \leqslant R_{p\,0{\cdot}2}$

Material of strut fails due to bending and compression

p 131 $$p_t = \frac{F}{A} = a - b\,\lambda + c\,\lambda^2 = p_c\,\nu_k$$

Material	GB-Standard	US : ASTM	a	b	c	valid for λ =
			N/mm²			
mild steel	BS 970, 050 A20	A 283 Gr. C	289	0·818	0	80 ... 100
mild steel	BS 970, 080 M30	A 440	589	3·918	0	60 ... 100
cast iron	BS 1452 – 220	A 48 A 258	776	12·000	0·053	5 ... 80
timber			30	0·20	0	2 ... 100
beech or oak			38	0·25	0	0 ... 100

Calculation method

First determine the minimum second moment of area using the Euler formula:

p 132 $$I_{min} = \frac{P_e\, l_k^2}{\pi^2 E} \quad ; \quad P_e = \nu_k \times F.$$

Then select a suitable cross section, e. g. circular tube, solid rectangle, etc., and find I and A.

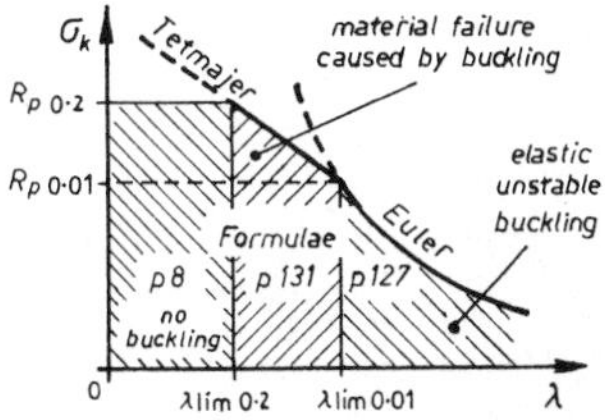

continued on P 23

Continued from P 22

$\nu_k = 3 \dots 5$ in the Tetmajer range

$\nu_k = 4 \dots 6$ | $\nu_k = 6 \dots 8$ in the Euler range | for large / for small | structures

Present / Limiting | slenderness ratio | λ / $\lambda_{\lim 0 \cdot 01}$ and $\lambda_{\lim 0 \cdot 2}$ | calculate from | p 129 / p 130

Determine buckling and compressive stress as follows:

If $\lambda \geq \lambda_{\lim 0 \cdot 01}$	use p 127
$\lambda_{\lim 0 \cdot 01} > \lambda \geq \lambda_{\lim 0 \cdot 2}$	use p 131
$\lambda < \lambda_{\lim 0 \cdot 2}$	use p 8.

If $P_e < F \nu_k$, redesign with larger cross section.

Method of buckling coefficient ω (DIN 4114)

Specified for building and bridge construction, steelwork and cranes.

p 133 Buckling coefficient $\omega = \dfrac{p_c}{p_k} = \dfrac{\text{permissible compressive stress}}{\text{buckling stress}}$

p 134 $\sigma_\omega = \omega \dfrac{F}{A} \leqslant p_c$ where $\omega = f(\lambda)$

λ	Standard	Buckling coefficient ω for			
	BS	mild steel 050 A 20	BS 970 080 M 30	alum. alloy BS L 102	cast iron BS 1452-220
	USA	mild steel A 283 Grade C	ASTM-A 440	alum. alloy AA 2017	cast iron ASTM-A48A25B
20		1·04	1·06	1·03	1·05
40		1·14	1·19	1·39	1·22
60		1·30	1·41	1·99	1·67
80		1·55	1·79	3·36	3·50
100		1·90	2·53	5·25	5·45
120		2·43	3·65	7·57	not valid in this range
140		3·31	4·96	10·30	
160		4·32	6·48	13·45	
180		5·47	8·21	17·03	
200		6·75	10·31	21·02	

Calculation method:

Estimate ω and choose cross section calculate A, $I_{\min}$ and λ from p 134. Then from table read off ω. Repeat the calculation with the appropriate new value of ω, until the initial and final values are identical.

Combination of direct stresses

Bending in two planes with end loads

The stresses σ arising from bending and end loads must be added together.

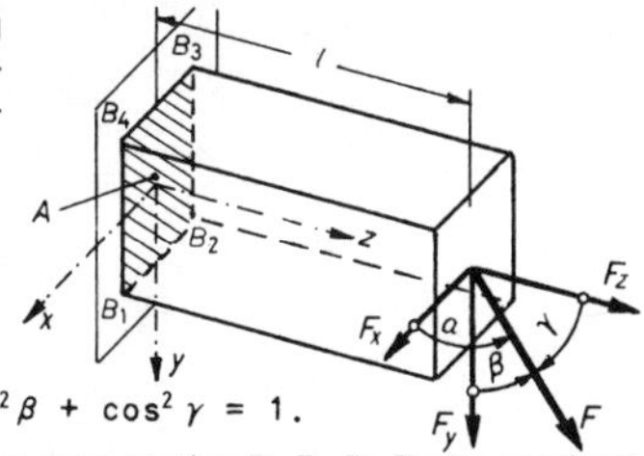

p 135 $$F_x = F \cos\alpha$$

p 136 $$F_y = F \cos\beta$$

p 137 $$F_z = F \cos\gamma$$

p 138 where $\cos^2\alpha + \cos^2\beta + \cos^2\gamma = 1.$

For any point $P(x, y)$ on the cross section $B_1 B_2 B_3 B_4$ the resultant normal stress is in the z-direction:

p 139 $$\sigma = \frac{F_z}{A} - \frac{F_y\, l}{I_x}\, y - \frac{F_x\, l}{I_y}\, x$$

Note the sign of x and y. If F_z is a compressive force, α, β and γ will be in different quadrants. For the sign of cosine functions see E 3.

Long beams in compression should be examined for buckling.

Neutral axis of $\sigma = 0$ is the straight line:

p 140 $$y = -\frac{F_x}{F_y} \times \frac{I_x}{I_y}\, x + \frac{F_z}{F_y} \times \frac{I_x}{A\, l}$$

which intercepts the axes at:

p 141 $$x_0 = \frac{F_z}{F_x} \times \frac{I_y}{A\, l}\,; \qquad y_0 = \frac{F_z}{F_y} \times \frac{I_x}{A\, l}$$

With asymmetrical cross section F resolve in the directions of the principal axes (see P 9).

Bending in one axis with end load

Either F_x or F_y in formulae p 139 ... p 141 is zero.

Bending with	tension / compression	displaces the neutral axis towards the	compression / tension	zone

Stress in curved beams $(R < 5\,h)$

The direct force F_n and bending moment M_x (see P 8) act at the most highly stressed cross section A.

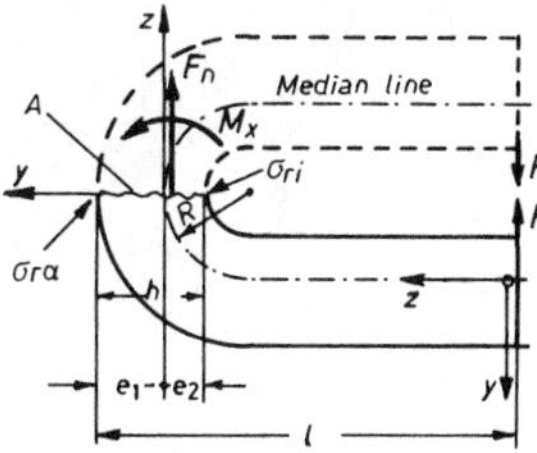

For the stress distribution over the cross section:

p 142 $$\sigma_t = \frac{F_n}{A} + \frac{M_x}{A\,R} + \frac{M_x\,R}{C} \times \frac{y}{R+y}$$

The stresses at the inner and outer surfaces are:

p 143 $$\sigma_{ra} = \frac{F_n}{A} + \frac{M_x}{A\,R} + \frac{M_x\,R}{C} \times \frac{|e_1|}{R+|e_1|} \leqslant p_t$$

p 144 $$\sigma_{ri} = \frac{F_n}{A} + \frac{M_x}{A\,R} - \frac{M_x\,R}{C} \times \frac{|e_2|}{R-|e_2|} \leqslant p_t$$

Formulae for coefficient C:

p 145 $$C = b\,R^3 \left(\ln \frac{1+\frac{d}{2R}}{1-\frac{d}{2R}} - \frac{d}{R} \right)$$

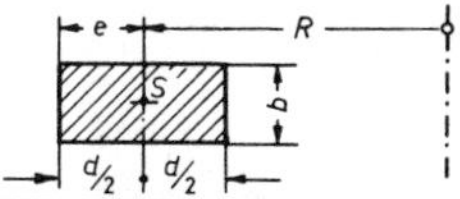

p 146 $$C = e^2\,\pi\,R^2 \frac{1-\sqrt{1-\left(\frac{e}{R}\right)^2}}{1+\sqrt{1-\left(\frac{e}{R}\right)^2}}$$

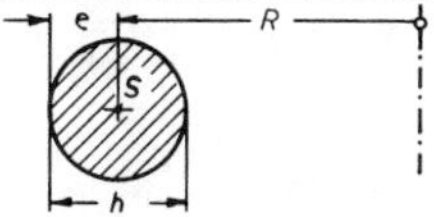

p 147 $$C = R^4 \left[\frac{a-b}{h} \left(1 + \frac{a e_1 + b e_2}{R(a-b)} \right) \times \ln \frac{1+\frac{e_1}{R}}{1-\frac{e_2}{R}} - \frac{a-b}{R} - \frac{(a+b)\,h}{2R^2} \right]$$

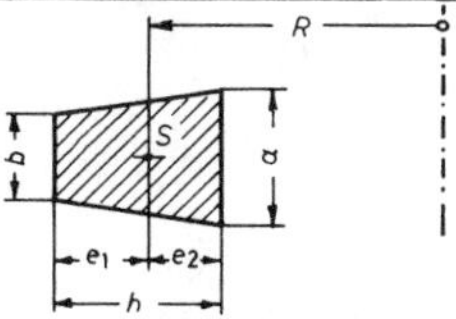

Position of the center of mass, see K 7

p 148 $$C = R^4 \left[\frac{b}{3h} \left(3+2\frac{h}{R} \right) \ln \frac{3+\frac{2h}{R}}{3-\frac{h}{3}} - \frac{b}{R} - \frac{b\,h}{2R^2} \right]$$

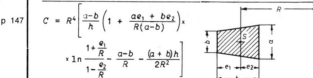

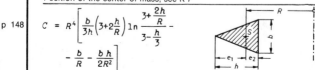

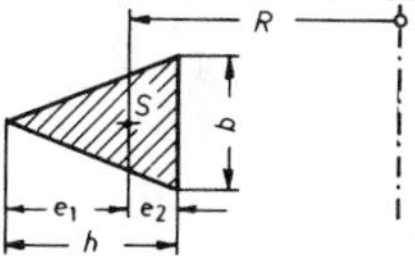

Combination of shear stresses

The stresses arising from shear and torsion at any cross section must be added vectorially.

The maximum shear stress τ_{res} occurs at point 1 and acts in the cross sectional plane. A complementary shear acts perpendicular to it.

Maximum torsional stress τ_{res} in point 1	cross section
p 149 $\quad \dfrac{5{\cdot}1\ T}{d^3} + \dfrac{1{\cdot}7\ F}{d^2} \leqslant p_{qt}$ where $T = F\dfrac{d}{2}$: p 150 $\quad 4{\cdot}244 \times \dfrac{F}{d^2} \leqslant p_{qt}$	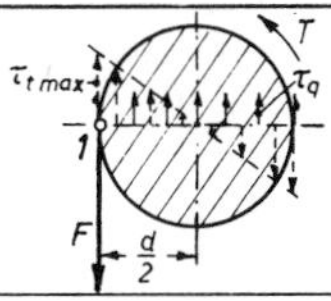
p 151 $\quad \dfrac{5{\cdot}1\ T D}{D^4 - d^4} + 1{\cdot}7\ F \times \dfrac{D^2 + Dd + d^2}{D^4 - d^4} \leqslant p_{qt}$ where $T = F\dfrac{D}{2}$: p 152 $\quad F\,\dfrac{4{\cdot}244\ D^2 + 1{\cdot}7\ d(D+d)}{D^4 - d^4} \leqslant p_{qt}$ For thin wall tubes: p 153 $\quad \dfrac{5{\cdot}1\ T D}{D^4 - d^4} + \dfrac{2{\cdot}55\ F}{D^2 - d^2} \leqslant p_{qt}$ p 154 $\quad 2{\cdot}55\ F \times \dfrac{2D^2 + d^2}{D^4 - d^4} \leqslant p_{qt}$	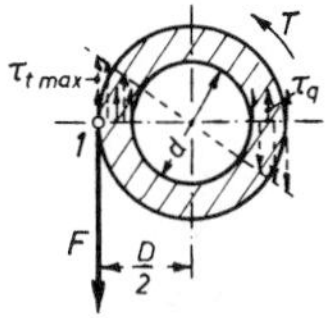
p 155 $\quad \dfrac{c_2}{c_1} \times \dfrac{T}{b^2 h} + \dfrac{1{\cdot}5\ F}{b\ h} \leqslant p_{qt}$ where $T = F\dfrac{b}{2}$: p 156 $\quad \dfrac{F}{2\ b\ h}\left(\dfrac{c_2}{c_1} + 3\right) \leqslant p_{qt}$	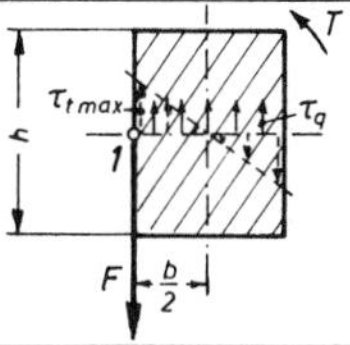

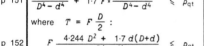

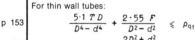

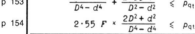

p_{qt} : permissible shear stress (see Z 16)
τ : shear stress
τ_q : calculated maximum torsional shear stress
F : force producing torsion
T : torque produced by F
for c_1 and c_2 see P 20

Combination of direct and shear stresses

Material strength values can only be determined for single-axis stress conditions. Therefore, multi-axis stresses are converted to single-axis equivalent stresses σ_V (see P 29). The following then applies, according to the type of load:

$$\sigma_V \lessgtr p_t \quad \text{or} \quad p_c \quad \text{or} \quad p_{bt}$$

Stresses in two dimensions

An element is subject to

direct stress	shear stress
σ_z in z direction	$\tau_{zy} = \tau$ in y–z plane
σ_y in y direction	$\tau_{yz} = \tau$ in y–z plane

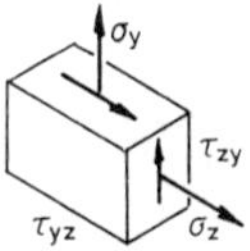

By rotating the element through the angle φ_σ the mixed stresses can be converted to tensile and compressure stresses only, which are called

Principal stresses

p 157 $$\sigma_1, \sigma_2 = 0{\cdot}5(\sigma_z + \sigma_y) \pm 0{\cdot}5\sqrt{(\sigma_z - \sigma_y)^2 + 4\tau^2}$$

Direction of the highest principal stress σ_1 at angle of rotation φ_σ from the original position is:

p 158 $$\tan 2\varphi_\sigma = \frac{2\tau}{\sigma_z - \sigma_y}\ ^{*)}$$ (where the shear stress is zero).

Rotating the element through the angle φ_τ gives the

Maximum shear stresses

p 159 $$\tau_{max}, \tau_{min} = \pm 0{\cdot}5\sqrt{(\sigma_z - \sigma_y)^2 + 4\tau^2} = \pm 0{\cdot}5(\sigma_1 - \sigma_2)$$

The direct stresses act simultaneously:

p 160 $$\sigma_M = 0{\cdot}5(\sigma_z + \sigma_y) = 0{\cdot}5(\sigma_1 + \sigma_2)$$

Direction of the maximum shear stress τ_{max} is:

p 161 $$\tan 2\varphi_\tau = -\frac{\sigma_z - \sigma_y}{2\tau}\ ^{*)}$$

The Principal stresses and Maximum shear stresses lie at 45° to each other.

*) The solution gives 2 angles. It means that both the Principal stresses and the Maximum shear stresses occur in 2 directions at right angles.

Stress in three dimensions

The stress pattern can be replaced by the

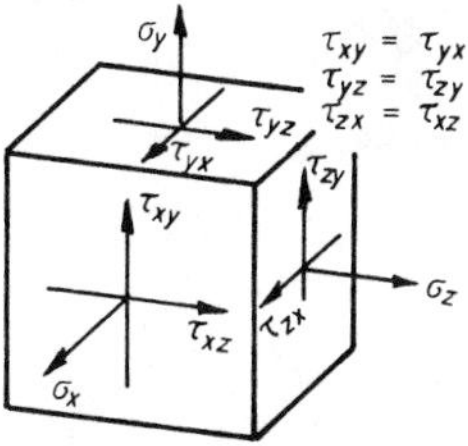

Principal stresses σ_1, σ_2, σ_3

They are the 3 solutions of the equation:

p 162 $$\sigma^3 - R\,\sigma^2 + S\,\sigma - T = 0$$

p 163 where $$R = \sigma_x + \sigma_y + \sigma_z$$

p 164 $$S = \sigma_x\,\sigma_y + \sigma_y\,\sigma_z + \sigma_z\,\sigma_x - \tau_{xy}^2 - \tau_{yz}^2 - \tau_{zx}^2$$

p 165 $$T = \sigma_x\sigma_y\sigma_z + 2\,\tau_{xy}\,\tau_{yz}\tau_{zx} - \sigma_x\,\tau_{yz}^2 - \sigma_y\,\tau_{zx}^2 - \sigma_z\,\tau_{xy}^2$$

Solve the cubic equation p 162 for σ_1, σ_2 and σ_3 as follows:

Put equation p 162 $= y$ (or substitute y for 0 on the right hand side), then plot $y = f(\sigma)$. The points of intersection with the zero axis give the solution. Substitute these values in p 162 and obtain more accurate values by trial and interpolation.

The case where $\sigma_1 > \sigma_2 > \sigma_3$ gives the Maximum shear stress $\tau_{max} = 0{\cdot}5\,(\sigma_1 - \sigma_3)$.

Bending and torsion in shafts of circular cross section

According to the theory of maximum strain energy:

p 166 Equivalent stress $$\sigma_E = \sqrt{f_{bt}^{\,2} + 3(\alpha_o\,f_{qt})^2} \leqslant p_{bt}$$

p 167 Equivalent moment $$M_E = \sqrt{M_b^{\,2} + 0{\cdot}75(\alpha_o\,T)^2}$$

To find the diameter of the shaft, determine the necessary section modulus Z from:

p 168 $$Z = \frac{M_E}{p_{bt}}$$

---

f_{bt} : tensile stress due to bending
f_{qt} : shear stress due to torsion
M : bending moment
T : torque
α_0 : according to P 29

		Based on theory of maximum strain energy: direct stress	shear stress	equivalent stress
three-dimensional stress		tension: $\sigma_1 > 0$: $\sigma_{vN} = \sigma_1 = \sigma_{max}$ compr.: $\sigma_3 < 0$: $\sigma_{vN} = \sigma_3 = \sigma_{min}$	$\sigma_{vS} = 2\,\tau_{max} = \sigma_1 - \sigma_3$	$\sigma_{vGe} = \sqrt{0{\cdot}5\left[(\sigma_1-\sigma_2)^2 + (\sigma_2-\sigma_3)^2 + (\sigma_3-\sigma_1)^2\right]}$
two-dimensional stress		tension: $\sigma_1 > 0$: $\sigma_{vN} = \sigma_1 = \sigma_{max}$ $= 0{\cdot}5\left[(\sigma_z+\sigma_y) + \sqrt{(\sigma_z-\sigma_y)^2 + 4(\alpha_0\,\tau)^2}\right]$ compr.: $\sigma_2 < 0$: $\sigma_{vN} = \sigma_2 = \sigma_{min}$ $= 0{\cdot}5\left[(\sigma_z+\sigma_y) - \sqrt{(\sigma_z-\sigma_y)^2 + 4(\alpha_0\,\tau)^2}\right]$	$\sigma_{vS} = 2\,\tau_{max} = \sigma_1 - \sigma_2$ $= \sqrt{(\sigma_z-\sigma_y)^2 + 4(\alpha_0\,\tau)^2}$	$\sigma_{vGe} = \sqrt{\sigma_1^2 + \sigma_2^2 - \sigma_1\,\sigma_2}$ $= \sqrt{\sigma_z^2 + \sigma_y^2 - \sigma_z\,\sigma_y + 3(\alpha_0\,\tau)^2}$
loads I, II, III for σ and τ	equal	$\alpha_0 = 1$	$\alpha_0 = 1$	$\alpha_0 = 1$
	un-equal	$\alpha_0 = \dfrac{p_t\ \text{I, II, III}}{p_q\ \text{I, II, III}} = \dfrac{\sigma_{lim}\ \text{I, II, III}}{\tau_{lim}\ \text{I, II, III}}$	$\alpha_0 = \dfrac{p_t\ \text{I, II, III}}{2\,p_q\ \text{I, II, III}} = \dfrac{\sigma_{lim}\ \text{I, II, III}}{2\,\tau_{lim}\ \text{I, II, III}}$	$\alpha_0 = \dfrac{p_t\ \text{I, II, III}}{1{\cdot}73\,p_q\ \text{I, II, III}} = \dfrac{\sigma_{lim}\ \text{I, II, III}}{1{\cdot}73\,\tau_{lim}\ \text{I, II, III}}$
used for	type of stress and material	tension. bending, torsion of brittle materials: cast iron glass stone	compression of brittle and ductile materials. Tension, bending, torsion of steel having pronounced yield point	all stressing of ductile materials: rolled, forged and cast steel; aluminium, bronze
	expected failure	cut off fracture	sliding fracture, flowing, deformation	permanent-, sliding-, cut off fraction; flowing, deformation

*) Give the best agreement with test results.

σ_{lim}, τ_{lim} are the typical values for the materials, e. g. R_m; q_s

σ_1; σ_2; σ_3 see P 27 + P 28

Lead screws see K 11

Fixing bolts

Bolted joints (approximate calculation)

Prestressed

axial working load F_A	shear load F_S
q 1 $A_3 \triangleq A_s = \dfrac{F_{max}}{p_t}$	calculation for friction effect:
q 2 $F_{max} = (1{\cdot}3 \ldots 1{\cdot}6)\, F_A$ (using load-extension diagram, see P 1)	$A_3 \triangleq A_s = \dfrac{F_{C\,req}}{(0{\cdot}25 \ldots 0{\cdot}5) R_{p\,0{\cdot}2}}$
q 3 $p_t = (0{\cdot}25 \ldots 0{\cdot}5)\, R_{p\,0{\cdot}2}$ (allowing for torsion and safety factor)	$F_{K\,erf} = \dfrac{\nu\, F_S}{\mu\; m\; n}$ (for values of μ see Z 7)

High-stress bolted joints see VDI 2230

Bracket attachment (precise calculation not possible)

Practical assumption: that the centre of pressure is the point of rotation,

q 4 e.g. $a \triangleq h/4$.

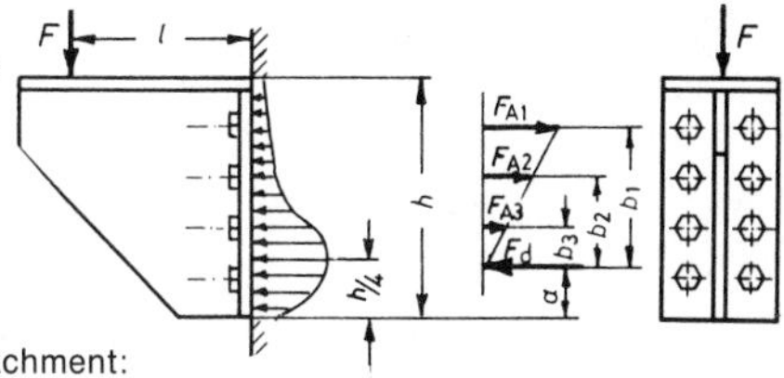

For a rigid attachment:

q 5 $$F\; l = F_{A1}\; b_1 + F_{A2}\; b_2 + \ldots F_{An}\; b_n$$

q 6 $$F_{A1} : F_{A2} : \ldots F_{An} = b_1 : b_2 \ldots b_n$$

Allow for the extra shear load $F_S = F$. There must be compressive stress over the whole attachment plane, when under load.

A_3 : core cross section

A_s : stress cross section $\left(A_s = \dfrac{\pi}{4}\left(\dfrac{d_2 + d_3}{2}\right)^2\right)$

$F_{C\,req}$: required clamping force

m : no. of bolts

n : no. of joint faces

e.g.

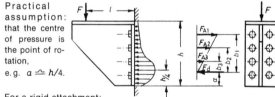

$m = 3;\ n = 1$

$m = 3;\ n = 2$

ν : safety factor against slipping $[\nu = 1{\cdot}5 \ldots (2)]$

$R_{p\,0{\cdot}2}$: proof stress

p_t : permissible stress

d_2 : outside diameter of bolt

d_3 : core diameter of bolt

Axles and shafts (approximate calculation)[1)]

Stability

	Axis	required section modulus for bending	solid axle of circular cross section ($Z \triangleq d^3/10$)	permissible bending stress[3)]
q 7	fixed[2)]			$p_{bt} = \frac{\sigma_{btU}}{(3 \ldots 5)}$
q 8		$Z = \frac{M}{p_{bt}}$	$d = \sqrt[3]{\frac{10\ M}{p_{bt}}}$	
q 9	rotating			$p_{bt} = \frac{\sigma_{btA}}{(3 \ldots 5)}$

	Shafts	diameter for solid shaft	permissible torsional stress[3)]
q 10	pure torsion		$p_{qt} = \frac{\tau_{tU}}{(3 \ldots 5)}$
q 11		$d = \sqrt[3]{\frac{5\ T}{p_{qt}}}$	
q 12	torsion and bending		$p_{qt} = \frac{\tau_{tU}}{(10 \ldots 15)}$

Bearing stress

q 13 on shaft extension $\left.\right\} f_{bm} = \frac{F}{d\ b} \leqslant p_b$

(p_b see Z 18)

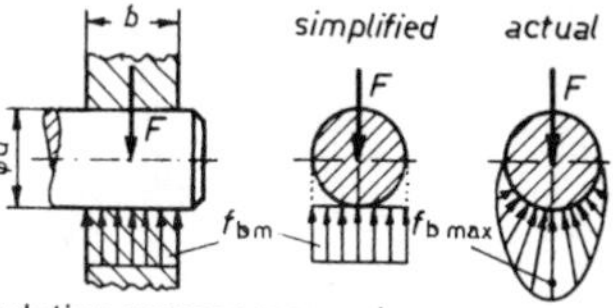

Shear due to lateral load: Calculation unnecessary when

$l > d/4$	for all shafts	with	circular	cross
$l > 0 \cdot 325\ h$	for fixed axles		rectangular	sections

Deflection due to bending see P 12
due to torsion see P 20

Vibrations see M 6.

[1)] For precise calculation see DIN 15017

[2)] Formulae are restricted to load classes I + II (P 2)

[3)] p_{bt} and p_{qt} allow for stress concentration-, roughness-, size-, (see DIN 15017), safety-factor and combined stresses

l : arm of force F

M, T : bending moment, torque

f_{bm} (p_b) : mean (permissible) bearing stress (see Z 18)

$f_{b\,max}$ see q 47 for hydrodynamically lubricated plain bearings, other cases see Z 18

σ_{btU}, σ_{btA}, τ_{tU} : values see Z 16.

Friction-locked joints

Proprietary devices (e. g. annular spring, Doko clamping device, Spieth sleeve, etc.): see manufacturer's literature.

For interference fits see DIN 7190 (graphical method).

Clamped joint

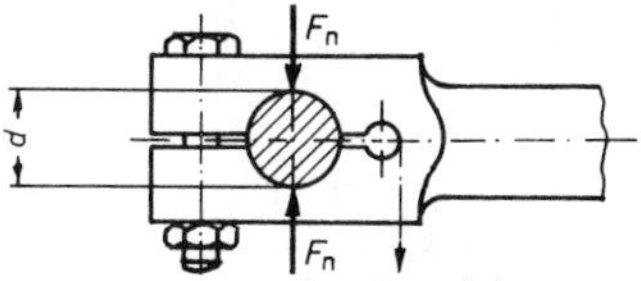

q 14 $$F_n = \frac{T\,\nu}{\mu\,d}$$

Taper joint

q 15 Taper from $\tan\alpha = \frac{D-d}{l}$

For tapered shaft extensions see DIN 1448, 1449.

Approximate formula for axial force F_A on the nut:

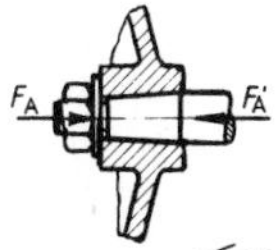

q 16 $$F_A = \frac{2\,T\,\nu}{\mu\,d_m}\tan\left(\frac{\alpha}{2}+\varrho\right)$$

q 17 $$d_m = \frac{D+d}{2}$$

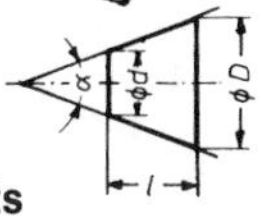

Specially machined joints

Proprietary splined fittings, hub to shaft e. g. polygon: see manufacturer's literature.

Plain key (approximate calculation)

Calculation is based on the bearing pressure on the side of the keyway in the weaker material. Allowing for the curvature of the shaft and the chamfer r_1, the bearing height of the key can be taken approximately as t_2.

The bearing length l is to transmit a torque T:

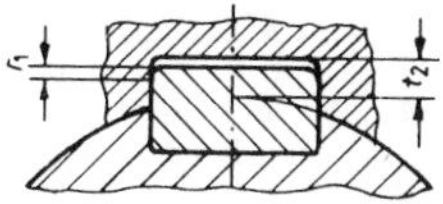

q 18 $$l = \frac{2\,T}{d\,t_2\,p_b}$$

Dimensions to DIN 6885, preferably Sheet 1. Allowance for fillets with form A.

For precise calculations refer to Mielitzer, Forschungsvereinigung Antriebstechnik e.V. Frankfurt/M., Forschungsheft 26, 1975.

continued on Q 4

continued from Q 3

Splined shaft

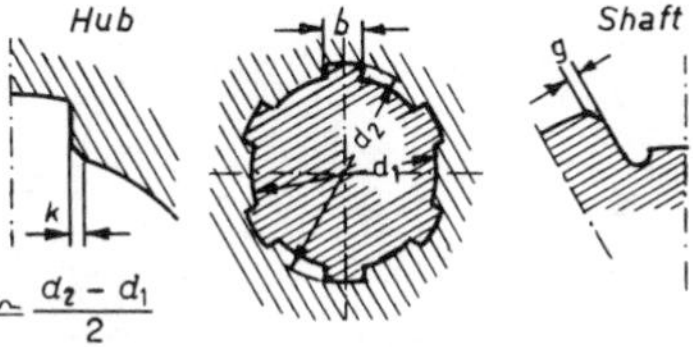

q 19 $$l = \frac{2\,T}{d_m\, h\, \varphi\, n\, p_b}$$

q 20 $$d_m = \frac{d_1 + d_2}{2}$$

q 21 $$h = \frac{d_2 - d_1}{2} - g - k \simeq \frac{d_2 - d_1}{2}$$

The load is not shared equally between the splines so allowance for unequal bearing is made with the factor φ:

Type of location	φ
shaft located	0·75
hub located	0·9

For cross section dimensions refer to DIN 5462 ... 5464.

Hub dimensions

Use diagram on page Q 5 to determine dimensions of hub.

Example: Find the length L and radial thickness s of a hub needed to transmit a torque T of 3000 N m, made in cast steel fitted with a plain key.

1. Determine the appropriate range "hub length L, CS/St, group e", follow the boundary lines to T = 3000 N m. Result: L = (110 ... 140) mm.
2. Determine the appropriate range "hub thickness s, CS/St, group 1", follow the boundary lines to T = 3000 N m. Result: s = (43 ... 56) mm.

F_n : normal force of transmitting surface
l : bearing length of joint
n : no. of splines
μ : coefficient of sliding friction (see Z 7)
ν : safety factor (see Q 1)
ϱ : angle of friction (ϱ = arc tan μ)
p_b : permissible bearing pressure. For approximate calculation:

material	p_b in N/mm²
CI (gray cast iron)	40... 50
CS (cast steel), St (steel)	90...100

(higher values possible in special cases)

Diagram to obtain hub geometry for Q 4

These empirical values are for steel shafts made of steel to ASTM A 572 Grade 42 – resp. to BS 4360 43 B –, but not for special cases (such as high centrifugal force, etc.). Increase L when there are other forces or moments being carried.

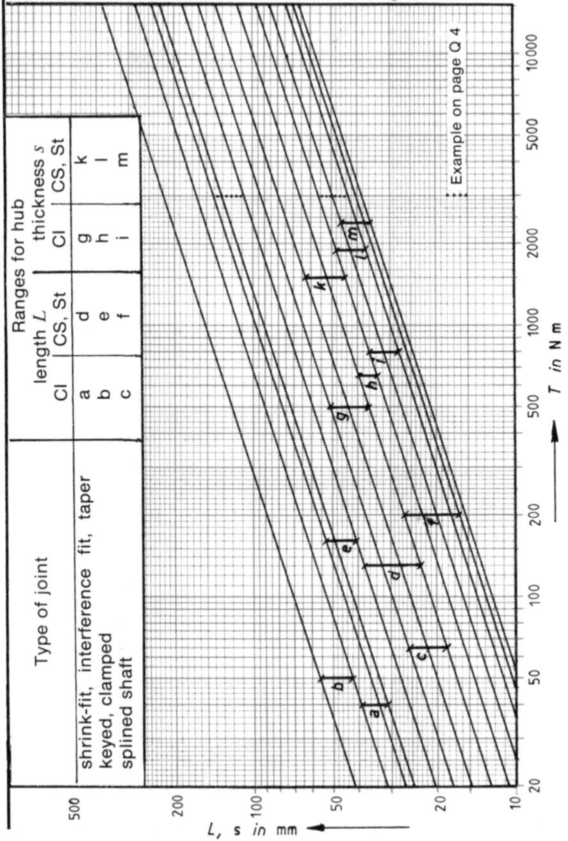

Spring rate R and spring work W (Strain energy)

	Characteristic general	const.
q 22 R	$\frac{dF}{ds}$	$\frac{F}{s}$
q 23 W_1	$\int_0^{s_1} F\, ds$	$F\,\frac{s_1}{2}$

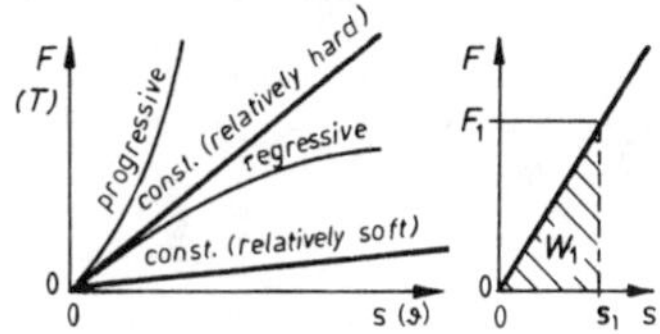

Springs in

	parallel	series
	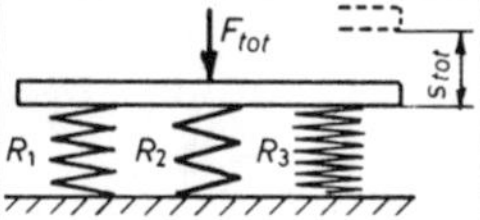	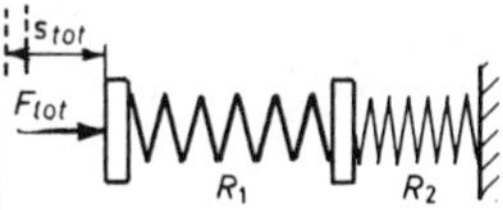
q 24 s_{tot} =	$s_1 = s_2 = s_3 \ldots = s_i$	$s_1 + s_2 + s_3 \ldots + s_i$
q 25 F_{tot} =	$F_1 + F_2 + F_3 \ldots + F_i$	$F_1 = F_2 = F_3 \ldots = F_i$
q 26 R	$R_{tot} = R_1 + R_2 + R_3 \ldots + R_i$	$\frac{1}{R_{tot}} = \frac{1}{R_1} + \frac{1}{R_2} + \ldots \frac{1}{R_i}$

Springs in tension and compression

e. g. ring spring (Belleville spring)

Springs in bending

Rectangular, trapezoidal, triangular springs

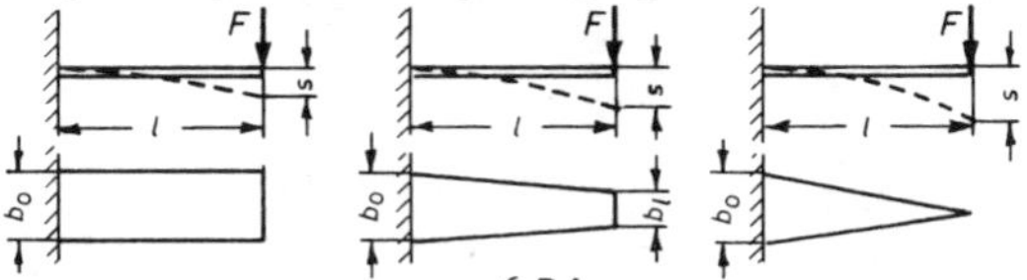

q 27 bending stress $f_{bt} = \frac{6\,F\,l}{b_0\,h^2}$

q 28 permissible load $F = \frac{b_0\,h^2\,p_{bt}}{6\,l}$

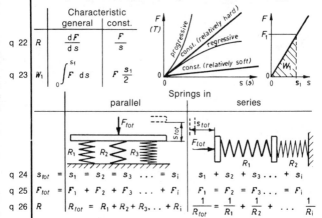

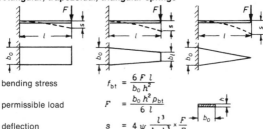

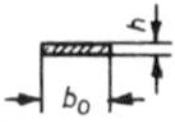

q 29 deflection $s = 4\,\psi\,\frac{l^3}{b_0\,h^3} \times \frac{F}{E}$

b_l/b_0	1 [1]	0·8	0·6	0·4	0·2	0 [2]
ψ	1·000	1·054	1·121	1·202	1·315	1·5

[1] Rectangular spring
[2] Triangular spring

continued on Q 7

continued from Q 6

Laminated leaf springs

Laminated leaf springs can be imagined as trapezoidal springs cut into strips and rearranged (spring in sketch can be replaced by two trapezoidal springs in parallel) of total spring width:

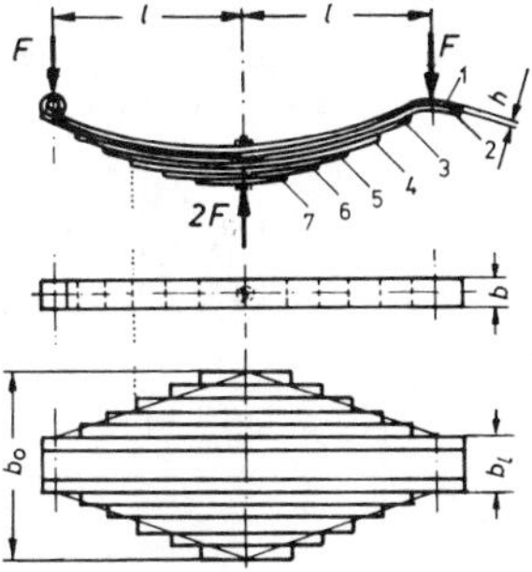

q 30 $$b_o = z\,b$$

z : no. of leaves.

Then (as q 28):

q 31 $$F \triangleq \frac{b\;h^2\,p_{bt}}{6\;l}$$

If leaves 1 and 2 are the same length (as in the sketch):

q 32 $$b_l = 2\,b$$

The calculation does not allow for friction. In practice, friction increases the carrying capacity by between 2 ... 12%.

Precise calculation according to Sheet 394, 1st edition 1974 Beratungsstelle für Stahlverwendung, Düsseldorf.

Disc springs (Ring springs)

Different characteristics can be obtained by combining n springs the same way and i springs the opposite way:

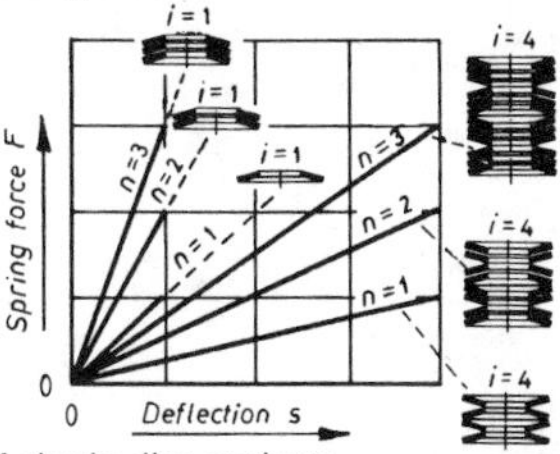

q 33 $$F_{tot} = n\;F_{single}$$

q 34 $$s_{tot} = i\;s_{single}$$

DIN 2092: Precise calculation of single disc springs.
DIN 2093: Dimensions and charact. of standard disc springs.

Material properties: Hot-worked steels for springs to ASTM A 322 e.g. for leaf springs 9255; 6150 – resp. to BS 970/5 e.g. 250 A 53; 735 A 50 – (Modulus of elasticity: $E = 200\,000$ N/mm²).

p_{bt} : static 910 N/mm²
oscillating (500 ± 225) N/mm² scale removed and tempered

continued on Q 8

continued from Q 7

Coiled torsion spring: The type shown in the sketch has both ends free and must be mounted on a guide post. Positively located arms are better.

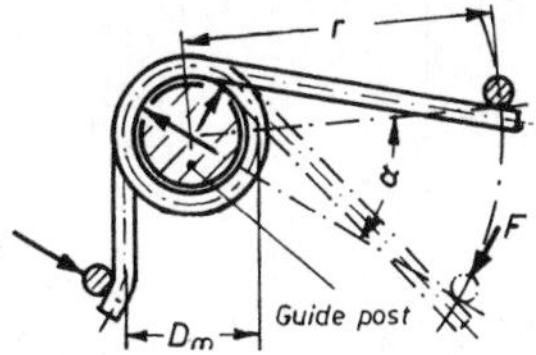

q 35 Perm. spring force[1] $F_p \mathrel{\hat=} \dfrac{Z\, p_{bt}}{r}$

q 36 Angle of deflection $\alpha \mathrel{\hat=} \dfrac{F\, r\, l}{I\, E}$

q 37 Spring coil length $l \mathrel{\hat=} D_m\, \pi\, i_f$

i_f : no. of coils.

(Additional correction is needed for deflection of long arms).

For precise calculation, see DIN 2088.

Springs in torsion

Torsion bar spring

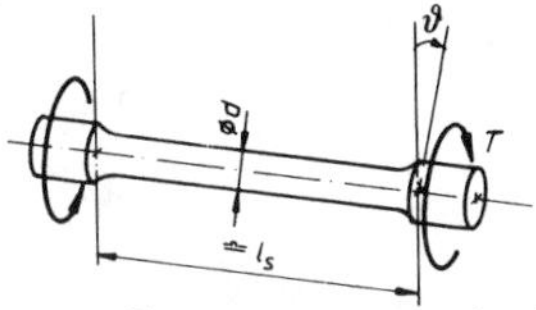

	Shear stress	Torque	Angle of twist
q 38	$\tau = \dfrac{5\,T}{d^3}$	$T = \dfrac{d^3}{5} p_{qt}$	$\vartheta = \dfrac{T\, l_s}{G\, I_p} \mathrel{\hat=} \dfrac{10\, T\, l_s}{G\, d^4}$

l_s : spring length as shown in sketch.

Stress p_{qt} and fatigue strength τ_f in N/mm²

		static		oscillating[2]		
q 39	p_{qt}	not preloaded	700	$\tau_f = \tau_m \pm \tau_A$	$d = 20$ mm	500 ± 350
		preloaded	1020		$d = 30$ mm	500 ± 240

τ_m : mean stress

τ_A : alternating stress amplitude of fatigue strength

Precise calculation see DIN 2091, especially for the spring length.

[1] Not allowing for the stress factor arising from the curvature of the wire.

[2] Surface ground and shot-blasted, preloaded.

continued on Q 9

continued from Q 8

Cylindrical helical spring (compression and tension)

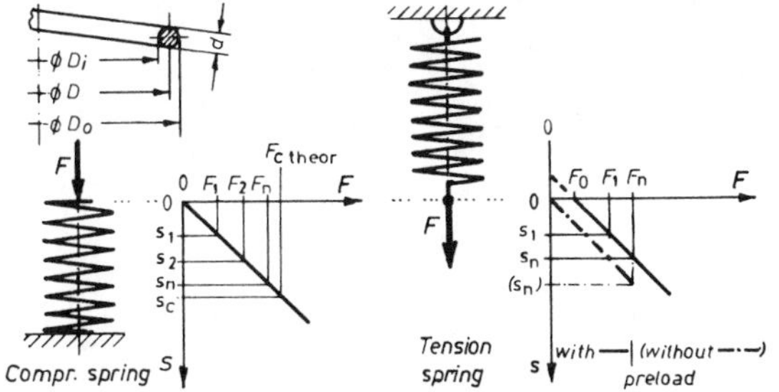

Normal coil ratio: $D/d = 5 \ldots 15; \quad D = (D_o + D_i)/2$

Static stress:

		compression spring	tension spring
q 40	D known	$d \geq \sqrt[3]{\dfrac{8\, F_{c\,theor}\, D}{\pi\, p_{qt}}}$	$d \geq \sqrt[3]{\dfrac{8\, F_n\, D}{\pi\, p_{qt}}}$
q 41	D unknown estimate D/d	$d \geq \sqrt{\dfrac{8\, F_{c\,theor}}{\pi\, p_{qt}} \times \dfrac{D}{d}}$	$d \geq \sqrt{\dfrac{8\, F_n}{\pi\, p_{qt}} \times \dfrac{D}{d}}$
q 42	max. perm. deflection	$s_n = s_c - s_A$	
q 43	sum of min. distances between coils	$s_A = x\, d\, n$ with $x = \lvert 0{\cdot}2 \rvert \ldots \lvert 0{\cdot}7 \rvert$ at $D/d = \lvert 4 \rvert \ldots \lvert 20 \rvert$	solid length
q 44	no. of effective coils	$n = \dfrac{s}{F} \times \dfrac{G\, d^4}{8\, D^3}$	
q 45	permissible shear stress *)	$p_{qt} = p_{qtc} = 0{\cdot}56 \times R_m$ see diagram	$p_{qt} = 0{\cdot}45 \times R_m$ diagram $\times\, 0{\cdot}8$

*) For higher relaxation requirement see DIN 2089.

Theoretical spring deflection | force | perm. stress when just solid (compression spring)

deflection	force	perm. stress
s_c	$F_{c\,theor}$	p_{qtc}

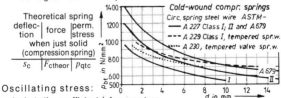

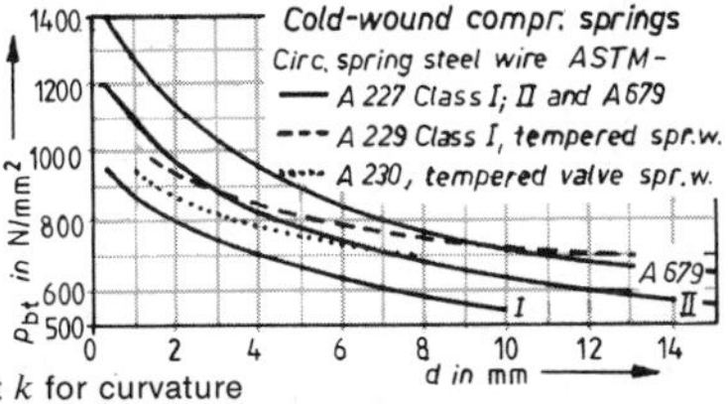

Oscillating stress:

Include the coefficient k for curvature of the wire and use the fatigue strength of spring steel (see DIN 2089) in the calculations.

Rolling bearings

Use the formulae from the manufacturer's literature which gives load capacities and dimensions, e. g. S.K.F., Timken.

Journal bearings

Hydrodynamically-lubricated plain journal bearing

Bearing must be running at proper temperatures and without excessive wear, i.e. separation of the journal and bearing by a film of lubricant.

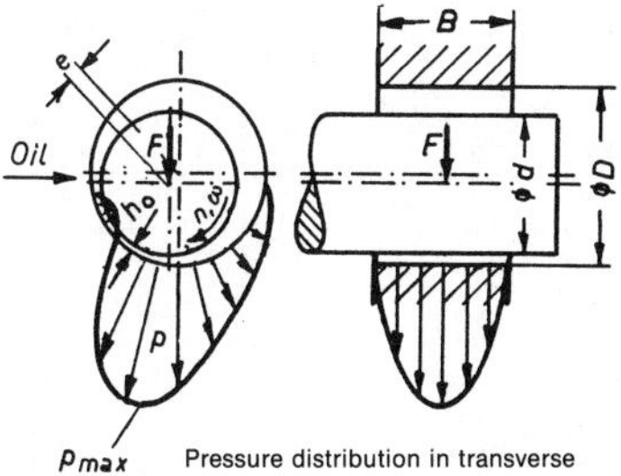

Pressure distribution in transverse and longitudinal sections

Length/diameter ratio *B/D*

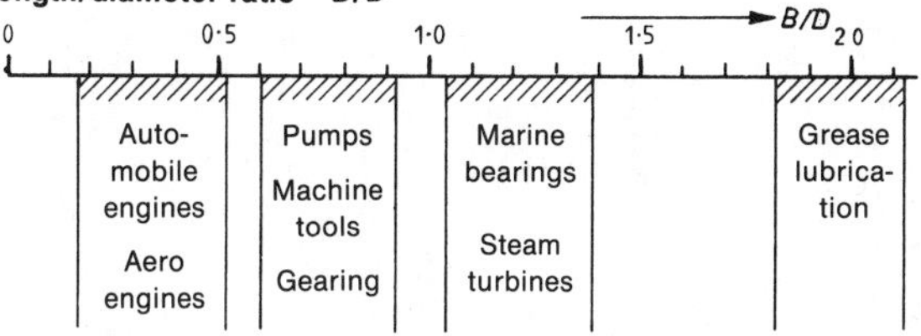

General properties

Short bearings	Long bearings
Large pressure drop at each end, therefore good cooling with adequate oil flow.	Small pressure drop at each end, therefore high load capacity
Excellent for high rotational speeds.	Good at low rotational speeds.
Low load capacity at low rotational speeds.	Poor cooling facilities.
	Danger of edge loading.

continued on Q 11

continued from Q 10 (journal bearings)

Bearing pressure $\overline{p}$, p_{max}

q 46 mean	bearing pressure	$\overline{p} = \dfrac{F}{D\ B}$
q 47 max.		$p_{max} \leq \dfrac{2}{3}\sigma_{dF}$

$f_{b\,max}$ depends mainly on the relative thickness δ of the lubricating film (see Sommerfeld number q 56).

The adjacent diagram shows the ratio of maximum pressure to mean bearing pressure in relation to the relative thickness of the lubrication film. (According to Bauer, VDI 2204).

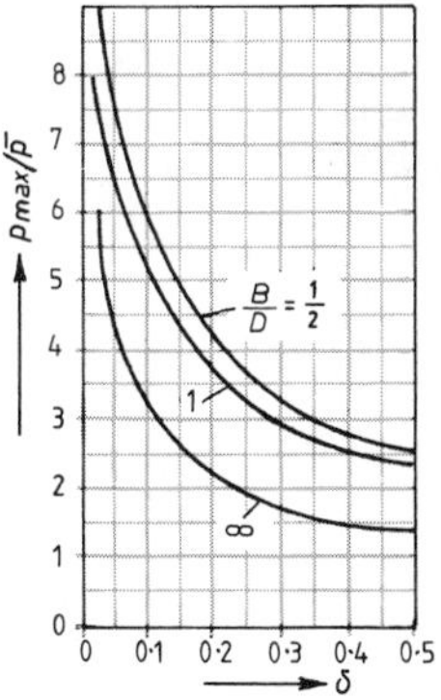

Bearing clearance s **relative bearing clearance** ψ

q 48 $s = D - d$; $\psi = s/D$

ψ is basically the relative bearing clearance established during operation (including thermal expansion and elastic deformation).

q 49 Typical values $\psi = (0{\cdot}3 \ldots 1 \ldots 3)\ 10^{-3}$ [1)]

Criteria for the choice of ψ:

	Lower value	Upper value
bearing material	soft (e. g. white metal)	hard (e. g. phosphor)
viscosity	relatively low	relatively high
peripheral speed	relatively low	relatively high
bearing pressure	relatively high	relatively low
length/diameter ratio	$l/d \leq 0{\cdot}8$	$l/d \geq 0{\cdot}8$
support	self-aligning	rigid

q 50 Minimum values for plastics $\psi \geq (3 \ldots 4)\ 10^{-3}$

q 51 sintered metals $\psi \geq (1 \cdot 5 \ldots 2)\ 10^{-3}$

q 52 [1)] Grease-lubricated plain bearings $\psi = (2 \ldots 3)\ 10^{-3}$

continued on Q 12

continued from Q 11 (journal bearings)

Minimum thickness of lubricating film h_{min} in μm

Theoretical	Actual
q 54 $h_o \geq h_{o\,lim}$ = shaft deflection + bearing distortion + sum of peak-to-valley heights ($R_{zB} + R_{zS}$)	$h_{o\,lim} = [(1) \ldots 3{\cdot}5 \ldots 10 \ldots (15)]$ (1) … 3·5: special cases, e.g. some automobile crankshaft bearings 10: small- | (15): large shaft diameters

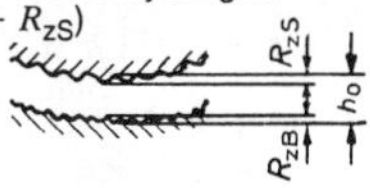

Relative thickness of lubricating film δ

q 55 $$\delta = \frac{h_o}{s/2} = \frac{2\ h_o}{\psi\ d}$$

For statically loaded bearings:
$\delta \leq 0{\cdot}35$
otherwise instability.

Relative eccentricity ε

$$\varepsilon = \frac{e}{s/2} = 1 - \delta$$

Sommerfeld number *So*
(Dimensionless)

q 56 $$So = \frac{\overline{p}\,\psi^2}{\eta_{eff}\ \omega}$$

Inserting *So* in the adjacent diagram gives δ and therefore also h_o from q 55.

To a first estimate η_{eff} viscosity is based on mean temperature of bearing.

A better estimate is:

$t_{eff} = 0{\cdot}5\,(t_1 + t_2)$

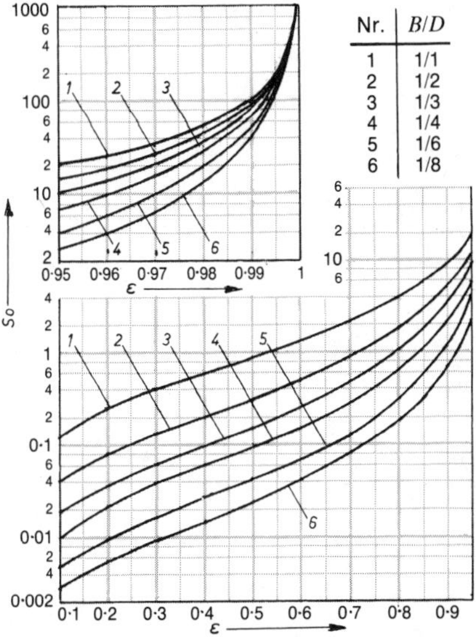

Nr.	B/D
1	1/1
2	1/2
3	1/3
4	1/4
5	1/6
6	1/8

continued on Q 13

continued from Q 12 (journal bearings)

Lubricant flow rate $\dot{Q}_1$

The theoretical flow rate required to maintain hydrodynamic lubrication is (exactly values see DIN 31652):

q 57 $\dot{Q}_1 = 0{\cdot}5\, B\, u(s - 2h_o)$

Rules: Oil inlet to the expanding part of the bearing.

Oil velocity v:

q 58 In supply lines: $v \triangleq 2$ m/s ; $p_r = 0{\cdot}05 \ldots 0{\cdot}2$ MPa

q 59 In return lines: $v \triangleq 0{\cdot}5$ m/s ; $p_r = 0$

Oilbags, oil slots (deepth $\triangleq 2 \cdot s$) never in loaded zones, no connection with the surface area of the bearings, of higher pressure only short oil slots or -bags; bigger oil bags only in special cases for greater heat removal.

Heat removal

Requirement:

q 60 Friction power $P_F = \mu F u = P_A$ (rate of heat removal)

(Calculate μ by the use of the following diagrams and So by q 56)

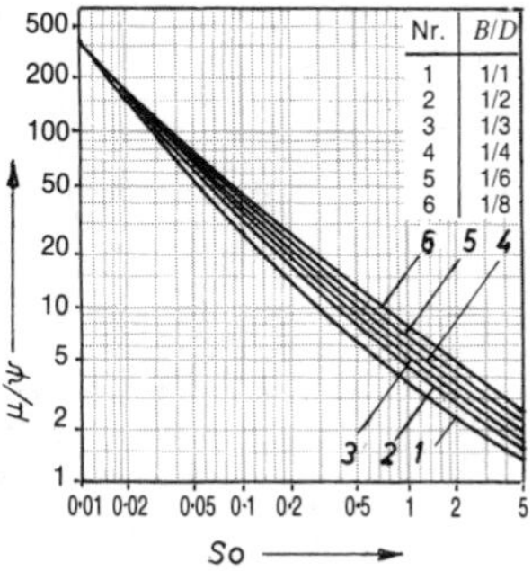

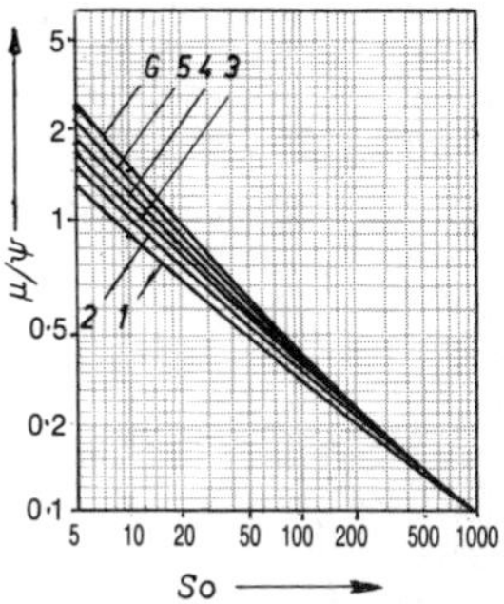

Heat removal by convection at the housing surface A.

q 61 $P_A = k\,A(t_h - t_o)$ with empirical formula for k:

$$\frac{k}{\mathrm{W/(m^2K)}} = 7 + 12\sqrt{\frac{w}{\mathrm{m\,s^{-1}}}}$$ (equation scaled for units)

continued on Q 14

Symbols see Q 14

continued from Q 13 (journal bearings)

If the area of the heat removal surface A is unknown, approximate value for step-bearings:

q 68 $$A \approx \pi H \left[B_H + \frac{H}{2} \right]$$

Since η is temperature-sensitive and the running oil temperature is initially unknown, use iteration method for preliminary and successively improved estimates of t_h by q 61 until, according to q 60, $P_A = P_F$.

Heat removed by lubricant P_Q.

Oil circulation, if necessary with oil cooler (convection, heat conduction are neglected):

q 69 $$P_Q = \dot{Q}\, c\, \varrho\, (t_2 - t_1)$$

Guide values for simple calculations for mineral oils:

q 70 $$c \cdot \varrho \approx 1{\cdot}8 \times 10^6 \quad \mathrm{J\,m^{-3}\,K^{-1}}$$

---

c : specifique heat (values see Z 5)
h_o : min. thickness of lubricating film at running speed (r.p.m.)
$h_{o\,lim}$: min. thickness of lubricating film at lowest running (min. permissible thickness of oil lubricating film)
k : heat transmission coefficient
p_r : pressure (relative to atmospheric) of oil flow
$\bar{p}$, p_{max} : mean, max. bearing pressure
s : radial clearence
t_h : operating temperature of housing surface
t_o : ambient temperature
t_1 , t_2 : inlet, outlet temperature of oil
u : peripheral speed of bearing journal
w : velocity of cooling air (m/s)
B_H : width of housing in axial direction
F : radial bearing force
H : pedestal bearing, total height
$\dot{Q}$: flow rate of lubricant
$\dot{Q}_1$: flow rate of lubricant, in a consequence of an increase of its self-pressure
η : dynamic or absolute viscosity (values see Z 14)
η_{eff} : dynamic viscosity based on mean temperature of bearing
μ : friction coefficient (values see Z 7)
ϱ : density of lubricant (values see Z 5)
ψ : relative bearing clearence (values see Q 11)
ω : angular velocity

Crosshead guide

Crosshead guide will operate smoothly only when

q 71 $$\tan \alpha < \frac{l}{(2h + l)\mu} \quad \text{or}$$

the length ratio is

q 72 $$\frac{l}{h} = \lambda > \frac{2\mu \tan \alpha}{1 - \mu \tan \alpha}$$

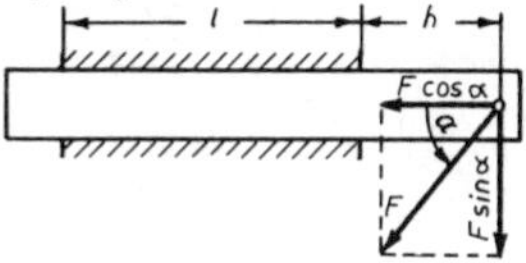

If the above conditions for $\tan \alpha$ are not satisfied there is a danger of tilting and jamming.

Friction clutches

Slip time and energy loss per operation

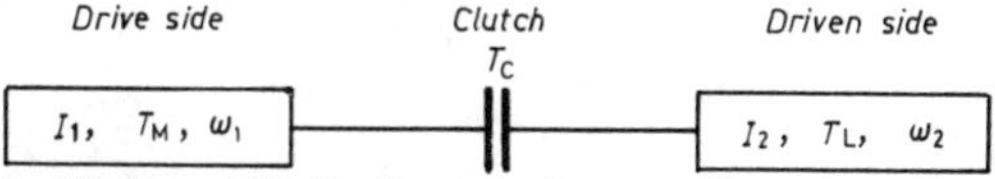

A simplified model with the following conditions is sufficent for approximate calculation:

Acceleration of driven side from $\omega_2 = 0$ to $\omega_2 = \omega_1$, $\omega_1 = \text{const.}$; $T_L = \text{const.}$; $T_C = \text{const.} > T_L$. Then, per operation:

q 73 energy loss $$W_l = I_2 \times \frac{\omega_1^2}{2} \times \left(1 + \frac{T_L}{T_C - T_L}\right)$$

q 74 slip time $$t_f = \frac{I_2\, \omega_1}{T_C - T_L}$$

Calculating the area of the friction surface

single- plate	twin- plate	multi-plate	cone	cylin- drical
flat clutches			clutches	

Number and size of friction surfaces depend on the permissible contact pressure p_b and the permissible thermal capacity per unit area q_p.

continued on Q 16

Symbols see Q 17

continued from Q 15 (friction clutches)

Calculation of contact pressure p_b
(for values see Z 19)

For all types of friction surfaces:

q 75 $$i\,A \geqslant \frac{T_c}{p_b \times \mu_{dyn} \times R_m}$$

q 76 where $$R_m = \frac{2}{3} \times \frac{R_a^3 - R_i^3}{R_a^2 - R_i^2} \triangleq \frac{R_a + R_i}{2}$$

		flat	conical	cylindrical
		friction surfaces		
q 77	operating force (axial)	$F_a = A\,p$	$F_a = A\,p \times \sin \alpha$	
q 78		for multiplate clutches usually:	to prevent locking:	$R_a = R_i = R_m$
q 79		$\frac{R_i}{R_a} = 0{\cdot}6 \ldots 0{\cdot}8$	$\tan \alpha > \mu_{stat}$	

Calculation for a shaft: $$T_t = T_c \cdot \frac{\mu_{stat}}{\mu_{dyn}}$$

Calculation permissible temperature rise

In HEAVY-LOAD STARTING the maximum temperature is reached in one operation. It depends on the energy loss, slip time, heat conduction, specific heat and cooling. These relationships cannot be incorporated in a general formula.

With CONTINUOUS OPERATION constant temperature is only established after several operations. There are empirical values for permissible thermal capacity per unit area q_p with continuous operation (see Z 19).

q 80 Friction power $$P_F = W_l\,z$$

q 81 Condition $$i\,A \geqslant \frac{W_l\,z}{q_p}$$

Friction brakes

All friction clutches can also be used as brakes. But there are also:

Disc brakes
with caliper and pads.

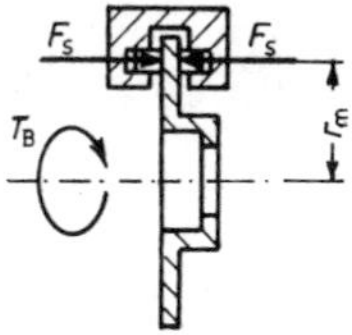

Braking torque T_B:

q 82 $$T_B = 2\,\mu\,F_s\,j\,r_m$$

Expanding-shoe drum brakes
(Drawing of simplex brake showing, simplified, the forces acting on the shoes).

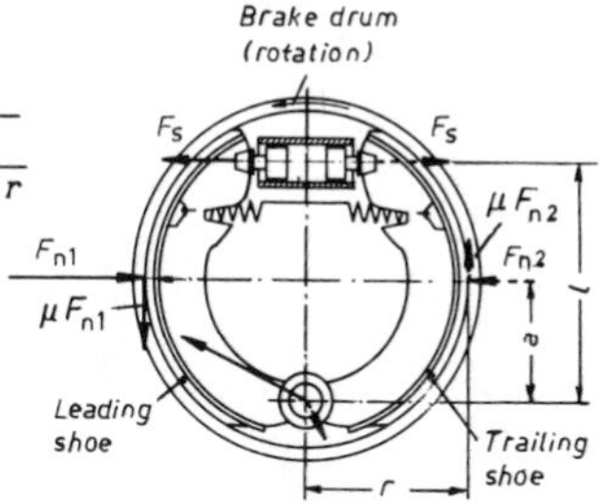

Leading shoes	Trailing shoes
q 83/84 $F_{n1} = \dfrac{F_s\, l}{a - \mu\, r}$	$F_{n2} = \dfrac{F_s\, l}{a + \mu\, r}$
(Servoaction)	

Braking torque T_B:

q 85 $$T_B = (F_{n1} + F_{n2})\,\mu\,r$$

Band brakes see K 13

Notation for friction clutches and brakes (Q 15 ... Q 17)

A : area of friction surface
T_C : operating torque of clutch
T_L : load torque
T_M : motor torque
T_t : transfer moment of clutch
W_l : energy loss per operation
μ, μ_{dyn}, μ_{stat} : friction-, sliding friction-, static coefficient of friction
i : no. of friction surfaces
j : no. of calipers on a disc brake
R : radius of friction surface
R_m, R_a, R_i : mean, outside, inside radius of friction surface
z : operating frequency (EU: s^{-1}, h^{-1})
ω : angular velocity

(For properties of friction materials see Z 19)

Involute-tooth gears

Spur gears, geometry

q 86 Gear ratio $$u = \frac{z_2}{z_1}$$

q 87 Transmission ratio $$i = \frac{\omega_a}{\omega_b} = \frac{n_a}{n_b} = -\frac{z_b}{z_a}$$ [1)]

Transmission ratio of multi-stage gearing:

q 88 $$i_{tot} = i_I \times i_{II} \times i_{III} \times \ldots \times i_n$$

q 89 Involute function $$\text{inv}\,\alpha = \tan\alpha - \hat{\alpha}$$

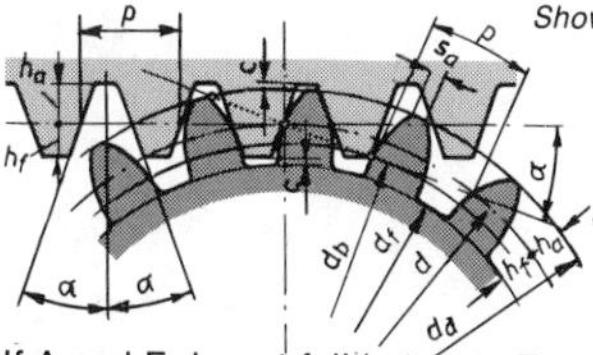

Showing the transverse path of contact (see ISO/R 1122)

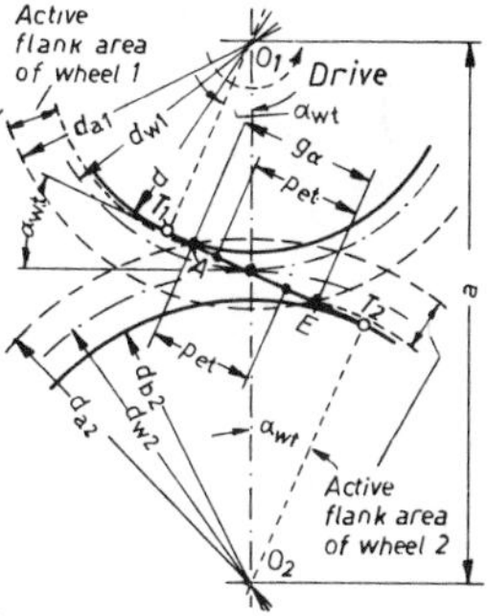

If A and E do not fall between T_1 and T_2, interference will occur and "modified" gears as in Q 20 should be used.

[1)] Negative for external gears because rotation is opposite. Positive for internal gears. The sign can normally be disregarded.

		Standard gears spur	Standard gears helical
q 90	normal pitch	$p = \frac{\pi d}{z} = m\pi$	$p_n = m_n \pi$
q 91, q 92	circular pitch		$p_t = \frac{m_n \pi}{\cos\beta}$
q 93	normal module	$m = \frac{p}{\pi} = \frac{d}{z}$	$m_n = \frac{p_n}{\pi} = \frac{d}{z}\cos\beta$
q 94, q 95	circular module		$m_t = \frac{m_n}{\cos\beta} = \frac{d}{z}$
q 96	addendum	$h_a = h_{aP} = m$	
q 97	dedendum	$h_f = h_{fP} = m + c$	
q 98	bottom clearance	$c = (0{\cdot}1 \ldots 0{\cdot}3)\,m \simeq 0{\cdot}2\,m$	

continued on Q 19

For symbols see Q 29, suffixes see Q 23

MACHINE PARTS

Involute-tooth gears

continued from Q 18 (spur gears)

		Standard gears: spur	Standard gears: helical
99/100	reference diameter	$d = m\ z$	$d = \frac{m_n\ z}{\cos\beta} = m_t\ z$
q 101	tip diameter	$d_a = d + 2\ h_a$	
q 102	root diameter	$d_f = d - 2\ h_f$	
q 103	pressure angle	$\alpha = \alpha_n = \alpha_t = \alpha_P$	$\alpha_n = \alpha_P$
q 104			$\tan\alpha = \frac{\tan\alpha_n}{\cos\beta}$
105/106	base diameter	$d_b = d\ \cos\alpha$	$d_b = d\ \cos\alpha_t$
q 107	equivalent no of teeth		$z_{nx} = z\frac{1}{\cos^2\beta \times \cos\beta}$ table see DIN 3960
q 108			$\approx \frac{z}{\cos^3\beta}$
	min. no. of teeth		
q 109	to avoid interference: theory	$z_{lim} = \frac{2}{\sin^2\alpha} \approx 17$ for $\alpha_P = 20^\circ$	
110/111	to avoid interference: practice	$z'_{lim} \approx 14$	$z'_{lim\,h} \approx 14\cos^3\beta$
q 112	spread		$U = b\ \tan\lvert\beta\rvert$

		Standard gearing: spur	Standard gearing: helical
113/114	centre distance	$a = \frac{d_1+d_2}{2} = m\frac{z_1+z_2}{2}$	$a = \frac{d_1+d_2}{2} = m_n\frac{z_1+z_2}{2\cos\beta}$
q 115	length of path of contact (total length)	$g_\alpha = \frac{1}{2}\left[\sqrt{d_{a1}^2 - d_{b1}^2} + \sqrt{d_{a2}^2 - d_{b2}^2} - (d_{b1} + d_{b2})\tan\alpha_t\right]$	
116/117	transverse contact ratio	$\varepsilon_\alpha = \frac{g_\alpha}{p\ \cos\alpha}$	$\varepsilon_\alpha = \frac{g_\alpha}{p\ \cos\alpha}$
q 118	overlap ratio		$\varepsilon_\beta = \frac{b\ \sin\lvert\beta\rvert}{m_n\ \pi}$
q 119	contact ratio		$\varepsilon_\gamma = \varepsilon_\alpha + \varepsilon_\beta$

continued on Q 20

For symbols see Q 29, suffixes see Q 23

Q 19	MACHINE PARTS Involute tooth gears

[illegible]

continued on Q 20

continued from Q 19 (gearing)

		modified gears: spur	modified gears: helical
	p, p_n, p_t, z, z_{nx}, m, m_n, m_t, d, d_b	see standard gears	
120/121	profile offset	$x\,m$	$x\,m_n$
122/123	profile offset factor: to avoid interference	$x_{min} = -\dfrac{z\,\sin^2\alpha}{2} + \dfrac{h_{a0} - \varrho_{a0}(1 - \sin\alpha)}{m}$	$x_{min} = -\dfrac{z\,\sin^2\alpha}{2\,\cos\beta} + \dfrac{h_{a0} - \varrho_{a0}(1 - \sin\alpha_n)}{m_n}$
		can be up to 0·17 mm	
124/125	profile offset factor: ditto[1]	$x \triangleq \dfrac{14 - z}{17}$	$x \triangleq \dfrac{14 - (z/\cos^3\beta)}{17}$
q 126	profile offset factor: to give a specific centre distance (total)	$x_1 + x_2 = \dfrac{(z_1 + z_2) \times (\mathrm{inv}\,\alpha_{wt} - \mathrm{inv}\,\alpha_t)}{2\,\tan\alpha_n}$	
q 127	α_{wt} calculated from	$\cos\alpha_{wt} = \dfrac{(z_1 + z_2)\,m_t}{2\,a}\cos\alpha_t$	
q 128	or	$\mathrm{inv}\,\alpha_{wt} = \mathrm{inv}\,\alpha_t + 2\,\dfrac{x_1 + x_2}{z_1 + z_2}\tan\alpha_n$	
q 129	centre distance	$a = a_d\,\dfrac{\cos\alpha_t}{\cos\alpha_{wt}}$	
q 130	addendum modification coefficient	$k\,m_n = a - a_d - m_n \times (x_1 + x_2)$ [2]	
q 131	addendum	$h_a = h_{aP} + x\,m_n + k\,m_n$	
q 132	dedendum	$h_f = h_{fP} - x\,m_n$	
q 133	outside diameter	$d_a = d + 2\,h_a$	
q 134	root diameter	$d_f = d - 2\,h_f$	
q 135	length of path of contact	$g_\alpha = \dfrac{1}{2}\left[\sqrt{d_{a1}^2 - d_{b1}^2} + \sqrt{d_{a2}^2 - d_{b2}^2} - (d_{b1} + d_{b2})\tan\alpha_{wt}\right]$	
136/137	transverse contact ratio	$\varepsilon_\alpha = g_\alpha/(p\,\cos\alpha)$	$\varepsilon_\alpha = g_\alpha/(p_t\,\cos\alpha_t)$
q 138	overlap ratio		$\varepsilon_\beta = b\,\sin\lvert\beta\rvert/(m_n\,\pi)$
q 139	contact ratio		$\varepsilon_\gamma = \varepsilon_\alpha + \varepsilon_\beta$

1) If tool data unknown take $\alpha_P = 20^\circ$.

2) Note the sign. With external gears $k \times m_n < 0$! When $k < 0{\cdot}1$ addendum modification can often be avoided

Spur gears, design

The dimensions are derived from

load-carrying capacity of the tooth root
load-carrying capacity of the tooth flank,

which must be maintained independently.

Gearing design is checked in accordance with DIN 3990. By conversion and rough grouping of various factors it is possible to derive some approximate formulae from DIN 3990.

Load capacity of tooth (approximate calculation)

Safety factor S_F against fatigue failure of tooth root:

q 140

$$S_F = \frac{\sigma_{F\,lim}}{\frac{F_t}{b\ m_n} \times Y_F \times Y_\varepsilon \times Y_\beta} \times \frac{Y_S \times K_{FX}}{K_I \times K_V \times K_{F\alpha} \times K_{F\beta}} \geqslant S_{F\,min}$$

Giving the approximate formulae:

q 141

$$m_n \geqslant \frac{F_t}{b}\ Y_F \times K_I \times K_V \times \underbrace{Y_\varepsilon \times Y_\beta \times K_{F\alpha}}_{\simeq 1} \times \underbrace{\frac{K_{F\beta}}{Y_S \times K_{FX}}}_{\simeq 1} \times \frac{S_{F\,min}}{\sigma_{F\,lim}}$$

Y_F : tooth form factor for external gearing (see diagram)

q 142 $K_I \times K_V$ = 1 ... 3, rarely more, (allowing for external shock and irregularities exceeding the rated torque, additional internal dynamic forces arising from tooth errors and circumferential velocity).

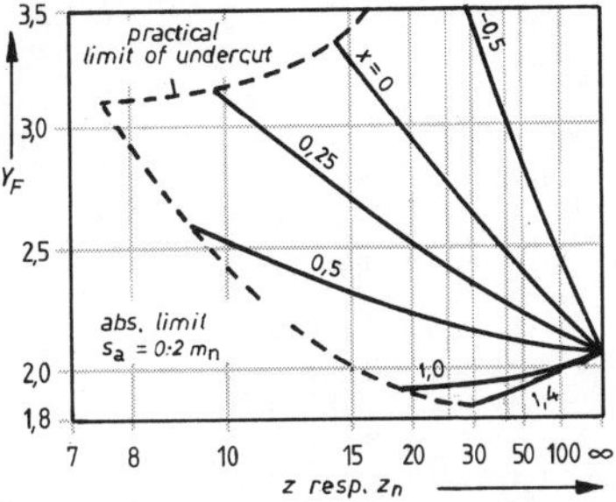

q 143 $S_{F\,min}$ = 1·7 (guide value)

q 144 $\sigma_{F\,lim}$: guide values see table on Q 22

continued on Q 22

For symbols see Q 29, suffixes see Q 23

continued from Q 21 (spur gears, design)

Load capacity of tooth flank (approximate calculation)

Safety factor S_H against pitting:

q 145
$$S_H = \frac{\sigma_{H\,lim}}{\sqrt{\frac{u+1}{u} \times \frac{F_t}{b\; d_1}} \times Z_H \times Z_M \times Z_\varepsilon} \times \frac{Z_V \times K_{HX} \times Z_R \times K_L}{\sqrt{K_I \times K_V \times K_{H\alpha} \times K_{H\beta}}} \geq S_{H\,min}$$

For metals the material factor Z_M is simplified to:

q 146
$$Z_M = \sqrt{0\cdot35\; E} \quad \text{where} \quad E = \frac{2\; E_1\; E_2}{E_1 + E_2}$$

Therefore, the approximate formula becomes:

q 147
$$d_1 \geq \sqrt{\frac{2\,T_1}{b} \times \frac{u+1}{u}\, 0\cdot35\; E} \times Z_H \times Z_\varepsilon \underbrace{\sqrt{K_{H\alpha}}}_{\approx 1} \frac{\overbrace{\sqrt{K_I \times K_V} \times \sqrt{K_{H\beta}}}^{\approx 1}}{Z_V \times K_{HX} \times Z_R \times K_L} \times \frac{S_{H\,min}}{\sigma_{H\,lim}}$$

Approximate values for strength
(Diagrams in DIN 3990, part 5)

Mat.	Specification to Norm	Grade	$\sigma_{F\,lim}$ N/mm²	$\sigma_{H\,lim}$ N/mm²
CI	ASTM	A 48-50 B	80	360
CS	ASTM	A 536-20-90-02	230	560
CS	ASTM	A 572 Gr. 65	200	400
CS	SAE	1064	220	620
AS	SAE	4140	290	670
ASCH	SAE	3240	500	1630

CI: cast iron
CS: carbon steel
AS: alloy steel
ASCH: case hardened alloy steel

In DIN 3990: Z_H is higher by a factor $\sqrt{2}$, but Z_M (new Z_E = factor for elasticity) is lower by a factor $\sqrt{2}$, so

$$(Z_{H\,old}) \cdot Z_M = (Z_{H\,new}) \cdot Z_E.$$

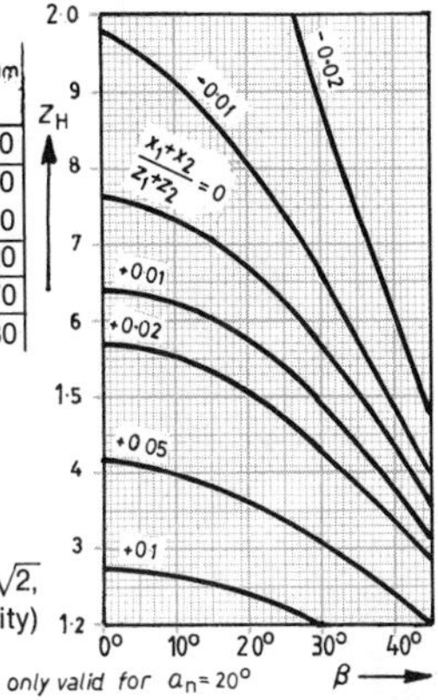

continued on Q 23

Z_H : flank form factor (see diagram)

q 148 $K_I \times K_V$: see load capacity of tooth root (q 142)

q 149 $S_{H\,min} \approx 1\cdot2$ (guide value)

q 150 $\sigma_{H\,lim}$: guide values see table

q 151 $Z_V \times K_{HX} \times Z_R \times K_L = 0\cdot5 \ldots 1$. Higher value for higher circumferential velocity, higher lubricating oil viscosity and lower roughness.

For symbols see Q 29, for suffixes see Q 23

MACHINE PARTS

Involute-tooth gears

continued from Q 22, (spur gears, design)

In q 141, q 145 and q 147 b or b and d must be known. The following ratios are for estimating purposes and should be used for the initial calculation:

Pinion dimensions

Either:

		$\frac{d_1}{d_{shaft\ 1}}$
q 152	pinion integral with shaft	1·2 ... 1·5
q 153	pinion free to turn on shaft	2

Or: from gear ratio i and a specified centre distance a (see q 113–114–129)

Tooth width ratios

	Tooth- and bearing-quality	$\frac{b}{m}$	$\frac{b}{d_1}$
q 154	teeth smoothly cast or flame cut	6 ... 10	
q 155	teeth machined; bearings supported each side on steel construction or pinion overhung	(6) ... 10 ... 15	
q 156	teeth well machined; bearings supported each side in gear casing	15 ... 25	
q 157	teeth precision machined; good bearings each side and lubrication in gear casing: $n_1 \leq 50\ s^{-1}$.	20 ... 40	
q 158	overhung gearwheel		$\leqslant$ 0·7
q 159	fully supported		$\leqslant$ 1·5

Suffixes for Q 18 ... 25

- $_a$: driving wheel
- $_b$: driven wheel
- $_1$: small wheel or pinion
- $_2$: large wheel or wheel
- $_t$: tangential
- $_n$: normal
- $_m$: tooth middle for bevel gears
- $_v$: on back cone (or virtual cylindrical gear)

For symbols see Q 29

Bevel gears

Bevel gears, geometry
Equations q 86...q 88 are applicable and also:

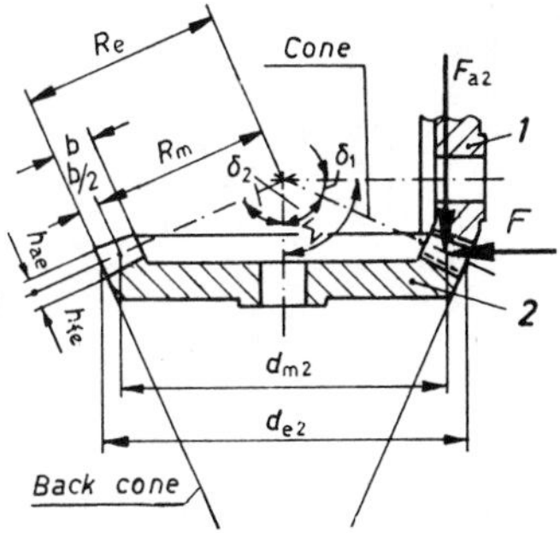

Only the axial and radial forces acting on mesh wheel 2 are shown

cone angle δ:

q 160 $\tan\delta_1 = \dfrac{\sin\Sigma}{\cos\Sigma + u}$;

q 161 $\left(\Sigma = 90^\circ \Rightarrow \tan\delta_1 = \dfrac{1}{u}\right)$

q 162 $\tan\delta_2 = \dfrac{\sin\Sigma}{\cos\Sigma + 1/u}$;

q 163 $(\Sigma = 90^\circ \Rightarrow \tan\delta_2 = u)$

q 164 angle between shafts } $\Sigma = \delta_1 + \delta_2$

q 165 external pitch cone distance } $R_e = \dfrac{d_e}{2\sin\delta}$

Development of the back cone to examine the meshing conditions gives the virtual cylindrical gear (suffix "v" = virtual) with the values:

q 166	straight	bevel gears	$z_v = \dfrac{z}{\cos\delta}$	$u_v = \dfrac{z_{v2}}{z_{v1}}$
q 167				
q 168	spiral		$z_v \approx \dfrac{z}{\cos\delta \times \cos^3\beta}$	

Formulae q 92, q 95 ... q 100 are also applicable to the surface of the back cone (suffix "e").

Bevel gears, design

The design is referred to the MID-POINT OF THE WIDTH b (suffix "m") with the values:

169/170 $R_m = R_e - \dfrac{b}{2}$ | $m_m = \dfrac{d_m}{z}$

171/172 $d_m = 2\,R_m \sin\delta$ | $F_{tm} = \dfrac{2\,T}{d_m}$

continued on Q 25

For symbols see Q 29, for suffixes see Q 23

continued from Q 24

Axial and radial forces in mesh

q 173 axial force $F_a = F_{tm} \tan \alpha_n \times \sin \delta$

q 174 radial force $F_r = F_{tm} \tan \alpha_n \times \cos \delta$

Load capacity of tooth root (approx. calculation)

Safety factor S_F against fatigue failure of tooth root:

q 175
$$S_F = \frac{\sigma_{F\,lim}}{\frac{F_{tm}}{b\ m_{nm}} \times Y_F \times Y_{\varepsilon V} \times Y_\beta} \times \frac{Y_S \times K_{FX}}{K_I \times K_V \times K_{F\alpha} \times K_{F\beta}} \geqslant S_{F\,min}$$

Giving the approximate formula:

q 176
$$m_{nm} \geqslant \frac{F_{tm}}{b} \times Y_F \times K_I \times K_V \times \underbrace{Y_{\varepsilon V} \times Y_\beta \times K_{F\alpha}}_{\doteq 1} \times \underbrace{\frac{K_{F\beta}}{Y_S \times K_{FX}}}_{\doteq 1} \times \frac{S_{F\,min}}{\sigma_{F\,lim}}$$

q 177 Y_F : substitute the number of teeth of the complementary spur gear z_v or, with spiral gears $z_{vn} \doteq z_v / \cos^3 \beta$. The graph for spur gears on page Q 21 is then also applicable to bevel gears.

For all other data see q 142, q 143 and q 144.

Load capacity of tooth flank (approximate calculation)

Safety factor S_H against pitting of tooth surface.

q 178
$$S_H = \frac{\sigma_{H\,lim}}{\sqrt{\frac{u+1}{u} \times \frac{F_{tm}}{b\ d_1} \times Z_H \times Z_M \times Z_{\varepsilon V}}} \times \frac{Z_V \times K_{HX} \times Z_R \times K_L}{\sqrt{K_I \times K_V \times K_{H\alpha} \times K_{H\beta}}} \geqslant S_{H\,min}$$

For metals the material factor Z_M is simplified to:

q 179
$$Z_M = \sqrt{0 \cdot 35\ E} \quad \text{with} \quad E = \frac{2\ E_1\ E_2}{E_1 + E_2}$$

Giving the approximate formula:

q 180
$$d_{vm1} \geqslant \sqrt{\frac{2\ T_1}{b} \times \frac{u_v+1}{u_v}\ 0 \cdot 35\ E \times Z_{HV} \times \underbrace{Z_{\varepsilon V} \times \sqrt{K_{H\alpha}}}_{\doteq 1}} \times \frac{\sqrt{K_I \times K_V} \times \overbrace{\sqrt{K_{H\beta}}}^{\doteq 1}}{Z_V \times K_{HX} \times Z_R \times K_L} \times \frac{S_{H\,min}}{\sigma_{H\,lim}}$$

q 181 Z_{Hv} : see diagram for Z_H (page Q 22), but only valid for $(x_1 + x_2)/(z_1 + z_2) = 0$ with $\beta = \beta_m$.

For all other data see q 148 … q 151.

For symbols see Q 29, for suffixes see Q 23

Velocity diagram and angular velocities
(referred to fixed space)

		arm fixed	wheel 1 fixed	wheel 3 fixed / wheel 4 fixed
q 182	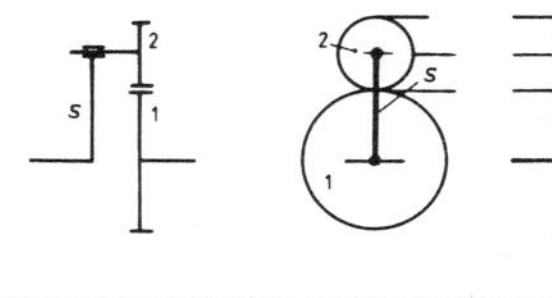	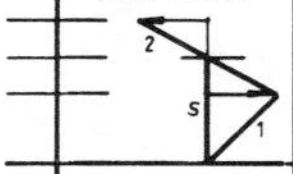 $\omega_2 = -\omega_1 \frac{r_1}{r_2}$	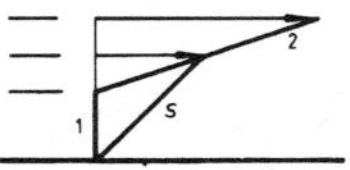 $\omega \cdot r \longrightarrow$ $\omega_2 = \omega_s \left(1 + \frac{r_1}{r_2}\right)$	
q 183	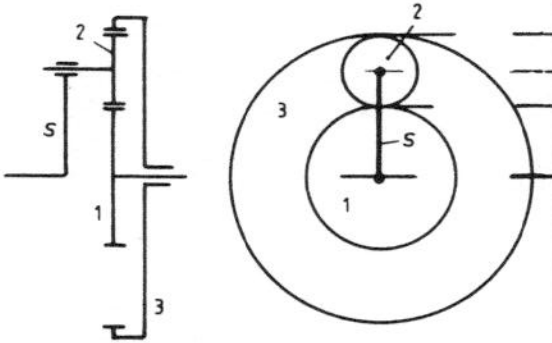	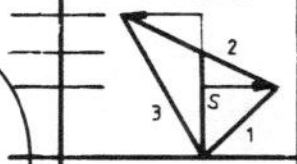 $\omega_3 = -\omega_1 \frac{r_1}{r_3}$	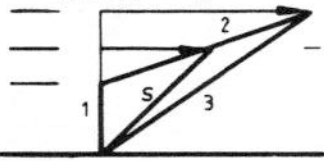 $\omega_3 = \omega_s \left(1 + \frac{r_1}{r_3}\right)$	wheel 3 fixed 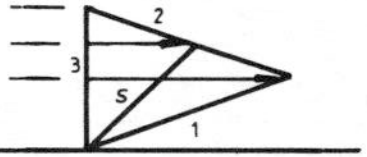 $\omega\, r \longrightarrow$ $\omega_1 = \omega_s \left(1 + \frac{r_3}{r_1}\right)$
q 184	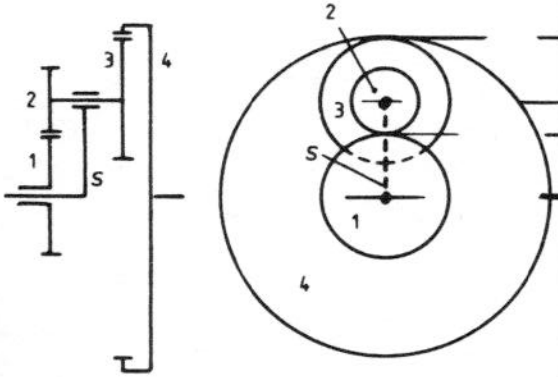	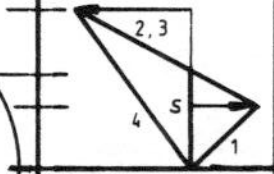 $\omega_4 = -\omega_1 \frac{r_1}{r_2} \frac{r_3}{r_4}$	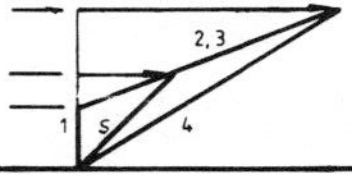 $\omega_4 = \omega_s \frac{(r_1+r_2)(r_2+r_3)}{r_2 \quad r_4}$	wheel 4 fixed 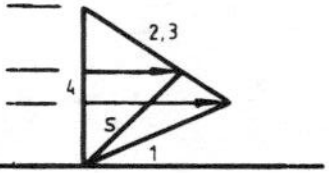 $\omega\, r \longrightarrow$ $\omega_1 = \omega_s \frac{(r_1+r_2)(r_2+r_3)}{r_1 \quad r_3}$

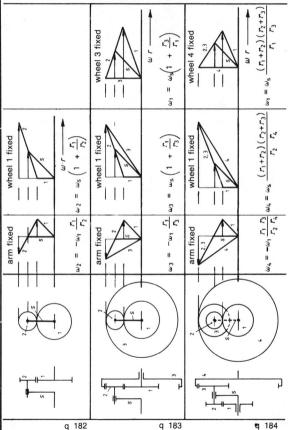

Worm gearing, geometry

(Cylindrical worm gearing, normal module in axial section, BS 2519, angle between shafts $\Sigma = 90^\circ$).

Drive worm

All the forces acting on the teeth in mesh are shown by the three arrows F_a, F_t and F_r.

In the example:
$z_1 = 2$, right-hand helix.

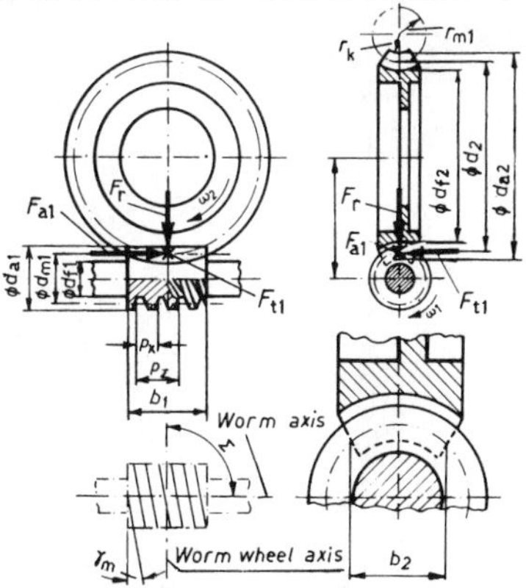

Gear tooth ratio and transmission ratio as q 86 ... 88

		Worm, suffix 1	Worm wheel, suffix 2
q 185	module	$m_x = m = m_t$	
q 186	pitch	$p_x = m\,\pi = p_z = d_2\,\pi/z_2$	
q 187	mean diameter (free to choose, for normal values see DIN 3976)	$d_{m1} = 2\,r_{m1}$	
q 188	form factor	$q = d_{m1}/m$	
q 189	centre helix angle	$\tan\gamma_m = \dfrac{m\,z_1}{d_{m1}} = \dfrac{z_1}{q}$	
q 190	pitch diameter		$d_2 = m\,z_2$
q 191	addendum	$h_{a1} = m$	$h_{a2} = m(1+x)$ [1)]
q 192	dedendum	$h_{f1} = m(1+c_1^*)$	$h_{f2} = m(1-x+c_2^*)$
q 193	tip clearance factor	$c_1^* = (0{\cdot}167\ldots\underline{0{\cdot}2}\ldots0{\cdot}3) = c_2^*$	
q 194	outside diameter	$d_{a1} = d_{m1} + 2\,h_{a1}$	$d_{a2} = d_2 + 2\,h_{a2}$
q 195	tip groove radius		$r_k = a - d_{a2}/2$
q 196	tooth width	$b_1 \geq \sqrt{d_{a2}^2 - d_2^2}$	$b_2 \approx 0{\cdot}9\,d_{m1} - 2\,m$
q 197	root diameter	$d_{f1} = d_{m1} - 2\,h_{f1}$	$d_{f2} = d_2 - h_{f2}$
q 198	centre distance	$a = (d_{m1} + d_2)/2 + x\,m$ [1)]	

[1)] Profile offset factor x for check of a pre-set centre distance, otherwise $x = 0$.

continued on Q 28

For symbols see Q 29, for suffixes see Q 23

continued from Q 27

Worm gearing, design (worm driving)

		Worm	Worm wheel
q 199	peripheral force	$F_{t1} = \frac{2\ T_1}{d_{m1}} K_I \times K_V$	$F_{t2} = F_{a1}$
q 200	axial force	$F_{a1} = F_{t1} \times \frac{1}{\tan(\gamma + \varrho)}$	$F_{a2} = F_{t1}$
q 201	radial force	$F_{r1} = F_{t1} \times \frac{\cos\varrho \times \tan\alpha_n}{\sin(\gamma + \varrho)}$	$= F_r = F_{r2}$
q 202	rubbing speed	$v_g = \frac{d_{m1}}{2} \times \frac{\omega_1}{\cos\gamma_m}$	

Efficiency

	Worm driving	Worm wheel driving
q 203	$\eta = \tan\gamma_m / \tan(\gamma_m + \varrho)$	$\eta' = \tan(\gamma_m - \varrho) / \tan\gamma_m$
		$(\gamma_m < \varrho) \Longrightarrow$ self-locking!

Coefficient of friction (typical values) $\mu = \tan\varrho$

	$v_g \triangleq 1$ m/s	$v_a \triangleq 10$ m/s
worm teeth hardened and ground	0·04	0·02
worm teeth tempered and machine cut	0·08	0·05

For calculation of worm shaft deflection see P 12

Calculation of module m

Load capacity of teeth root and flanks and temperature rise are combined in the approximate formula:

q 204 $F_{t2} = C\ b_2\ p_2$; where $b_2 \triangleq 0{\cdot}8\ d_{m1}$; $p_2 = m\,\pi$.

q 205 $m \triangleq \sqrt[3]{\frac{0{\cdot}8\ T_2}{C_{perm}\ q\ z_2}}$ | $F_{t2} = 2\ T_2/d_2 = 2\ T_2/(m\ z_2)$

q 206 $q \triangleq 10$ for $i = 10, 20, 40$

$q \triangleq 17$ for $i = 80$, self-locking

Assumed values for normal, naturally-cooled worm gears (worm hardened and ground steel, worm wheel of bronze):

v_g	m s^{-1}	1	2	5	10	15	20
C_{perm}	N mm^{-2}	8	8	5	3·5	2·4	2·2

When cooling is adequate this value can be used for all speeds:

q 207 $C_{perm} \geq 8$ N mm^{-2}

For all symbols see Q 29, for suffixes see Q 23

Notation for Q 18 ... Q 28 (suffixes see Q 23)

a : standard centre distance
b : tooth width
h_{a0} : addendum of cutting tool
h_{aP} : addendum of reference profile (e.g. DIN 867)
h_{fP} : dedendum of reference profile
k : change of addendum factor
p_e : normal pitch ($p_e = p \cos\alpha$, $p_{et} = p_t \cos\alpha_t$)
z : no. of teeth
z_{nx} : equivalent no. of teeth
(C_{perm}), C : (permissible) load coefficient
F_t : peripheral force on pitch cylinder (plane section)
K_I : operating factor (external shock)
K_V : dynamic factor (internal shock)
$K_{F\alpha}$: end load distribution factor } for root stress
$K_{F\beta}$: face load distribution factor } for root stress
K_{FX} : size factor } for root stress
$K_{H\alpha}$: end load distribution factor } for flank stress
$K_{H\beta}$: face load distribution } for flank stress
R_e : total pitch cone length (bevel gears)
R_m : mean pitch cone length (bevel gears)
T : torque
$Y_{F'}$ (Y_S) : form factor, (stress concentration factor)
Y_β : skew factor
Y_ε : load proportion factor
Z_H : flank form factor
Z_ε : engagement factor
Z_R : roughness factor
Z_V : velocity factor
α_P : reference profile angle (DIN 867 : $\alpha_P = 20^\circ$)
α_W : operating angle
β | β_b : skew angle for helical gears | pitch cylinder | base cylinder
ϱ : sliding friction angle ($\tan\varrho = \mu$)
ϱ_{a0} : tip edge radius of tool
$\sigma_{F\,lim}$: fatigue strength
$\sigma_{H\,lim}$: Hertz pressure (contact pressure)

Precise calculations for spur and bevel gears: DIN 3990.
Terms and definitions for

spur gears and gearing : DIN 3960 } or BS 2519
bevel gears and gearing : DIN 3971 } or BS 2519
straight worm gearing : DIN 3975 } or BS 2519

Machine tool design: general considerations

Components of machine tools which are subjected to working stresses (frames with mating and guide surfaces, slides and tables, work spindles with bearings) are designed to give high accuracy over long periods of time. They are made with generous bearing areas and the means to readjust or replace worn surfaces should become necessary.

The maximum permissible deflection at the cutting edge (point of chip formation) is approximately 0·03 mm. For spindle deflection refer to formula P13 and for cutting forces see r 4.

Cutting drives (main drives) with v = const. over the entire working range (max. and min. workpiece or tool-diameter) are obtainable with output speeds in geometric progression:

r 1 $$n_k = n_1 \varphi^{k-1}$$

The progressive ratio φ for the speeds $n_1 \ldots n_k$ with k number output speeds

r 2 are calculated by: $$\varphi = \sqrt[k-1]{\frac{n_k}{n_1}}$$

and the preferred series is selected.

Standardized progressive ratio φ: 1·12–1·25–1·4–1·6–2·0

Speed basic series R_{20} where $\varphi = \sqrt[20]{10} = 1{\cdot}12$:

...100–112–125–140–160–180–200–224–250–280–315–355–400–
–450–500–560–630–710–800–900–1000–... rpm.

Cutting gears are designated by the number of shafts and steps.

Example: A III/6 gear drive incorporates 3 shafts and provides 6 output speeds. Representation of gear unit as shown (for k = 6; φ =1·4; n_1 = 180; n_k = 1000):

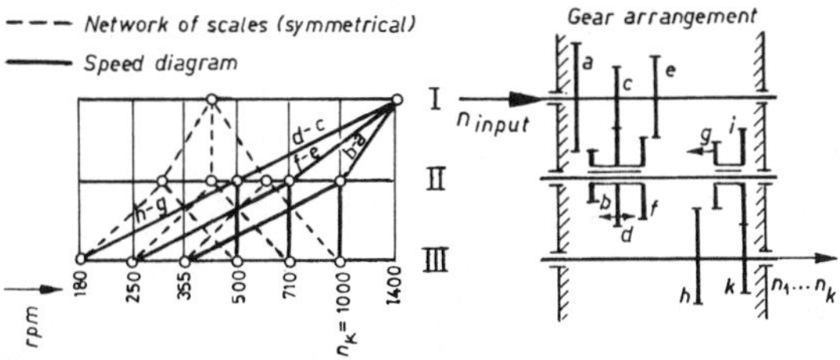

For explanation of symbols refer to R 5

PRODUCTION ENGINEERING

Cutting power P_s

			General	Drilling
r 3	Cutting power	P_s =	$\dfrac{F_s \times v}{\eta_{mech} \times \eta_{electr}}$	$\dfrac{F_s\,(D+d)\pi \times n}{2 \times \eta_{mech} \times \eta_{electr}}$
r 4	Cutting force	F_s =	$K \times k_{c1.1} \times b \left(\dfrac{h}{\text{mm}}\right)^{1-mc} \text{mm} \times z_e$	

Table of values for K, b, h, z_e ($k_{c1.1}$; $1-mc$ see Z 17)

No	Method	Sketch	K	b	h	z_e	Notes
r5	Turning external longitud.		1 HM + HSS	$\dfrac{a}{\sin \varkappa}$	$s \times \sin \varkappa$	1	
r6	internal	analogous to r 5	1·2				
r7	Planing and shaping	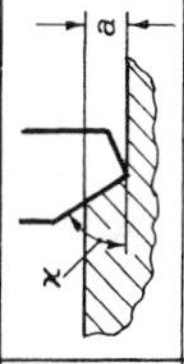	1·1 HM ····· 1·2 HSS				
r8	Drilling and boring	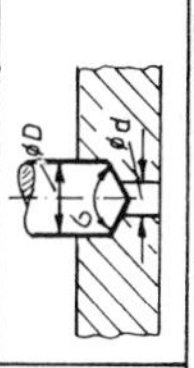	0·85 HM 1 HSS	$\dfrac{D-d}{2\sin\frac{\sigma}{2}}$	$s_z \times \sin\dfrac{\sigma}{2}$ $s_z = 0{\cdot}5\ s$	2 for twist drill	$d = 0$ when drilling $\sigma = 118°$ for steel
r9	Plane milling down-cut and up-cut	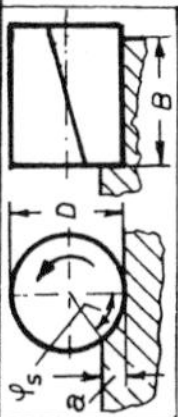	1·1 HM 1·2 HSS	B	$\dfrac{2a}{D\,\varphi_s} s_z$	$\dfrac{\varphi_s\, z_s}{360°}$	$\cos\varphi_s = 1 - 2a/D$
r 10	End milling down-cut and up-cut	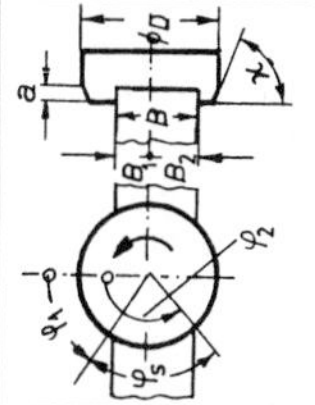		$\dfrac{a}{\sin\varkappa}$	$(\cos\varphi_1 - \cos\varphi_2) \times \dfrac{1}{\varphi_s} s_z \times \sin\varkappa$		$\varphi_s = \varphi_2 - \varphi_1$ $\cos\varphi_1 = \dfrac{2\,B_1}{D}$ $\cos\varphi_2 = \dfrac{2\,B_2}{D}$ calculate φ_1 and φ_2 in direction of rotation

For explanation or symbols refer to R 5

Nr.	Method	Sketch	$K =$	$b =$	$h =$	$z_e =$	Comments
r 11	Circular grinding, flat	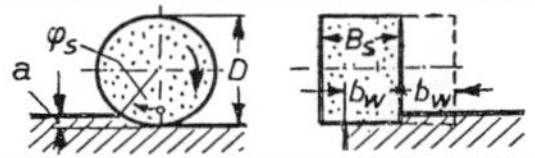			$\frac{l_K\ u}{v}\sqrt{\frac{a}{D}}$		$\cos\varphi_s = 1 - 2a/D$
r 12	round-grinding, outside inside	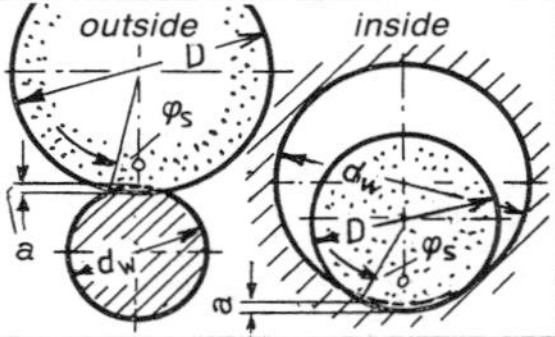	Tab. 1	b_W	$\frac{l_K\ u}{v}\sqrt{a\left(\frac{1}{D} \pm \frac{1}{d_W}\right)}$ + outside − inside	$\frac{D\ \varphi_s}{2\ l_K}$	$\varphi_s = \sqrt{\frac{4a}{D(1 \pm D/d_W)}}$ + outside − inside
r 13	cutting round						
r 14	front-grinding	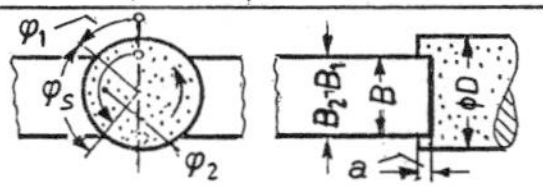		a	$(\cos\varphi_1 - \cos\varphi_2) \times \frac{1}{\varphi_s} \times \frac{l_K\ u}{v}$		angle φ_s as in r 10

Table 1

grain size \ h in mm	0·001	0·002	0·003	0·004
	Correction factor K			
40	5·1	4·3	4·0	3·6
60	4·5	3·9	3·5	3·2
80	4·0	3·6	3·2	3·0
120	3·4	3·0	2·8	2·5
150	3·2	2·8	2·6	2·3
180	3·0	2·6	2·4	2·2

Table 2

grain size \ a in mm	Roughplane 0·01	0·02	0·03	Smoothing 0·003	0·004	0·005	0·006
	effective grain distance l_K						
40	24	14	9	–	–	–	–
60	32	23	15	39	38	37	36
80	40	31	24	47	46	45	44
120	53	44	37	60	59	58	57
150	56	48	40	64	63	62	61
180	58	50	42	66	65	64	63

For explanation or symbols refer to R 5

Feed drives

Feeds in geometrical progression with progressive ratio

$$\varphi = 1{\cdot}12 - 1{\cdot}25 - 1{\cdot}4 - 1{\cdot}6 - 2{\cdot}0.$$

Feed rate

	Method	Feed rate	Notes
r 15	Turning, longitudinal (external and internal)	$u = n \times s$	
r 16	Drilling	$u = n \times s_Z \times z_S$	for twist drills $z_e = z_S = 2$; $s_Z = 0{\cdot}5\, s$
r 17	Planing, shaping	$u = v$	
r 18	Milling, plane milling and end milling	$u = s_Z \times n \times z_S$	

Cutting times t_S

r 19 $$t_S = \frac{l_1}{u} \quad ; \quad \text{where} \quad l_1 = l + l'.$$

When calculating the cycle and machining times for each workpiece, the feed and infeed travels and also the lengths covered during non-cutting motions, divided by the corresponding speeds, must be taken into account.

Feed power P_V

r 20 Feed power $$P_V = \frac{u(F_R + F_V)}{\eta_{mech} \times \eta_{electr}}$$

r 21 Feed force $F_V \approx 0{\cdot}2\, F_S$; (F_S from r 4)

r 22 Friction force $F_R = m_b \times g \times \mu$

where m_b is the mass moved, e.g. in the case of milling machines the sum of the table and workpiece masses.

It must be determined whether the feed power as calculated under r 20 is sufficient to accelerate the moving components to rapid motion speed u_E within a given time t_b (in production machines $u_E \approx 0{\cdot}2$ m/s).
Otherwise the following applies:

r 23 $$P_V = u_E\, m_b \left(\mu\, g + \frac{u_E}{t_b}\right) \frac{1}{\eta_{mech} \times \eta_{electr}}$$

For explanation or symbols refer to R 5

Explanation of symbols
used on pages R 1 ... R 4

- a : infeed
- b : width of chip
- b_w : effective width
  rough plane grind. $b_w = B_s/1{,}4$
  smooth grinding $b_w = B_s/3$
- B : milling width
- B_1, B_2: milling width measured from tool centre
- B_s : width of the buffing disk
- d : diameter of pre-drilled hole
- d_w : working-part – outside, inside diameter resp.
- D : tool diameter
- F_R : friction force
- F_c : cutting force
- F_v : feed force
- g : gravitational acceleration
- h : chip thickness
- k : number of output speeds
- $k_{c1\cdot1}$: basic cutting force related to area
- K : method factor
- l : cutting travel
- l_1 : work travel
- l' : overrun travel at both ends with feed rate u
- z_s : number of cutting edges per tool
- l_K : effective grain distance see table 2
- n : speed
- n_1 : minimum output speed
- n_k : maximum output speed
- P_c : cutting power
- P_v : feed power
- s : feed
- s_z : feed per cutting edge
- t_b : acceleration time
- t_s : cutting time
- u : feed rate
- u_E : rapid traverse speed
- v : cutting speed
- z_e : number of cutting edges in action
- ε_s : slenderness ratio ($\varepsilon_s = a/s$)
- η_{electr}: electrical efficiency
- η_{mech}: mechanical efficiency
- $\varkappa$: setting angle
- μ : friction coefficient, see Z 7
- σ : drill tip angle
- φ : progressive ratio
- φ_s : incident angle for milling, cutting or grinding resp.

HM : carbide tip
HSS : high-speed tip

Cold working of sheet

Deep drawing

Initial blank diameter D

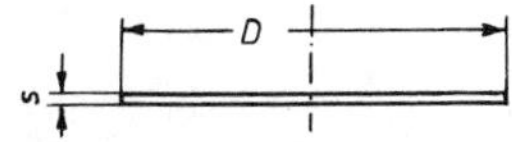

r 24 $$D = \sqrt{\frac{4}{\pi} \times \Sigma A_{mi}}$$

A_{mi} are the surface areas of the finished item which can be calculated from the following formulae b 30, c 12, c 16, c 21, c 25, c 27 or c 30. The surface areas at the transition radii for both drawing and stamping dies are calculated as follows:

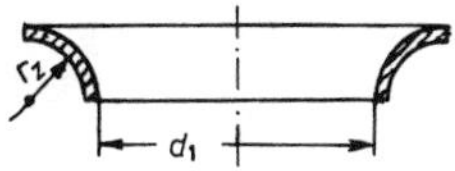

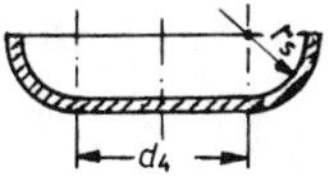

r 25 $$A_m = \frac{\pi}{4}\left[2\pi d_1 r_z + 4(\pi - 2) r_z^2\right] \qquad A_m = \frac{\pi}{4}(2\pi d_4 + 8 r_s) r_s + \frac{\pi}{4} d_4^2$$

Example: (assume $r_s = r_z = r$)

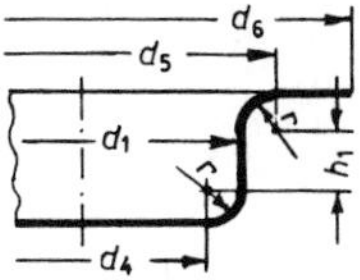

r 26 $$D = \sqrt{d_4^2 + d_6^2 - d_5^2 + 4 d_1 h + 2\pi r(d_1 + d_4) + 4\pi r^2}$$

1st and 2nd stages

	1st stage	2nd stage
	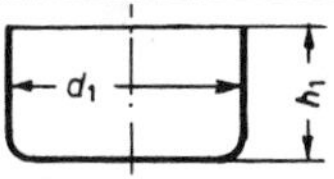	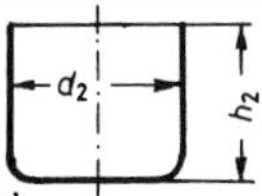
r 27	$\beta_1 = \frac{D}{d_1}$	$\beta_2 = \frac{d_1}{d_2}$
r 28	$\beta_{1max} = \beta_{100} + 0{\cdot}1 - \left(\frac{d_1}{s}\, 0{\cdot}001\right)$	$\beta_{2max} = \beta_{100} + 0{\cdot}1 - \left(\frac{d_2}{s}\, 0{\cdot}001\right)$
r 29	$F_{D1} = \pi d_1 s k_{fm1} \varphi_1 \frac{1}{\eta_{E1}}$	$F_{D2} = \frac{F_{D1}}{2} + \pi d_2 s k_{fm2} \varphi_2 \frac{1}{\eta_{E2}}$
r 30	$\varphi_1 = \left\lvert \ln\sqrt{0{\cdot}6\, \beta_1^2 - 0{\cdot}4} \right\rvert$	$\varphi_2 = \left\lvert \ln\sqrt{0{\cdot}6\, \beta_2^2 - 0{\cdot}4} \right\rvert$
r 31 r 32	$k_{fm1} = \frac{w}{\varphi_1}$	without intermediate annealing: $k_{fm2} = \frac{k_{f1} + k_{f2}}{2}$
r 33		with intermediate annealing: $k_{fm2} = \frac{w}{\varphi_2}$

continued on R 7

continued from R 6

The work w, related to the volume and the yield strength k_f is obtained from the deformation curves for the appropriate value of logarithmic deformation ratio φ (see Z 20).

Blank holding forces F_{B1} and F_{B2}

1st stage	2nd stage
r 34 $F_{B1} = (D^2 - d_1^2)\frac{\pi}{4}\,\frac{R_m}{400}\left[(\beta_1^2 - 1) + \frac{d_1}{s}\right]$	$F_{B2} = (d_1^2 - d_2^2)\frac{\pi}{4}\,\frac{R_m}{400}\left[(\beta_2^2 - 1) + \frac{d_2}{s}\right]$

Bottom tearing occurs if

r 35 $R_m \leqslant \frac{F_{D1} + 0{\cdot}1\,F_{B1}}{\pi\, d_1\, s}$	$R_m \leqslant \frac{F_{D2} + 0{\cdot}1\,F_{B2}}{\pi\, d_2\, s}$

r 36 **Maximum drawing conditions** β and R_m

Material – Sheet metal	Specification to U.S.A.		BS	Grade	β_{100}	without intermediate annealing β_{2max}	with intermediate annealing β_{2max}	R_m N/mm²
Carbon steel	ASTM-	366-79	1449 P. 1	15	1·7	1·2	1·5	390
		619-75		4	1·8	1·2	1·6	360
		–		3	1·9	1·25	1·65	350
		620-75		2	2·0	1·3	1·7	340
		283 Gr. C		1	1·7	–	–	410
Stainless steel (18% Cr; 9% Ni)	SAE 3310		970	321 S 20	2·0	1·2	1·8	600
Al Mg Si soft	AA 6004 soft		1490	LM 5	2·05	1·4	1·9	150

Notation for R 6 and R 7

A_{mi} : surface area
F_{D1}, F_{D2} : drawing force in 1st and 2nd stage
k_{fm1} or k_{fm2} : mean yield strength, 1st or 2nd stage
k_{f1}, k_{f2} : yield strength for φ_1 and φ_2
r : radius
r_s : radius of stamping die
r_d : radius of drawing die
w : work per unit volume $= \frac{\text{work of deformation}}{\text{forminged volume}}$
β_1, β_2 : drawing ratio, 1st and 2nd stage
β_{100} : max. drawing ratio for s = 1 mm and d = 100 mm
β_{1max}, β_{2max} : max. drawing ratio, 1st and 2nd stage
η_{E1}, η_{E2} : process efficiency, 1st and 2nd stage
φ_1, φ_2 : logarithmic deformation ratio, 1st and 2nd stage

Extrusion

r 37 Remodel force, pressure force $F = A\ k_{fm}\ \varphi_A \frac{1}{\eta_F}$

r 38 Remodel work $W = V\ k_{fm}\ \varphi_A \frac{1}{\eta_F}$

r 39 Mean strength for deformation $k_{fm} = \frac{w}{\varphi_A}$

	Extrusion forward: full body	Extrusion forward: hollow body	Extrusion backward

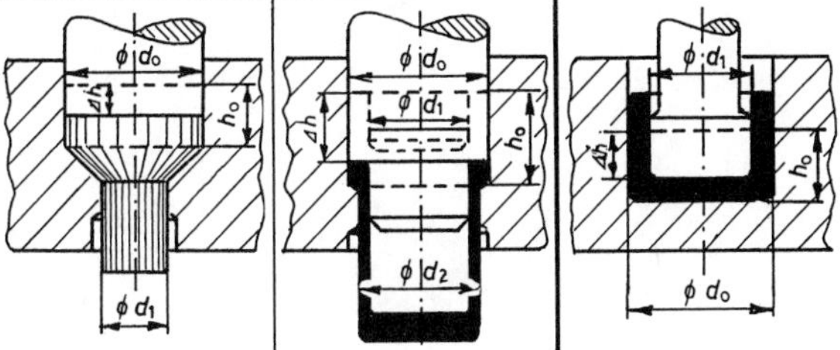

	full body	hollow body	backward
r 40	$A = \frac{\pi}{4}\ d_0^2$	$A = \frac{\pi}{4}\ (d_0^2 - d_1^2)$	$A = \frac{\pi}{4}\ d_0^2$
r 41	$\varphi_A = \ln \frac{d_0^2}{d_1^2}$	$\varphi_A = \ln \frac{d_0^2 - d_1^2}{d_2^2 - d_1^2}$	$\varphi_A = \ln \frac{d_0^2}{d_0^2 - d_1^2}$
r 42	$V = \frac{\pi}{4}\ h_0\ d_0^2$	$V = \frac{\pi}{4}\ h_0\ (d_0^2 - d_1^2)$	$V = \frac{\pi}{4}\ d_0^2\ h_0$
r 43	$\eta_F = 0{\cdot}7 \ldots 0{\cdot}8$	$\eta_F = 0{\cdot}6 \ldots 0{\cdot}7$	$\eta_F = 0{\cdot}5 \ldots 0{\cdot}6$

Maximum logarithmic ratio of deformation $\varphi_{A\,max}$ without intermediate annealing

material / method	Al 99,5	Al Mg Si soft	Steel C<0,1%	C<0,15%	C>0,15%	low alloyed	alloyed
forward	3·9	3·0	1·4	1·2	0·9	0·8	0·7
backward	4·5	4·0	1·2	1·1	1·1	0·95	0·8

A : used area
φ_A: logarithmic deformation ratio
η_F : deformation efficiency
V : remodelled volume
w : volume related remodel-work according to curves Z 20
Δh: depth of stroke

The most important electrical quantities and their units. – Basic rules

s 1 **Note regarding capital and small letters used as symbols**

Electrical engineering quantities that are independent of time are mainly denoted by capital letters. Quantities that vary with time are denoted by small letters or by capital letters provided with the subscript t

Examples: formulae s 8, s 9, s 13

Exceptions: f, ω, $\hat{\imath}$, $\hat{u}$, p_{Fe10}

Electrical work W

Electrical work W is equivalent to mechanical work W as explained on M 1. Energy conversion, however, is subject to losses.

Units: J; W s (wattsecond); kW h; MW h

$$1\ \text{W s} = 1\ \text{Joule} = 1\ \text{J} = 1\ \text{N m}$$

Further the following relation applies, using quantities explained on S1 and S2:

s 2 $$W = I\,V\,t = \frac{V^2}{R}\,t = I^2 R\,t$$

Electrical power P

Electrical power P is equivalent to mechanical power P, as explained on M 1. Energy conversion, however, is subject to losses.

Units: W (Watt); kW; MW

$$1\ \text{W} = 1\ \frac{\text{J}}{\text{s}} = 1\ \frac{\text{N m}}{\text{s}}$$

Further the following relation applies, using quantities explained on S1 and S2:

s 3 $$P = \frac{V^2}{R} = I^2 R$$

Frequency f see L 1

Period T see L 1

Angular frequency ω, **angular velocity** ω see L 1

Current I

Is a base quantity (see preface and instructions)

Units: A (ampere); mA; kA

The current of 1 A has been defined by means of the attracting force which two parallel current-carrying conductors exert on each other.

continued on S 2

continued from S 1

Current density J

s 4 $$J = \frac{I}{A}$$

Applicable only where distribution of current I is uniform over cross section A.

Units: A/m^2; A/mm^2

Potential difference V

s 5 $$V = \frac{P}{I}$$

Units: V (volt); mV; kV

Where a direct current of 1 A through a conductor converts energy at a rate of 1 W, the voltage across this conductor is 1 V.

$$1\,V = 1\,\frac{W}{A} = 1\frac{J}{s\,A} = 1\,A\,\Omega = 1\frac{N\,m}{s\,A}$$

Resistance R

s 6 $$R = \frac{V}{I} \qquad \text{(Ohm's law)}$$

Units: Ω (ohm); kΩ; MΩ

Where a voltage of 1 V across a conductor causes a current of 1 A to the flow through it, resistance is 1 Ω.

$$1\,\Omega = \frac{1\,V}{1\,A} = 1\,\frac{W}{A^2} = 1\,\frac{W}{s\,A^2} = 1\,\frac{N\,m}{s\,A^2}$$

Conductance G

Conductance G is the reciprocal of resistance R.

s 7 $$G = 1/R$$

Unit: 1/Ω

$$1/\Omega = [1\ \text{Mho}]$$

Quantity of electricity, charge Q

s 8 $$q = \int i\ dt \qquad \text{(see s 1)}$$

For direct current:

s 9 $$Q = I\,t$$

Unit: C (coulomb);

$$1\,C = 1\,A\,s$$

continued on S 3

continued from S 2

Capacitance C

The capacitance C of a capacitor is the ratio of quantity of electricity Q stored in it and voltage V across it:

s 10 $$C = \frac{Q}{V}$$

Units: F (farad); μF; nF; pF

Where a capacitor requires a charge of 1 C to be charged to a voltage of 1 V, its capacitance is 1 F.

$$1\,\text{F} = 1\frac{\text{C}}{\text{V}} = 1\frac{\text{A s}}{\text{V}} = 1\frac{\text{A}^2\,\text{s}}{\text{W}} = 1\frac{\text{A}^2\,\text{s}^2}{\text{J}} = 1\frac{\text{A}^2\,\text{s}^2}{\text{N m}}$$

Magnetic flux Φ

s 11 $$\Phi = \frac{1}{N}\int v \; dt \qquad \text{(see s 1)}$$

Here N is the number of turns of a coil and v the voltage induced, when the magnetic flux Φ linked with the coil varies with time.

Units: Wb (weber) = V s = 10^8 M (maxwell)

1 Wb is the magnetic flux which, linking a circuit of 1 turn, induces in it a voltage of 1 V as it is reduced to zero at a uniform rate in 1 s.

Magnetic induction (flux density) B

The magnetic induction in a cross section A is:

s 12 $$B = \frac{\Phi}{A}$$

Here A is the cross-sectional area traversed perpendicularly by the homogeneous magnetic flux Φ.

Units: T (tesla); μT; nT; V s/m^2; G (gauss)

$$1\,\text{T} = 1\frac{\text{V s}}{\text{m}^2} = 10^{-4}\frac{\text{V s}}{\text{cm}^2} = \left[10^4\text{G} = 10^4\frac{\text{M}}{\text{cm}^2}\right]$$

Where a homogeneous magnetic flux of 1 Wb perpendicularly traverses an area of 1 m^2, its magnetic induction is 1 T.

continued on S 4

continued from S 3

Inductance L

s 13 $$L = N\frac{\Phi}{I} = N\frac{\Phi_t}{i}$$ (see s 1)

Here I is the current flowing through a coil of N turns and Φ the magnetic flux linked with this coil.

Units: H (henry); mH

1 H is the inductance of a closed loop of one turn which, positioned in vacuum and passed through by a current of 1 A, enclosed a magnetic flux of 1 Wb.

$$1\ \text{H} = 1\ \frac{\text{Wb}}{\text{A}} = 1\ \frac{\text{V s}}{\text{A}}$$

Magnetic field strength H

s 14 $$H = \frac{B}{\mu_0\ \mu_r}$$

Units: A/m; A/cm; A/mm; (Ampere Turn/m)

Magnetomotive force F

s 15 $$F = N\ I$$

Units: A; kA; mA; (Ampere Turn)

Magnetomotive force F_i in the i-th section of a magnetic circuit:

s 16 $$F_i = H_i\ l_i$$

here l_i is the length of this section.

s 17 $$\sum_{i=1}^{n} F_i = F$$

Reluctance S of a homogeneous section of a magnetic circuit:

s 18 $$S = \frac{F}{\Phi}$$ (equivalent to Ohm's law for magnetic circuits)

Units: 1/H; = A/V s; (Ampere Turn/Wb)

for symbols see S 16

Basic properties of electric circuits

Directions of currents, voltages, arrows representing them

s 20 s 21 Direction of the current and of arrows representing positive currents in	generator	$-\rightarrow+$
	load	$+\rightarrow-$

s 22 Direction of the potential difference and of arrows representing positive voltages always | $+\rightarrow-$

Directions of arrows representing currents or voltages

	where function of element (generator or load) as well as polarity is	determine directions of arrows	where calculation results in a positive value, direction with respect to arrow of current or voltage is	where calculation results in a negative value, direction with respect to arrow of current or voltage is
s 23	known	as stated above	—	—
s 24	unknown	at random	equal	opposite

Special rule

Arrows representing voltage drop across a resistor and current causing it, should always be determined in same direction (as $R > 0$).

Ohm's Law

Current through a resistor:

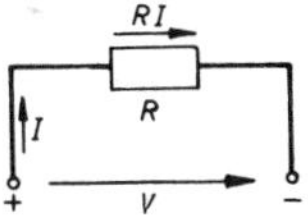

s 25 $$I = \frac{V}{R}$$ (see also s 6)

Resistance R of conductor

s 26 $$R = \frac{\rho\, l}{A} = \frac{l}{\gamma\, A}$$

Resistance R of conductor at temperature ϑ (in degrees centigrade)

s 27 $$R = R_{20}\left[1 + \alpha\,(\vartheta - 20^{\circ}\mathrm{C})\right]$$

Electric heating of a mass m

s 28 $$V\,I\,t\,\eta = c\,m\,\Delta\vartheta$$

α : temperature coefficient (see Z 21)
γ : conductivity (see Z 21)
ρ : resistivity (see Z 21)
c : specific heat (see Z 1 ... 4 and O 2)
η : efficiency
$\Delta\vartheta$: temperature change
t : time
R_{20} : resistance at $\vartheta = 20^{\circ}$C

continued on S 6

continued from S 5

1st Kirchhoff Law

The algebraic sum of all currents entering a branch point (node) is zero.

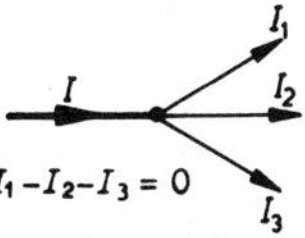

s 29 $$\Sigma I = 0 \qquad I - I_1 - I_2 - I_3 = 0$$

Here currents	into / out of	the node are considered	positive / negative

Ratio of currents

Where several resistors are connected in parallel, total current and partial currents are inversely proportional to their respective resistances.

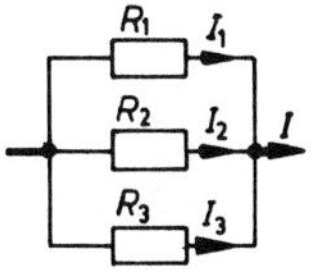

s 30 $$I : I_1 : I_2 : I_3 = \frac{1}{R} : \frac{1}{R_1} : \frac{1}{R_2} : \frac{1}{R_3}$$

Current division

Partial currents of 2 resistors connected in parallel:

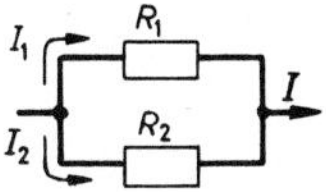

s 31 $$I_1 = I \frac{G_1}{G_1 + G_2} = I \frac{R_2}{R_1 + R_2}$$

2nd Kirchhoff Law

The algebraic sum of all voltages around a closed mesh (loop) is zero.

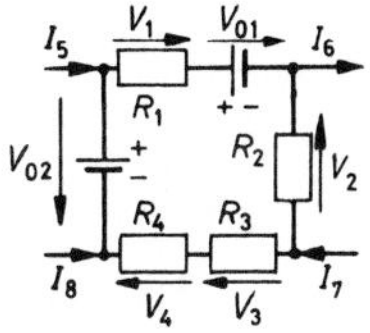

s 32 $$\Sigma V = 0$$

Here voltages traversed in accordance with (opposite to) direction of arrow are considered positive (negative).

$$V_1 + V_{01} - V_2 + V_3 + V_4 - V_{02} = 0$$

Ratio of voltages

Where several resistors are connected in series, the ratio of partial voltages is equal to the ratio of the respective resistances.

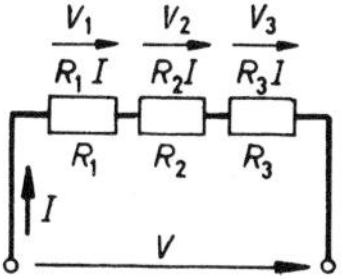

s 33 $$V_1 : V_2 : V_3 = R_1 : R_2 : R_3$$

Voltage divider

Partial voltages across 2 resistors connected in series:

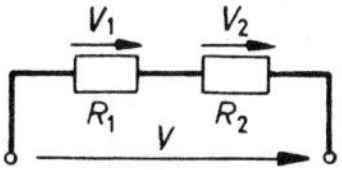

s 34 $$V_1 = V \frac{G_2}{G_1 + G_2} = V \frac{R_1}{R_1 + R_2}$$

Series connection

Total resistance R (according to s 26)

generally:

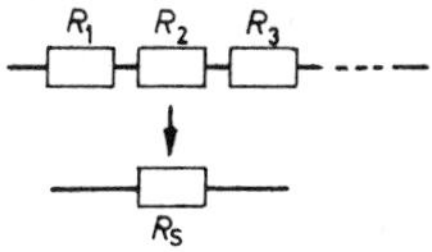

s 35 $$R_S = R_1 + R_2 + R_3 + \dots$$

for n equal resistances R:

s 36 $$R_S = n\,R$$

Parallel connection

Total resistance R (according to s 30)

generally:

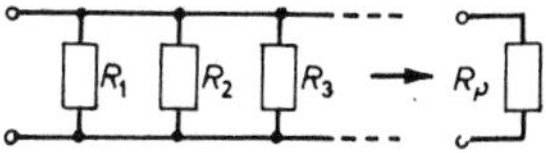

s 37 $$\frac{1}{R_p} = \frac{1}{R_1} + \frac{1}{R_2} + \frac{1}{R_3} + \dots$$

s 38 $$G_p = G_1 + G_2 + G_3 + \dots$$

	for 2 several resistances	for 3 several resistances	for n equal resistances
s 39	$R_p = \dfrac{R_1 R_2}{R_1 + R_2}$	$R_p = \dfrac{R_1 R_2 R_3}{R_1 R_2 + R_2 R_3 + R_1 R_3}$	$R_p = \dfrac{R}{n}$
s 40	$= \dfrac{1}{G_1 + G_2}$	$= \dfrac{1}{G_1 + G_2 + G_3}$	$= \dfrac{1}{n\,G}$

Multiple connection

A multiple connection of several known resistances is subdivided into parallel and series connections, proceeding outwards. These are separately converted as to be conveniently combined again, e.g.:

s 41 $$I = \frac{R_2 + R_3}{R_1 R_2 + R_1 R_3 + R_2 R_3} V = \frac{G_1 (G_2 + G_3)}{G_1 + G_2 + G_3} V$$

s 42 $$I_3 = \frac{R_2}{R_1 R_2 + R_1 R_3 + R_2 R_3} V = \frac{G_1 G_3}{G_1 + G_2 + G_3} V$$

s 43 $$V_2 = \frac{R_2 R_3}{R_1 R_2 + R_1 R_3 + R_2 R_3} V = \frac{G_1}{G_1 + G_2 + G_3} V$$

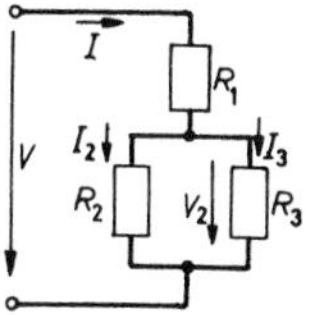

Solutions of Linear Networks

General: There are special methods which allow the calculation of unknown voltages and currents in a network more easily than mesh or node analysis, e.g.:

Use of the Superposition Theorem: In a general network, let all voltage[1] and current[2] sources be successively applied to the network, compute the voltages and currents caused by each source acting alone.

- The remaining voltage sources are short-circuited.
- The remaining current sources are open-circuited.

The complete solution is the sum of these partial solutions.

General procedure to compute V_x in a general network with voltage sources $V_o \dots V_\nu$ and current sources $I_o \dots I_\mu$:

s 44 $$V_x = a_o\, V_o + a_1 V_1 + \dots + a_\nu\, V_\nu + b_o\, I_o + b_1\, I_1 + \dots + b_\mu\, I_\mu$$

s 45 $$= V_{xao} + V_{xa1} + \dots + V_{xa\nu} + V_{xbo} + V_{xb1} + \dots + V_{xb\mu}$$

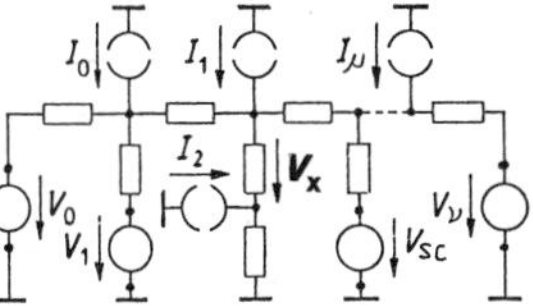

Computation of the partial solut.:

s 46 $V_x = V_{xaq}$, where $V_o \dots V_\nu = 0$, with $V_q = 0$, and $I_o \dots I_\mu = 0$

s 47 $V_x = V_{xbq}$, where $I_o \dots I_\mu = 0$, with $I_q = 0$, and $V_o \dots V_\nu = 0$

Example:

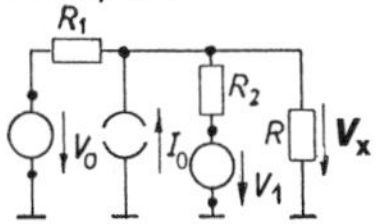

s 48 $$V_x = a_o\, V_o + a_1 V_1 + b_o\, I_o = V_{xao} + V_{xa1} + V_{xbo}$$

Equivalent networks for computation of each partial solution:

s 49	$V_o \neq 0;\ V_1 = 0;\ I_o = 0$	$V_o = 0;\ V_1 \neq 0;\ I_o = 0$	$V_o = 0;\ V_1 = 0;\ I_o \neq 0$
	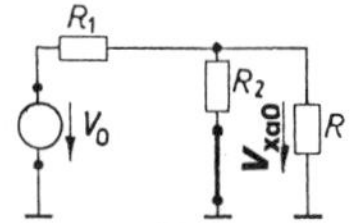	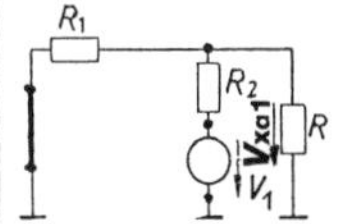	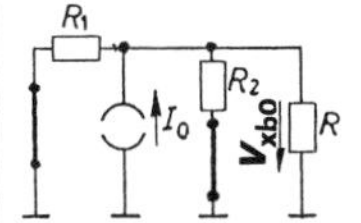
s 50	$V_{xa0} = \frac{V_o}{R_1} \times \frac{1}{\frac{1}{R_1}+\frac{1}{R_2}+\frac{1}{R}}$	$V_{xa1} = \frac{V_1}{R_2} \times \frac{1}{\frac{1}{R_1}+\frac{1}{R_2}+\frac{1}{R}}$	$V_{xb0} = I_o \times \frac{1}{\frac{1}{R_1}+\frac{1}{R_2}+\frac{1}{R}}$

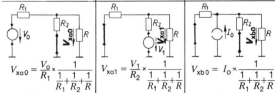

s 51 Required voltage (cf. s 48) $$V_x = \left(\frac{V_o}{R_1} + \frac{V_1}{R_2} + I_o\right) \times \frac{1}{1/R_1 + 1/R_2 + 1/R}$$

[1], [2] Explanations cf. S 9

continued on S 9

Use of Thévenin's theorem: Consider a general network containing voltage[1] and current[2] sources. It is required to compute the voltage V_x, across resistance R in branch AA′. This may be achieved by replacing the rest of the network by an equivalent voltage-source V_l and resistance R_i.

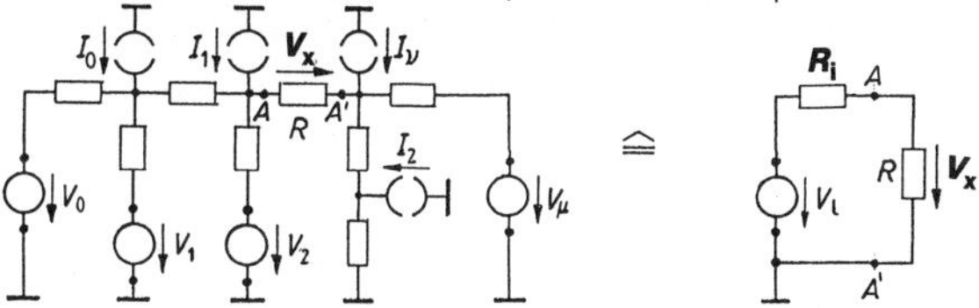

Produce to compute R_i and V_l:

Remove the branch AA′ from the network.
R_i is the resistance between A and A′.
V_l is the voltage at AA′.

Note: If R_i is known, V may be computed using $V_l = I_{sc} R_i$. I_{sc} is the current which flows when A and A′ are connected. Hence:

s 52 $$V_x = V_l \frac{R}{R+R_i} = I_{sc} R_i \frac{R}{R+R_i} = I_{sc} \frac{1}{1/R_i + 1/R}$$

Example:

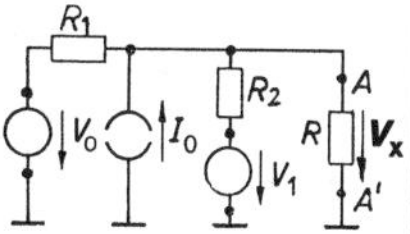

Equivalent circuit

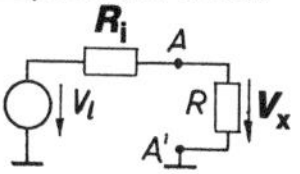

Computation of I_{sc}	Computation of R_i
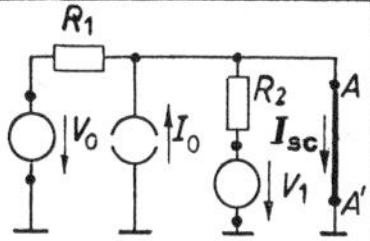	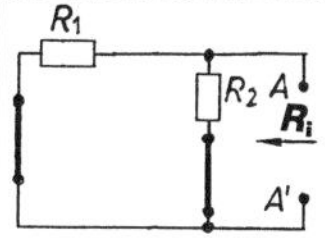
$I_{sc} = \frac{V_0}{R_1} + \frac{V_1}{R_2} + I_0$	$R_i = \frac{1}{1/R_1 + 1/R_2}$

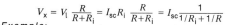

Hence: $V_l = I_{sc} R_i$

s 53 Using s 52: $V_x = \left(\frac{V_0}{R_1} + \frac{V_1}{R_2} + I_0\right) \times \frac{1}{1/R_1 + 1/R_2 + 1/R}$; cf. result of S 8, s 51

Explanations:

$a_0 \ldots a_v$	coefficients of	voltages	which are determined by the resistors in the network
$b_0 \ldots b_v$		currents	

[1] voltage source	with internal resistance	$R_i = 0$
[2] current source		$R_i \rightarrow \infty$

Transformation of a delta to a star-circuit and vice versa

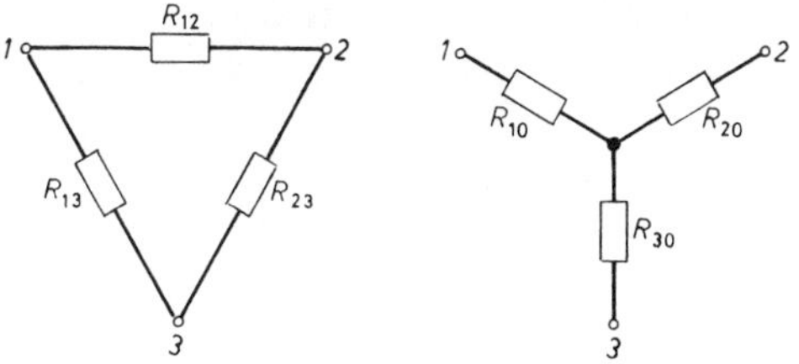

s 54 $$R_{12} = \frac{R_{10} \times R_{20} + R_{10} \times R_{30} + R_{20} \times R_{30}}{R_{30}}$$ $$R_{10} = \frac{R_{12} \times R_{13}}{R_{23} + R_{12} + R_{13}}$$

s 55 $$R_{13} = \frac{R_{10} \times R_{20} + R_{10} \times R_{30} + R_{20} \times R_{30}}{R_{20}}$$ $$R_{20} = \frac{R_{23} \times R_{12}}{R_{23} + R_{12} + R_{13}}$$

s 56 $$R_{23} = \frac{R_{10} \times R_{20} + R_{10} \times R_{30} + R_{20} \times R_{30}}{R_{10}}$$ $$R_{30} = \frac{R_{23} \times R_{13}}{R_{23} + R_{12} + R_{13}}$$

Potential divider

Potential dividers are used to provide reduced voltages.

s 57 $$V_V = \frac{R_2 R_V}{R_1 R_2 + R_1 R_V + R_2 R_V} V$$

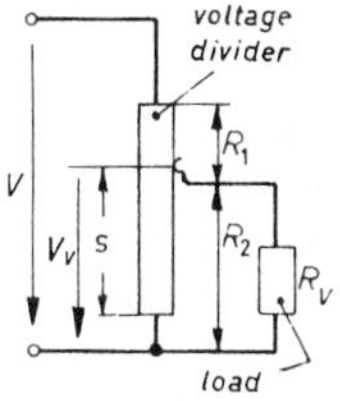

s 58 For applications, where V_V has to be approximately proportional to s, the condition $R_V \geq 10\,(R_1 + R_2)$ has to be satisfied.

s : distance of sliding contact from zero position

Applications in electrical measurements

Extending the range of a voltmeter

s 59 $$R_V = R_M\left(\frac{V_{max}}{V_{M\,max}} - 1\right)$$

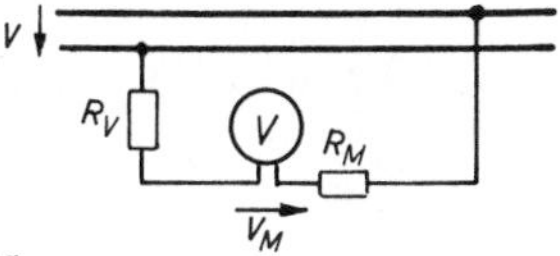

Extending the range of an ammeter

s 60 $$R_N = R_M \frac{I_{M\,max}}{I_{max} - I_{M\,max}}$$

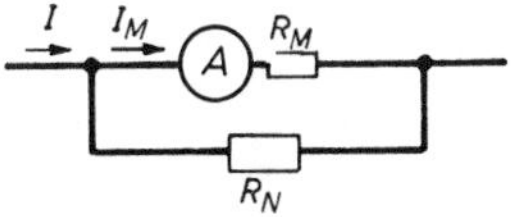

Wheatstone bridge for measuring an unknown resistance R_X

A slide-wire Wheatstone bridge may be used for measuring resistances of between 0·1 and 10^6 ohms. The calibrated slide wire is provided with a scale reading $a/(l-a)$. The sliding contact is adjusted, until the detector current I_B is zero. Then

s 61 $$\frac{R_x}{R} = \frac{a}{l-a}$$

s 62 and hence $$R_x = R\,\frac{a}{l-a}$$

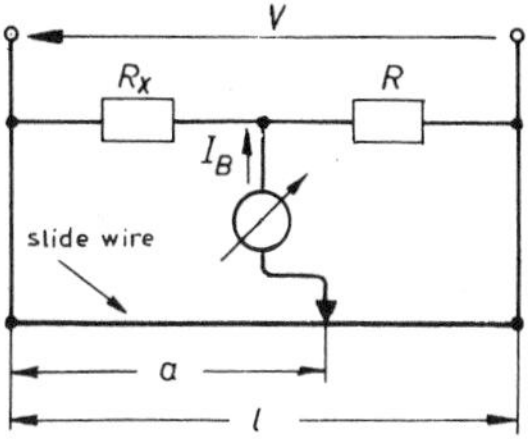

Wheatstone bridge used as a primary element

In many types of measuring equipment Wheatstone bridges serve as comparators for evaluating voltage differences.

R_1 : sensor resistor, the variation of which is proportional to the quantity x to be measured (e.g. temperature, distance, angle etc.)

R_2 : zero value of R_1

Approx. the relation applies

s 63 $$V_M \sim \Delta R \sim x$$

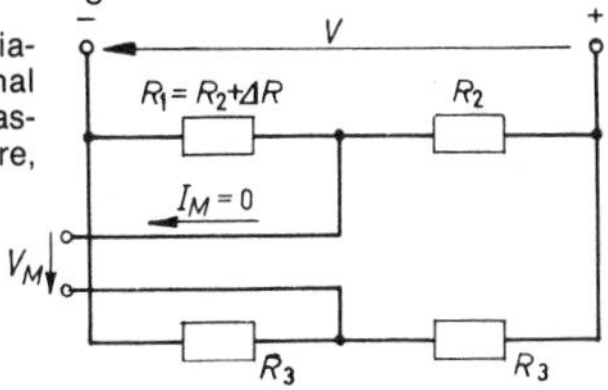

R_M : internal resistance of the measurement

Capacitance C of a capacitor

s 64 $$C = \frac{\varepsilon_0 \varepsilon_r A}{a}$$

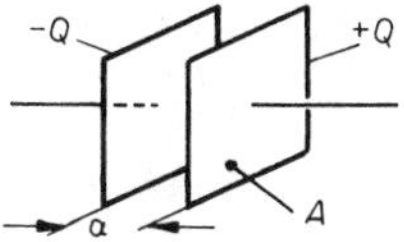

Quantity of electricity Q (see s 8)

Electrical work W_C stored in an electric field

s 65 $$W_C = \frac{1}{2} C V^2$$

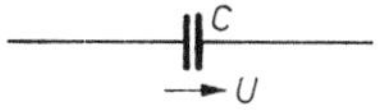

Capacitors connected in parallel

Where capacitors are added in parallel, the total capacitance C increases.

s 66 $$C = C_1 + C_2 + C_3$$

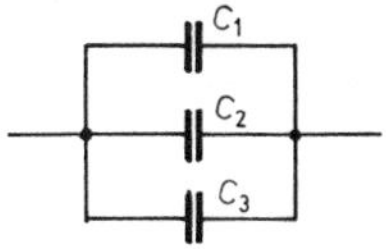

Capacitors connected in series

Where capacitors are added in series, the total capacitance C decreases.

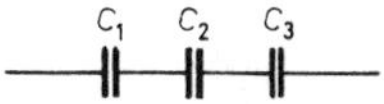

s 67 $$\frac{1}{C} = \frac{1}{C_1} + \frac{1}{C_2} + \frac{1}{C_3}$$

Capacitance of two coaxial cylinders

s 68 $$C = 2 \pi \varepsilon_0 \varepsilon_r \frac{l}{\ln \frac{r_2}{r_1}}$$

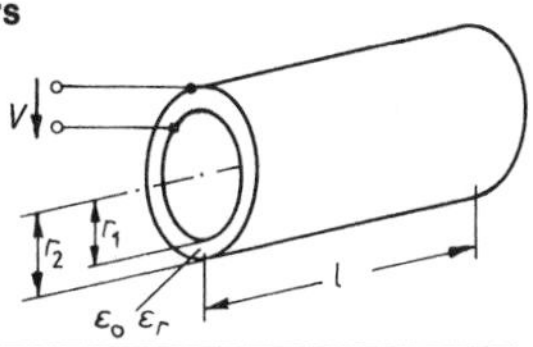

s 69
- ε_r : relative permittivity
- ε_0 : absolute permittivity
- A : plate area (one side)
- a : thickness of dielectric
- r_1 : radius of inner cylinder
- r_2 : radius of outer cylinder
- l : length of cylinders

(see Z 22)
$\varepsilon_0 = 8{\cdot}85 \times 10^{-12}$ A s/(V m)

s 70
Deflection of a magnetic needle

The N-pole of a magnetic needle is attracted by a magnetic S-pole and repelled by a magnetic N-pole.

Fixed conductors and coils

s 71 **Magnetic flux about a current-carrying conductor**

Assuming a corkscrew were screwed in the direction of the current, its direction of rotation would indicate the direction of the lines of magnetic flux.

s 72 **Magnetic flux within a current-carrying coil**

Assuming a corkscrew were rotated in the direction of the current through the coil, the direction of its axial motion would indicate the direction of the lines of magnetic flux through the coil.

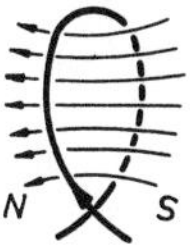

Movable conductors and coils

s 73 **Parallel conductors**

Two parallel conductors carrying currents of the same direction attract. Two carrying currents of opposite direction repel each other.

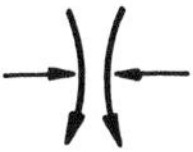

s 74 **Two coils facing each other**

Where two coils positioned face to face carry currents of the same direction, they attract, where they carry currents of opposite directions, they repel each other.

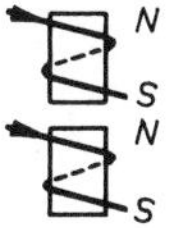

Machines

s 75 **Right Hand Rule** (generator)

Where the thumb points in the direction of the magnetic flux and the middle finger in the direction of motion, the index finger indicates the direction of current flow.

s 76 **Left Hand Rule** (motor)

Where the thumb points in the direction of the magnetic flux and the index finger in the direction of current flow, the middle finger indicates the direction of motion.

Quantities of magnetic circuits

Magnetic flux Φ

s 77 $$\Phi = \frac{F}{S} = \frac{N\,I}{R_m}$$ (see also s 11)

Magnetic induction (flux density) B

s 78 $$B = \frac{\Phi}{A} = \mu_r \mu_0 H$$ (see also s 12)

Inductance L

s 79 $$L = N\frac{\Phi}{I} = N^2\Lambda = \frac{N^2}{R_m}$$ (see also s 13)

For calculation of L see also s 150 through s 156

Magnetic field strength H (Magnetising force)

s 80 $$H = \frac{B}{\mu_r \mu_0} = \frac{F_i}{l_i}$$ (see also s 14)

Magnetomotive force F

s 81 $$F = N\,I = \sum_{i=1}^{n} F_i$$ (see also s 15)

Magnetomotive force F_i

s 82 $$F_i = H_i\, l_i$$ (see also s 16)

Reluctance S

s 83 $$S = \frac{F}{\Phi} = \frac{l}{\mu_r \mu_0 A}$$ (see also s 18)

Energy W_m stored in a magnetic field

s 85 $$W_m = \frac{1}{2} N\,I\,\Phi = \frac{1}{2} L\,I^2$$

Leakage flux Φ_L

Part of the total magnetic flux Φ leaks through the air and is thus lost for the desired effect. Φ_L is related to the useful flux Φ_u. Hence the leakage

s 86 coefficient is: $$\sigma = \frac{\Phi}{\Phi_u} = \frac{\text{total flux}}{\text{useful flux}}$$ (1·15 ... 1·25)

For symbols see S 18

The magnetic field and its forces

Force F_m acting between magnetic poles

In the direction of the magnetic flux a tensile force F_m occurs:

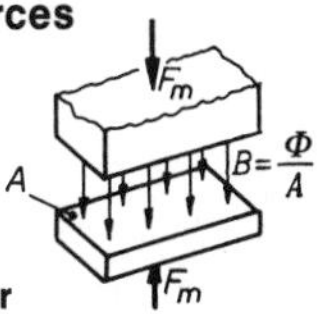

s 88 $$F_m = \frac{1}{2} \frac{B^2 A}{\mu_0}$$

Forces F_l acting on a current-carrying conductor

A conductor carrying a current I encounters a transverse force F_l over its length l perpendicular to the lines of magnetic flux:

s 89 $$F_l = B\, l\, I$$

When applied to the armature of a DC-machine, the moment is:

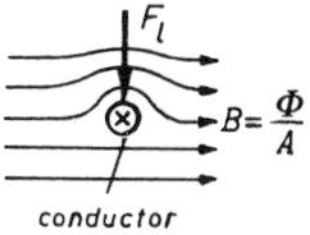

s 90 $$M_i = \frac{1}{2\pi} \Phi I \frac{p}{a} z$$

Φ : flux per pole

Induced voltage V_i (induction law)

Where a coil of N turns and resistance R_i is threaded by a magnetic flux Φ that varies with time, an open-circuit voltage

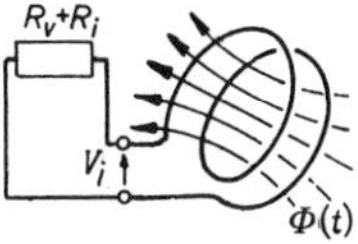

s 92 $$V_i = N \frac{d\Phi_t}{dt}$$ (see also s 11)

is induced across its terminals. This voltage causes a current through an external load resistor R_v.

	voltage induced by		
	motion of conduction perpendicular to flux Φ	rotation of conductor loop (in magnetic field)	rotation of generator armature (in magnetic field)
s 93	$V_i = B\, l\, v$	$V_i = \omega \Phi_{max} \times \sin(\omega t)$	$V_i = \Phi\, n\, z \frac{p}{a}$
s 94		$\Phi_{max} = l\, d\, B$	$= l\, d\, B \frac{z\, p}{2\pi a} \omega$
	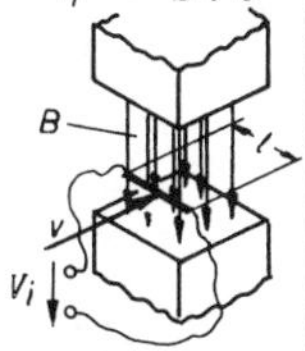		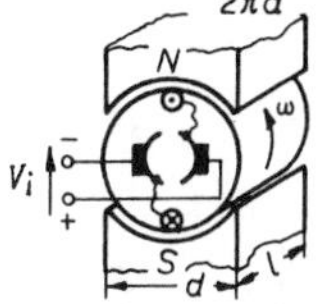

s 95 **Voltage V_i due to self-induction:** $V_i = L \times di/dt$

Continued on page S 18

For symbols see S 18

General terms relating to alternating-current circuit

Sense of phase angles

In vector diagrams arrows are sometimes used to represent phase angles. Here counter-clockwise arrows are taken positive, clockwise arrows negative.

Example:

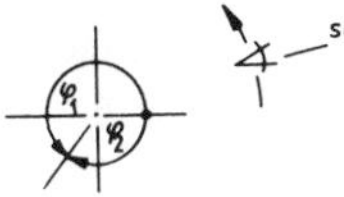

s 96 $$\varphi_1 - \varphi_2 = 360^\circ = 0$$

s 97 $$\varphi_1 = \varphi_2$$

Peak values (see also s 1)

Current i and voltage v of an alternating current vary periodically with time t, usually sinusoidally. The maximum values $\hat{\imath}$ and $\hat{v}$ are called peak values. At an angular frequency $\omega = 2\pi f$ the angle covered in time t is:

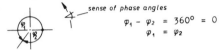

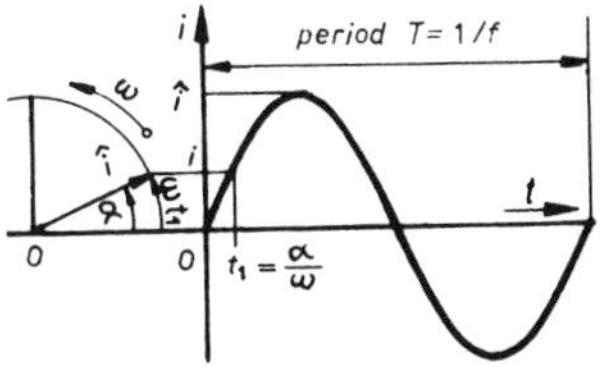

s 98 $$\alpha = \omega t = 2\pi f t$$

Hence at this time

s 99 the current is $i = \hat{\imath}\sin(\omega t) = \hat{\imath}\sin\alpha$

s 100 the voltage is $v = \hat{v}\sin(\omega t) = \hat{v}\sin\alpha$

Root-mean-square (rms) **values**

These are used for practical calculations and are usually indicated by meters

	generally	for sine waves
s 101	$I = I_{eff} = \sqrt{\frac{1}{T}\int_0^T i^2\,dt}$	$I = I_{eff} = \frac{\hat{\imath}}{\sqrt{2}}$
s 102	$V = V_{eff} = \sqrt{\frac{1}{T}\int_0^T v^2\,dt}$	$V = V_{eff} = \frac{\hat{v}}{\sqrt{2}}$

With these values the relation $P = V\,I$ also applies for al-
s 103 ternating current, if $\cos\varphi = 1$ (see s 115).

continued on S 17

continued from S 16

Phase shift, phase angle φ

Where different kinds of load (resistance, inductance and/or capacitance) are present in an alternating-current circuit, a phase shift between current and voltage occurs.

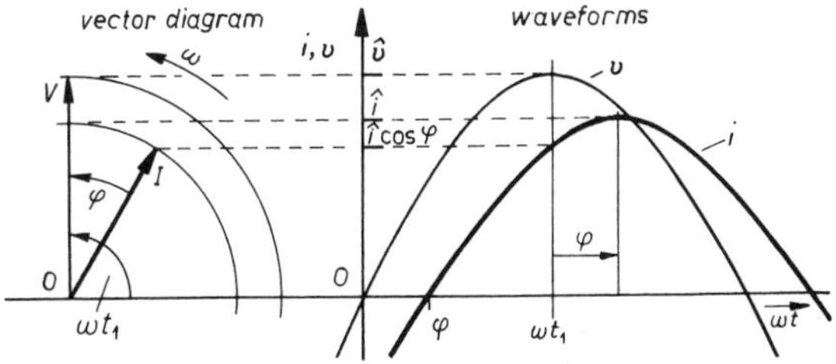

s 104 $\upsilon = \hat{\upsilon} \sin(\omega t)$ $\qquad i = \hat{i} \sin(\omega t - \varphi)$

Q factor, damping factor tan δ, **loss angle** δ

The Q factor of a circuit has been defined by:

s 106 $$Q = \frac{2\pi \hat{w}}{W_{VP}}$$

Here $\hat{w}$ is the peak value of the energy stored in the circuit and W_{VP} the loss energy dissipated in one period.

The reciprocal of Q factor is called damping factor

s 105 $\tan\delta = 1/Q$ (δ is the loss angle)

For a choke (s 125 and s 128) and for a capacitor-resistor combination (s 126 and s 129) this definition results in the simple relations:

s 107 $Q = \tan\varphi$ | $\tan\delta = 1/Q = 1/\tan\varphi$
s 108 $\delta = 90^{\circ} - \varphi$ | $= V_w/V_b$ (for series connection)
s 109 | $= I_w/I_b$ (for parallel connection)

For formulae regarding tan φ see S 19 and S 20.
Formulae s 138 and s 139 applicable for resonant circuits are not so simple.

Basic equations for single phase alternating current

	Impedance	Z see S 19 and S 20
s 110	Admittance	$Y = 1/Z$
s 111	Voltage across impedance	$V = I\,Z$
s 112	Current through impedance	$I = \frac{V}{Z}$
s 113	Reactance	$X = Z \sin\varphi$
s 114	Active power	$P = V\,I \cos\varphi = I^2 R$
s 115	Reactive power	$Q = V\,I \sin\varphi = I^2 X$
s 116	Apparent power	$S = V\,I = \sqrt{P^2 + Q^2} = I^2 Z$
s 117	Power factor	$\cos\varphi = \frac{P}{V\,I} = \frac{P}{S}$
s 118	Alternating magnetic flux in a coil	$\hat{\Phi} = \frac{V_L}{4 \cdot 44\, N f}$

s 95 continued (V_i due to self-induction)
Where the current i flowing through a coil changes with time, the magnetic field caused by this current also changes. Thereby a voltage V_i is induced in the coil. Its direction is such that it counteracts the instantaneous change of current (Lenz's law).

Symbols used on page S 15 :

s 119 μ_o : absolute permeability ($\mu_o = 4\,\pi \times 10^{-7}$ V s/A m)

μ_r : relative permeability

for vacuum, gases, fluids, and most solids: $\mu_r = 1$,
for magnetic materials take μ_r from Z 23

a : number of parallel paths through winding

l : length of magnetic circuit

N : number of turns of coil

p : number of pole pairs

z : number of conductors

R_R	resistance in	series	equivalent circuit of choke
R_P		parallel	
L_R	inductance in	series	
L_P		parallel	

Components, series- and parallel connections carrying alternating current

No.	component		symbol	vector diagram	phase relation	phase angle	impedance	tan φ =
s 120	resistive lamp, coil with bifilar winding		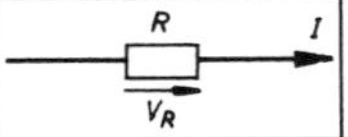		I and V in phase	$\varphi = 0°$	$Z = R$	0
s 121	inductive ideal inductance		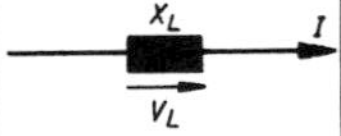	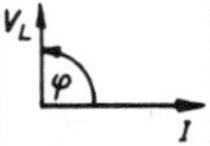	I lags V by 90°	$\varphi = 90°$	$Z = X_L = \omega L$	∞
s 122	capacitive capacitor		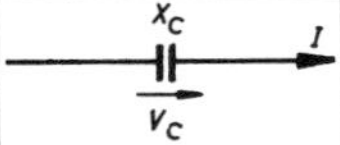	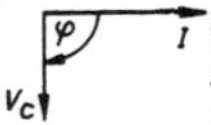	I leads V by 90°	$\varphi = -90°$	$Z = X_C = -\frac{1}{\omega C}$	∞
s 123	resistive inductive capacitive connected in series	$\omega L_R < \frac{1}{\omega C}$	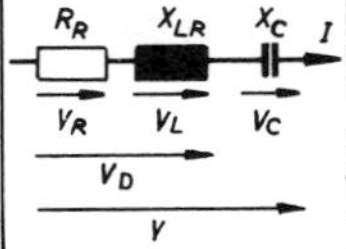	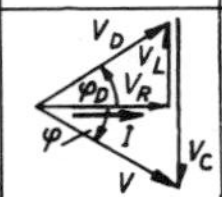	I leads V	$-90° < \varphi < 0°$	$Z = \sqrt{R_R^2 + \left(\omega L_R - \frac{1}{\omega C}\right)^2}$	$\frac{\omega L_R - \frac{1}{\omega C}}{R_R}$
s 124	choke in series with capacitor	$\omega L_R > \frac{1}{\omega C}$		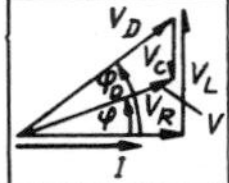	I lags V	$0° < \varphi < 90°$		
s 125	resistive and inductive in series choke		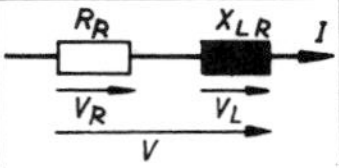	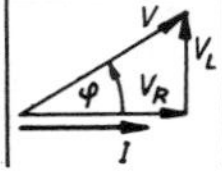	I lags V by less than 90°	$0° < \varphi < 90°$	$Z = \sqrt{R_R^2 + (\omega L_R)^2}$	$\frac{\omega L_R}{R_R}$

continued on S 20

continued from S 19

	kind of load	symbol	vector diagram	phase relation	phase angle	impedance	tan φ =
s 126	resistive + capacitive in series resistor in series with capacitor	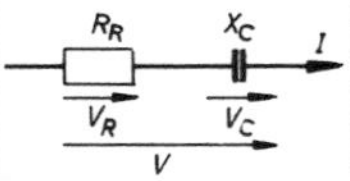	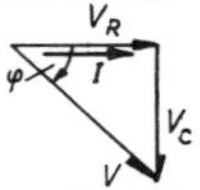	I leads V by less than 90°	$-90° < \varphi < 0°$	$Z = \sqrt{R_R^2 + \left(\frac{1}{\omega C}\right)^2}$	$-\frac{1}{R_R \omega C}$
s 127	resist. + induct. + capacitive in parallel choke + capacitor in parallel	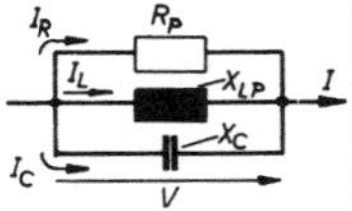	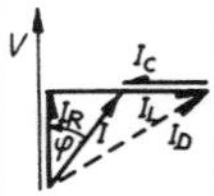	I leads or lags V depending on values	$-90° < \varphi < 90°$	$Z = \frac{1}{\sqrt{\left(\frac{1}{R_P}\right)^2 + \left(\frac{1}{\omega L_P} - \omega C\right)^2}}$	$R_P\left(\frac{1}{\omega L_P} - \omega C\right)$
s 128	resistive + inductive in parallel choke	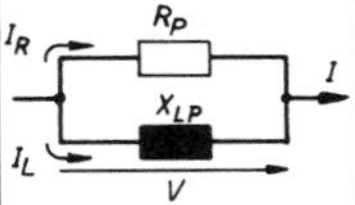	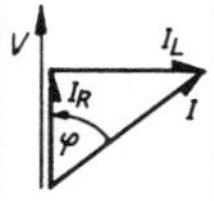	I lags V	$0° < \varphi < 90°$	$Z = \frac{1}{\sqrt{\left(\frac{1}{R_P}\right)^2 + \left(\frac{1}{\omega L_P}\right)^2}}$	$\frac{R_P}{\omega L_P}$
s 129	resist. + capac. in parallel resistor + capac. in parallel	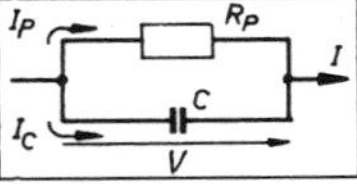	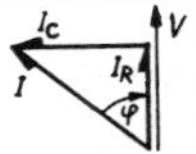	I leads V	$-90° < \varphi < 0°$	$Z = \frac{1}{\sqrt{\left(\frac{1}{R_P}\right)^2 + (\omega C)^2}}$	$-R_P \omega C$

s 130/131 The given values R and L of a choke always are the values R_R and L_R of the series equivalent circuit (see s 125). However, where a choke is connected in parallel with a capacitor, it is desirable to use the parallel equivalent circuit of a choke (see s 128). The values R_P and L_P contained therein may be calculated by:

$$R_P = R_R + \frac{(\omega L_R)^2}{R_R}$$

$$L_P = L_R + \frac{R_R^2}{\omega^2 L_R}$$

Resonant circuits

		series-	parallel-
		resonant circuit	resonant circuit
	symbol and general vector diagram	see s 123	see s 127
	vector diagram at resonance	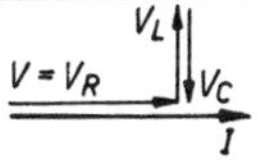	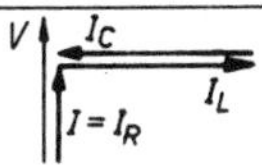
s 132	resonance condition	$V_L = V_C$	$I_L = I_C$
s 133		$\omega_r L_R - \frac{1}{\omega_r C} = 0$	$\frac{1}{\omega_r L_P} - \omega_r C = 0$
s 134		$\omega_r^2 L_R C = 1$	$\omega_r^2 L_P C = 1$
s 135	resonant frequency	$f_r = \frac{1}{2\pi\sqrt{L_R C}}$	$f_r = \frac{1}{2\pi\sqrt{L_P C}}$
		where line frequency $f = f_r$, resonance occurs	
s 136	current at resonance	$I_r = \frac{V}{R_R}$	$I_r = \frac{V}{R_P} = \frac{R_R C V}{L_R}$
s 137		at $V_b = V_L - V_C = 0$, $\varphi = 0$	at $I_b = I_L - I_C = 0$, $\varphi = 0$
s 138	Q factor	$Q_R = \frac{\omega_r L_R}{R_R} = \frac{1}{\omega_r C R_R}$	$Q_P = \omega_r C R_P = \frac{R_P}{\omega_r L_P}$
s 139	loss angle δ from	$\tan\delta_R = \frac{1}{Q_R} = \frac{R_R}{\omega_r L_R}$	$\tan\delta_P = \frac{1}{Q_P} = \frac{1}{\omega_r C R_P}$
s 140	wavelength	$\lambda = \frac{c}{f_r} = \frac{300 \times 10^6\ \mathrm{m}}{f_r\, \mathrm{s}}$	
s 141	resonant period	$T_r = 2\pi\sqrt{L_R C}$	$T_r = 2\pi\sqrt{L_P C}$

Tank Circuit

A parallel resonant circuit has its maximum impedance Z_{max} at its resonant frequency. Therefore it acts as a rejector for currents of this frequency.

s 142 $$Z_{max} = R_P = \frac{L_R}{R_R C} \qquad \text{and current} \qquad I = \frac{V}{Z_{max}}$$

for symbols see S 18

Alternating-current bridge

AC bridges are used to determine capacitances and inductances. For balancing the bridge variable capacitor C_2 and resistor R_2 are adjusted until the sound in the low resistance headphone K reaches its minimum or vanishes. The following circuits are independent of frequency.

measurement of

capacitance	inductance
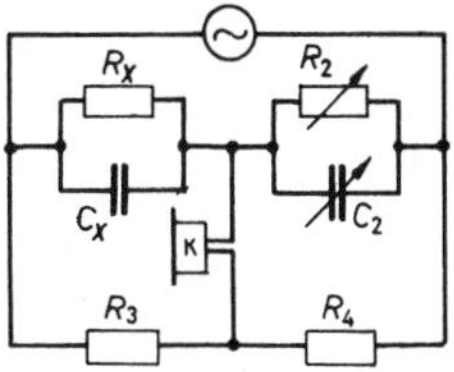	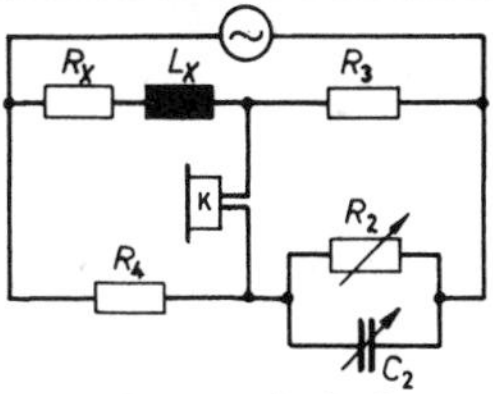

	capacitance	inductance
s 143	$C_X = \frac{R_4}{R_3} C_2$	$L_X = C_2 R_3 R_4$
s 144	$R_X = \frac{R_3}{R_4} R_2$	$R_X = \frac{R_3 R_4}{R_2}$
s 145	$\tan \delta_X = \frac{1}{R_X \omega C_X}$	$\tan \delta_X = \frac{R_X}{\omega L_X}$

Determination of an unknown impedance by measuring the voltages across this impedance and an auxiliary resistor:

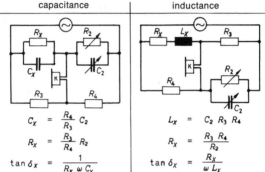

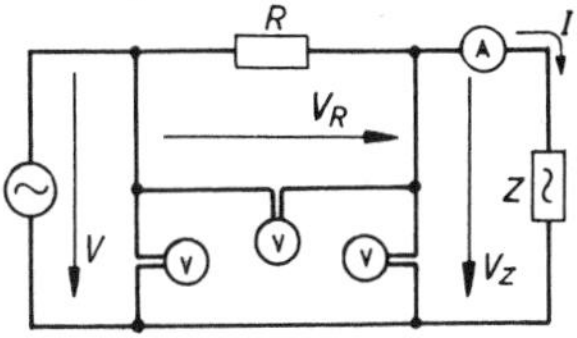

s 146 $$P_{WZ} = \frac{V^2 - V_R^2 - V_Z^2}{2R}$$

s 147 $$\cos \varphi_Z = \frac{P_{WZ}}{V_Z I}$$

s 148 $$Z = \frac{V_Z}{I}$$

s 149 select auxiliary resistor R such that $V_R \approx |V_Z|$

C_X : unknown capacitance
L_X : unknown inductance
δ_X : loss angle, see S 17
$R_{2 \ldots 4}$: known resistances
R_X : unknown resistance of coil or capacitor
C_2 : calibrated adjustable capacitance
Z : unknown impedance (inductive or capacitive)

Inductance L from impedance and resistance

Calculating L from impedance and resistance

s 150 Pass an alternating current ($J = I/A \approx 3$ A/mm²) through a coil and measure the terminal voltage V, current I, active power P:

s 151 impedance $Z = \frac{V}{I}$; resistance $R = \frac{P}{I^2}$

s 152 $L = \frac{1}{\omega}\sqrt{Z^2 - R^2}$

Calculating L for a toroidal coil

s 153 $L = \frac{\mu_0 h N^2}{2\pi} \ln\frac{r_2}{r_1}$

Calculating L for a square coil

armatures must be circular

	$\frac{D}{u}$	inductance
s 154	< 1	$L = 1{\cdot}05 \frac{D}{m} N^2 \sqrt[4]{\left(\frac{D}{u}\right)^3}$ µH
s 155	> 1	$L = 1{\cdot}05 \frac{D}{m} N^2 \sqrt{\frac{D}{u}}$ µH
s 156	$\geqq 3$	values become unreliable

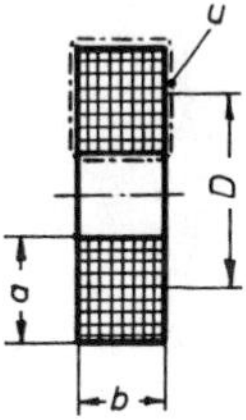

$1\ \mu\text{H} = 10^{-6} \frac{\text{V s}}{\text{A}}$

---

a : thickness of winding
A : cross section of wire
b : width of coil
d_a : external diameter of wire and insulation
D : mean diameter of coil
l_0 : internal length of armature winding
s 157 l_m : mean length of armature winding ($l_m = l_0 + \pi a$)
N : number of turns
u : circumference of coil cross section
α : ratio $a : b$
s 158 β : degree of loosening of turns $\left(\beta = \frac{a\,b}{N d_a^2}\right)$

Non-magnetic coils with specified inductance L

High frequency coils

	$\frac{D}{u}$	formula	
s 159	< 1	$\left(\frac{D}{\text{m}}\right)^{3.5} N^{3.25} \approx \frac{1}{39}\left(\frac{d_o}{\text{m}}\right)^{1.5}\left(\frac{L}{\text{H}}\right)^2 \times 10^{14}$	here: $d_o = \frac{u}{2\sqrt{N}}$
s 160	> 1	$\left(\frac{D}{\text{m}}\right)^{3} N^{3.5} \approx \frac{1}{55}\left(\frac{d_o}{\text{m}}\right)\left(\frac{L}{\text{H}}\right)^2 \times 10^{14}$	$= d_a(1+\alpha)\sqrt{\frac{\beta}{\alpha}}$

Low frequency coils

Assuming that

s 161 $\beta = 1$ and $D = u$, then

s 162 $$N \approx 975\sqrt{\frac{L}{\text{H}}\frac{\text{m}}{D}}$$

s 163 $$a = \frac{1}{4}\left(u \pm \sqrt{u^2 - 16\,N\,d_a^{\,2}}\right); \qquad b = \frac{u}{2} - a$$

Calculation of number of turns N of a coil

From cross section

s 164 $$N \approx \frac{a\,b}{d_a^{\,2}}$$

From resistance

s 165 $$N \approx \frac{R\,A}{\varrho\, l_m}$$

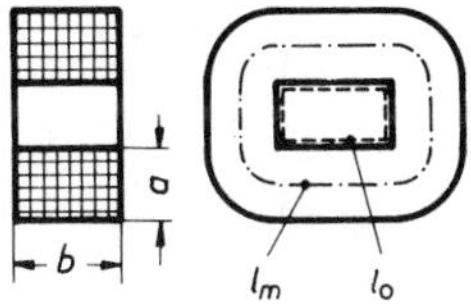

Using reference coil

Position unknown coil of N_x turns and reference coil of N_o turns at short distance on closed iron core. Magnetize core by alternating voltage V_e applied to magnetizing coil N_e. Measure voltages V_x and V_o using high impedance voltmeter. Then

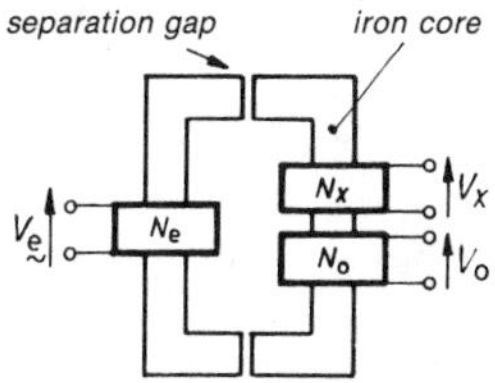

s 166 $$N_x = N_o\,\frac{V_x}{V_o}$$

For explanation of symbols see S 23

Hysteresis

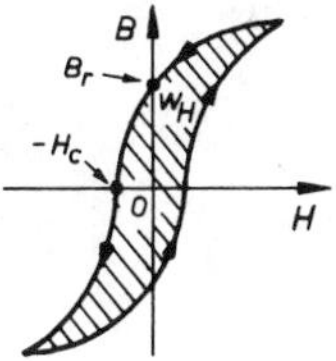

Remanent-flux density B_r

A residual magnetism of flux density B_r remains in the iron core, after the external magnetic field strength H has been removed.

Coercive force H_C

The coercive force H_C has to be applied to reduce the flux density B to zero.

Hysteresis work W_H

The energy W_H dissipated during a single cycle of the hysteresis loop is equal to the product of area of the hysteresis loop w_H and core volume V_{Fe}:

s 167
$$W_H = w_H V_{Fe}$$

Hysteresis power P_{VH}

s 168
$$P_{VH} = W_H f = w_H V_{Fe} f$$

Eddy currents

According to the induction law alternating voltages are also induced inside an iron core. Depending on the resistivity of the core iron these voltages cause induction currents called eddy currents. They are kept small by lamination (making up the core of thin metal sheets, which are insulated from each other).

Core losses (iron losses)

Core losses per unit mass p_{Fe}

They are the combined hysteresis and eddy-current losses per unit mass. They are measured at a peak induction $\hat{B} = 1\ \text{T} = 10\ \text{kG}$ or $1{\cdot}5\ \text{T} = 15\ \text{kG}$ and at a frequency $f = 50\ \text{Hz}$ and are then denoted P 1,0 or P 1,5 respectively. For values see Z 24.

Total core losses P_{Fe}

s 169
$$P_{Fe} = P\,1{\cdot}0 \left(\frac{\hat{B}}{\text{T}} \times \frac{f}{50\ \text{Hz}} \right)^2 m_{Fe} (1 + \varkappa)$$

m_{Fe}: mass of core | $\varkappa$: addition for punching ridges etc. (0·1 . . . 1·0)

Choke coil

Choke coil used as a dropping impedance

It is used in an ac circuit to reduce the line voltage V down to a value V_V for a restitive load with minimum losses.

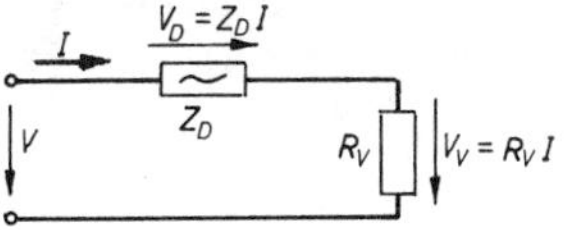

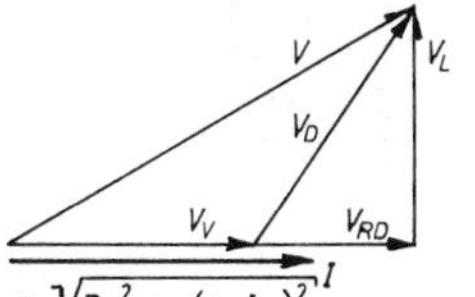

s 170 impedance of | choke $\quad Z_D = \sqrt{R_R^2 + (\omega L_R)^2}$

s 171 impedance of | total circuit $\quad Z = \sqrt{(R_R + R_V)^2 + (\omega L_R)^2}$

s 172 inductance $\quad L_R = \frac{1}{\omega}\sqrt{\left(\frac{V R_V}{V_V}\right)^2 - (R_V + R_R)^2}$

In a rough calculation of L_R neglect the unknown resistance R_R of the choke. After dimensioning the choke R_R is known, and Z may be determined exactly. Check V_V by

s 173 $$V_V = \frac{V R_R}{Z}$$

and repeat procedure, if necessary.

Choke of constant inductance without core

Dimension according to S 23. Make preliminary assumptions regarding values r_2/r_1 (toroid coil) or D/u (straight coil). In case of unfavourable results repeat procedure. Determine resistance of choke according to s 26.

Choke coil of constant inductance with iron core

The iron core essentially serves for guiding the magnetic flux and should incorporate as many single air gaps δ_1 as possible. These should be filled with insulating layers and should not exceed 1 cm in length. The m.m.f. required to magnetize the core is neglected. Peak values of H and B are used for calculations. The variation of inductance L_R my be expressed

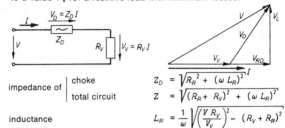

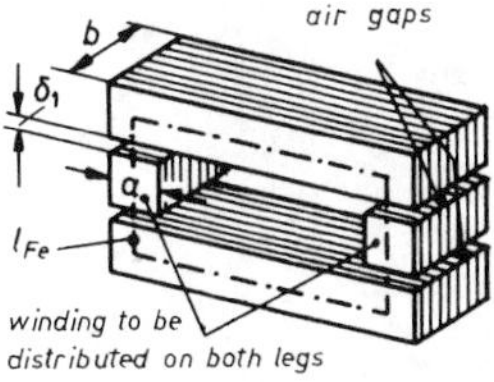

continued on S 27

continued from S 26

in terms of the maximum relative current-depending variation of inductance

s 174 $$g_L = \frac{|L_{R\,tot} - L_R|}{L_R}\ ; \qquad \frac{1}{g_L} = \frac{A_{Fe}\,\hat{B}_{Fe}\,\delta}{\hat{H}_{Fe}\,l_{Fe}\,\mu_0\,A_L} + 1$$

If $g_L > g_{L\,requ}$, repeat dimensioning with greater A_{Fe} and smaller $\hat{B}_{Fe}$ at unchanged product $A_{Fe} \times \hat{B}_{Fe}$.

Dimensioning. Given: L_R, f, $g_{L\,requ}$, $V_{L\,eff}$ or I_{eff}, then the

			preliminary dimensions are	final dimensions are
s 175	effective cross section of core		$A_{Fe}' = \sqrt{K\,I_{eff}\,V_{L\,eff}}$	take A_{Fe} from standards or determine a and b by
s 176			with $I_{eff} = \frac{V_{L\,eff}}{2\pi f L_R}$	$A_{Fe} = 0{\cdot}9\,a\,b \approx A_{Fe}'$
s 177	number of turns		$N = \frac{V_{L\,eff}}{4{\cdot}44\,f\,\hat{B}_{Fe}\,A_{Fe}}$	
s 178	cross section of air gap		$A_L' = ab + 5\text{cm}(a+b)$	$[A_L = a\,b + 5(a+b)\delta_1]$
s 179	total	length of air gap	$\delta' = \frac{N^2\,\mu_0\,A_L'}{L_R}$	$\delta = \frac{a\,b\,n\,N^2\,\mu_0}{n\,L_R - 5\,N^2\,\mu_0(a+b)}$
s 180	single		$\delta_1' = \delta'/n < 1$ cm	$\delta_1 = \delta/n < 1$ cm
s 181	diameter of wire		$d' = 2\sqrt{\frac{I}{J'\,\pi}}$	Use next standard values for d, d_a including insulation
s 182	cross section of winding		$A_W = 1{\cdot}12\,d_a^2\,N$	
	length of limb		l_s to be determined from dimensions of core section and A_W	

Choke coil of current-depending inductance

This type of choke employs an iron core without an air gap. It is used only for special purposes, e. g. as a magnetic amplifier.

K : power coefficient of choke
≈ 0·24 cm^4/VA for air-cooled chokes } core section see S 26
≈ 0·15 cm^4/VA for oil chokes }
for ⊡⊡ core section increase values by 75%

J' : preliminary current density for air-cooled choke $J' = 2$ A/mm^2; for oil choke $J' \triangleq 3 \ldots 4$ A/mm^2

$\hat{B}_{Fe}$: core induction (take approx. 1 ... 1·2 T)

$\hat{H}_{Fe}$: field strength in core corr. to $\hat{B}_{Fe}$ to be taken from Z 23 according to material employed

n : number of single air gaps, increase reduces stray flux

R_{Cu} : resistance of winding according to s 26

R_R : resistance of choke includ. core losses ($R_R \approx 1{\cdot}3\,R_{Cu}$)

l_{Fe} : mean length of magnetic path through iron

Transformer

Designation of windings

distinction by

nominal voltages		function in circuit (direction of power transfer)	
winding with higher nominal voltage	winding with lower nominal voltage	input winding	output winding
high-end winding	low-end winding	primary (index 1) winding	secondary (index 2) winding

Nominal values (index N)

s 183 rated power $S_N = V_{1N} \times I_{1N} = V_{2N} \times I_{2N}$

s 184 nominal transformation ratio $\ddot{u} = V_{1N} / V_{20} = I_{2N} / I_{1N}$

By the rated secondary voltage V_{2N} we mean the open-circuit secondary voltage ($V_{2N} = V_{2O}$), not the one at nominal load.

Core losses P_{Fe} and open-circuit measurements

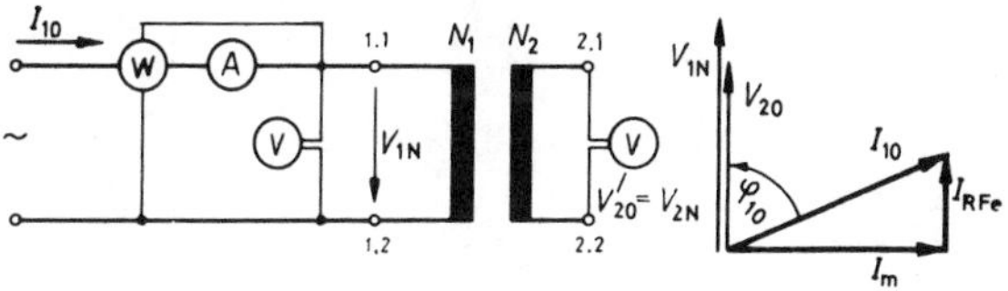

The core losses P_{Fe} only depend on primary voltage V_1 and frequency f, not on the kind of load.

s 185 $P_{10} = P_{Fe}$

Core losses P_{Fe} and nominal transformation ratio $\ddot{u}$ are determined by open-circuit measurements (see circuit diagram: secondary open, values provided with index O). The primary current's resistive component I_{RFe} covers the core losses, its reactive component is the magnetizing current I_m. The copper losses are negligibly small. The core losses P_{Fe} are required for calculating operational power dissipation and efficiency.

continued on S 29

continued from S 28

Copper losses P_{Cu} and short-circuit measurements

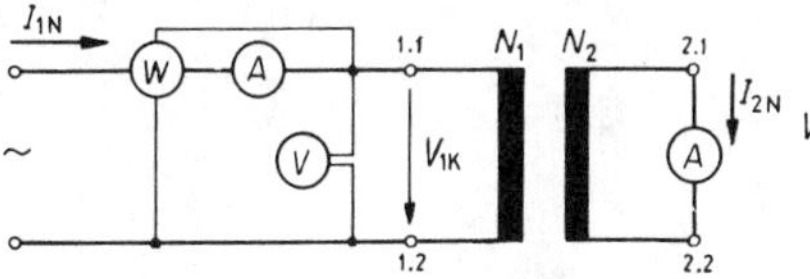

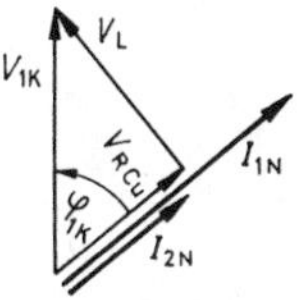

P_{Cu} depends only on the primary current I_1 and is determined by short-circuit measurements (see circuit diagram, values provided with index K). With the secondary shorted, the primary voltage is adjusted to a value V_{1K}, which causes the rated currents to flow. V_{1K} is so small that I_{RFe} and I_m are negligible. Then the short-circuit primary power P_{1K} is equal to the rated copper losses P_{CuN} of the total transformer at rated currents. P_{1K} is required for calculating operational power dissipation and efficiency.

s 186 $$P_{1K} = P_{CuN}$$

The values measured are used for calculating the relative short-circuit voltage v_K, which, for bigger transformers, is always indicated on the name plate:

s 187 $$v_K = 100(V_{1K}/V_{1N})\,\%$$

The following quantities may be determined using the vector diagram:

s 188 $$R_{Cu} = V_R/I_{1N}\,; \quad L = V_L/\omega I_{1N}\,; \quad \cos\varphi_{1K} = V_R/V_{1K} = \frac{P_{CuN}}{v_K\,P_{SN}}$$

Operating conditions

For calculating the operational secondary voltage V_2 for a given load all secondary quantities are first computed into those of an equivalent transformer having a transformer ratio of $i = 1$ (index '):

simplified equivalent circuit

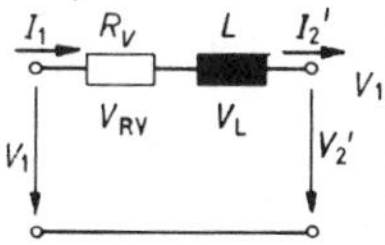

simplified vector diagram

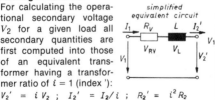

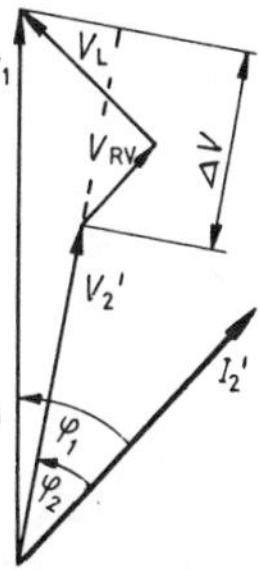

s 189 $$V_2' = i\,V_2\,; \quad I_2' = I_2/i\,; \quad R_2' = i^2 R_2$$

Load-dependent variation ΔV of V_2'

(approximation for $v_K \leqq 4\%$)

s 190 $$\Delta V \approx V_{1K}(\cos\varphi_{1K}\cos\varphi_2 + \sin\varphi_{1K}\sin\varphi_2)\,I_2/I_{2N}$$
$$\approx V_{1K}\cos(\varphi_{1K} - \varphi_2)\,I_2/I_{2N}$$

Secondary voltage V_2

s 191 $$V_2' \approx V_1 - \Delta V\,; \qquad V_2 = V_2'/i$$

Basic connections

Star

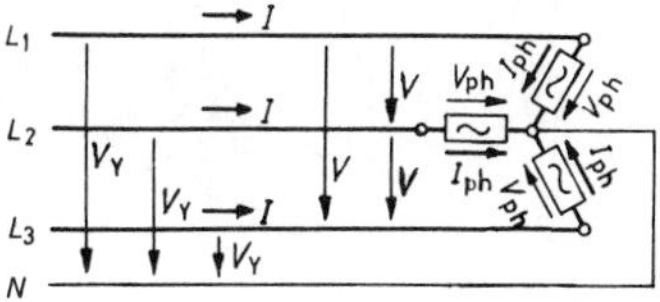

s 192 $V = V_{ph}\sqrt{3}$

s 193 $I = I_{ph}$

Delta

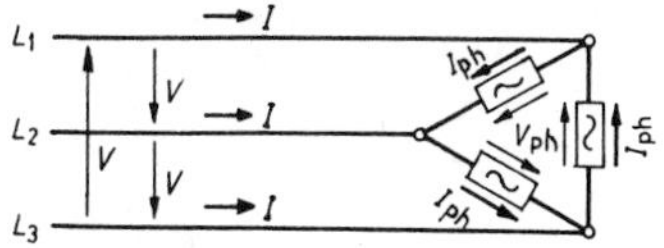

s 194 $V = V_{ph}$

s 195 $I = I_{ph}\sqrt{3}$

Measuring three-phase power

Load balanced

connection

with neutral point (star connected)	without neutral point (delta connected)
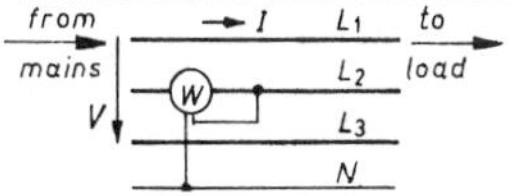	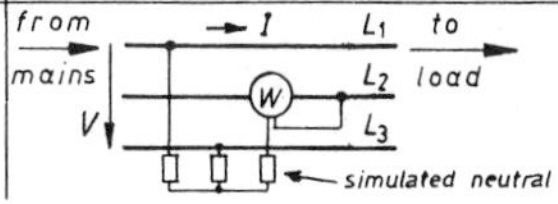

s 196 total power $P = 3\,P_{w\,ph} = \sqrt{3}\,V\,I\,\cos\varphi$

Load unbalanced (Two wattmeter method)

For delta connected without neutral point. (Also for balanced load without neutral point).

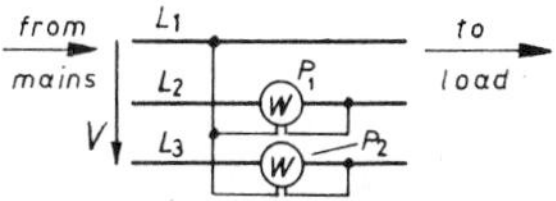

s 197 total power $P = P_1 + P_2$

I_{ph}	: phase current	V_{ph}	: phase voltage
I	: line current	V	: line voltage

L_1, L_2, L_3 : outer conductors
N : neutral conductor
$P_{w\,ph}$: active power of one phase

Reactive and active power, power factor

(for symmetrical load)

s 198 reactive power $Q = \sqrt{3}\, V I \sin\varphi$

s 199 active power $P = \sqrt{3}\, V I \cos\varphi$

s 200 power factor $\cos\varphi = \dfrac{P}{\sqrt{3}\, V I}$

Power factor correction

(for inductive consumers)

General

Adjust to power factor according to current rate, usually $\cos\varphi = 0{\cdot}8 \ldots 0{\cdot}9$. Adjust large consumers separately and directly and small consumers centrally to main or subdistributors.

Calculating the required capacitor power

Calculate power factor $\cos\varphi$ as above, use wattmeter (see connection in S30) or a current meter to determine P.

s 201 capacitor power $Q = (\tan\varphi_1 - \tan\varphi_2)P$

s 202 inherent consumption of condenser $P_c \approx 0{\cdot}003\, Q$

Table (numerical)

cos φ	tan φ	cos φ	tan φ	cos φ	tan φ	cos φ	tan φ
0·42	2·161	0·62	1·265	0·81	0·724	0·91	0·456
0·44	2·041	0·64	1·201	0·82	0·698	0·92	0·426
0·46	1·930	0·66	1·138	0·83	0·672	0·93	0·395
0·48	1·828	0·68	1·078	0·84	0·646	0·94	0·363
0·50	1·732	0·70	1·020	0·85	0·620	0·95	0·329
0·52	1·643	0·72	0·964	0·86	0·593	0·96	0·292
0·54	1·559	0·74	0·909	0·87	0·567	0·97	0·251
0·56	1·479	0·76	0·855	0·88	0·540	0·98	0·203
0·58	1·405	0·78	0·802	0·89	0·512	0·99	0·142
0·60	1·333	0·80	0·750	0·90	0·484	1	0·000

$\tan\varphi_1$ or $\tan\varphi_2$ can be calculated from the above table, $\cos\varphi_1$ representing the required power factor and $\cos\varphi_2$ the consumer power factor.

Direct-current machine
(motor and generator)

General

s 203	moment constant	$C_M = \frac{p\,z}{2\,\pi\,a}$
s 204	rotational source voltage	$V_q = C_M \Phi\, \omega = 2\,\pi\, C_M \Phi\, n$
s 205	torque	$M = C_M \Phi\, I_a$
s 206	armature current	$I_a = \frac{\pm(V - V_q)}{R_a}$ *)
s 207	terminal voltage	$V = V_q \pm I_a R_a$ *)
s 208	speed	$n = \frac{V \mp I_a R_a}{2\,\pi\, C_M \Phi}$ **)
s 209	internal power	$P_i = M_i\, \omega = V_q\, I_a$
s 210	mechanical power supplied to generator	$P_G = \frac{1}{\eta}\, V\, I_{tot}$
s 211	mechanical power supplied by motor	$P_M = \eta\, V\, I_{tot}$

Shunt motor (for circuit diagram see S 33)
Easy starting, speed is fairly independent of load and, within certain limits, easy to regulate.

Series motor (for circuit diagram see S 33)
Easy starting with powerful starting torque. Speed depends greatly on load. When running free may become unstable.

Compound wound motor (for circuit diagram see S 33)
Operates almost like a shunt motor. Main circuit winding ensures a powerful starting torque.

a : number of armature pairs
p : number of pole pairs
Φ : magnetic flux
z : number of conductors
R_a : armature resistance

*) + motor
− generator

**) − motor
+ generator

ELECTRICAL ENGINEERING

D. C. machines with commutating poles

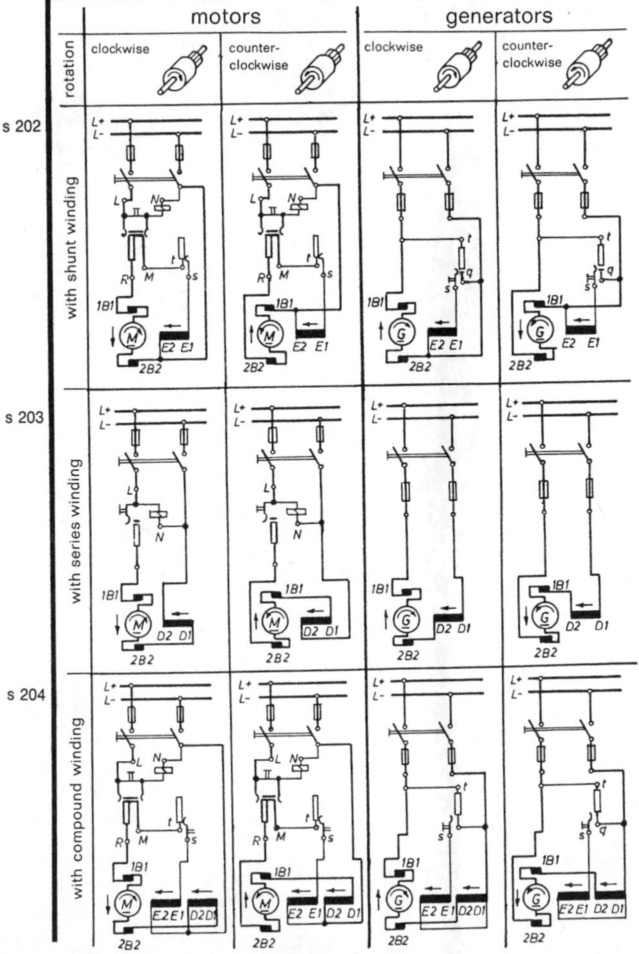

Three-phase motor

Speed

At a given frequency f the speed is determined by the number of pole pairs p.

s 215 $$\text{speed} \quad n = \frac{f}{p} = \frac{60\, f\, \text{s}}{p} \times \frac{1}{\text{min}}$$

Switching

Where both terminals of each winding are accessible on the switchboard, the three-phase motor can be connected either in star or in delta.

phase voltage in star	phase voltage in delta
s 216 $V_{ph} = \frac{V}{\sqrt{3}}$	$V_{ph} = V$

A 400/230 volt motor operates with its nominal values of current, torque and power, when connected to

s 217 $V = 230$ V in delta, meaning $V_{ph} = V = 230$ V

s 218 $V = 400$ V in star, meaning $V_{ph} = \frac{V}{\sqrt{3}} = \frac{400\text{ V}}{\sqrt{3}} = 230$ V

Star-delta connection

Higher powered motors usually operate in delta. To avoid excessive inrush currents, particularly in relatively low current networks, the motor is started in star and then switched over into delta. If, for instance, a 400/230 volt motor is connected in star to a 230/135 volt network, it is supplied with only $1/\sqrt{3}$ times its nominal voltage.

Induction motor

The rotating field of the stator causes voltage and current to be induced in the armature winding. Due to slip the rotational speed of the armature is about 3 to 5% lower than that of the rotating field; it remains almost constant under load.

Synchronous motor

Requires direct current for excitation and is synchronized with the speed of the rotating field by means of an auxiliary squirrel-cage armature. Can be used directly as a generator.

Switch groups generally used for transformers

	type: key-numb.	type: switch group	sign (PV, SV) and switch diagram (PV, SV)	ratio $V_1 : V_2$
			Threephase-output transformers	
s 219	0	D d 0	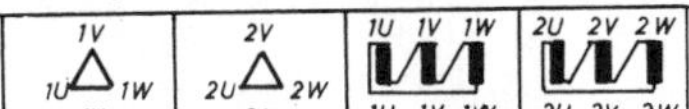	$\frac{N_1}{N_2}$
s 220	0	Y y 0		$\frac{N_1}{N_2}$
s 221	0	D z 0		$\frac{2\,N_1}{3\,N_2}$
s 222	5	D y 5		$\frac{N_1}{\sqrt{3}\,N_2}$
s 223	5	Y d 5	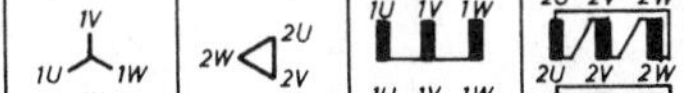	$\frac{\sqrt{3}\,N_1}{N_2}$
s 224	5	Y z 5	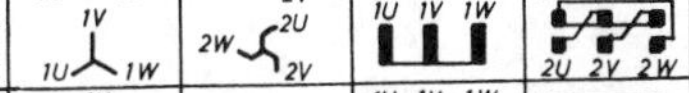	$\frac{2\,N_1}{\sqrt{3}\,N_2}$
s 225	6	D d 6		$\frac{N_1}{N_2}$
s 226	6	Y y 6	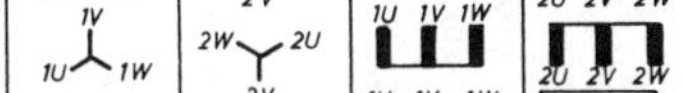	$\frac{N_1}{N_2}$
s 227	6	D z 6		$\frac{2\,N_1}{3\,N_2}$
s 228	11	D y 11		$\frac{N_1}{\sqrt{3}\,N_2}$
s 229	11	Y d 11		$\frac{\sqrt{3}\,N_1}{N_2}$
s 230	11	Y z 11	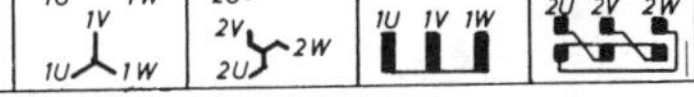	$\frac{2\,N_1}{\sqrt{3}\,N_2}$
			Single phase-output transformers	
s 231	0	I i 0	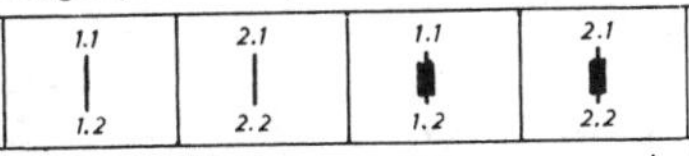	$\frac{N_1}{N_2}$

PV: primary voltage
SV: secondary voltage

| D / d | delta | Y / y | star | – / z | zig-zag |

Key numbers are used to calculate the phase angle (= key number × 30°) between the primary and secondary voltage, e. g. for Dy5 the phase angle is 5 × 30 = 150°.

Note: Use the framed switch groups for preference.

The most important measuring instruments

Symbol	Type of instrument	Construction	Basically measured quantity	Scale	Used for measuring	
	moving coil	moving coil in uniform radial field of a permanent magnet. 2 spiral springs or tension bands serve for counter moment and application of current	dc value (arithmetical mean)	linear	—	I and V
	moving coil with rectifier		arithmetical mean of rectified value	~ linear	~ [1]	I and V
	cross coil	2 coils fixed to each other moving in non-uniform field of a permanent magnet 2 current leads without counter moment	$\frac{I_1}{I_2}$	almost square-law	— ~ [2]	$\frac{I_1}{I_2}$
	moving coil with thermo-couple	Thermocouple in close thermal contact with heater. Thermocouple feeds moving coil instruments	root-mean-square value	almost square-law	— ~ [2) 3)]	I and V
	soft iron	1 moving and 1 fixed piece of soft iron, fixed coil, spiral spring as counter moment	root-mean-square value	non-linear	— ~ [2) 4)]	I and V
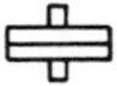	electro-dynamic	moving coil in uniform field of fixed coil, 2 spiral springs or tension bands as counter moment and current leads, magnetic screen	$I_1 \times I_2 \times \cos \varphi$	square-law for I and V, linear for P	— ~ [2]	I, V, P and $\cos \varphi$
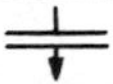	electro-static	1 fixed and 1 moving capacitor plate	root-mean-square value	non-linear	— ~ [2]	V 100 V

1) for sinusoidal waveforms only
2) also for non-sinusoidal waveform
3) also for rf currents and voltages
4) $f < 500$ Hz

Current rating I_z

PVC insulation, unburied copper conductors including overload protection devices at an ambient temperature of 30 °C [1)]

Nominal cross section mm²		1	1·5	2·5	4	6	10	16	25	35	50
Group 1	I_z in A	11	15	20	25	33	45	61	83	103	132
	I_n in A	6	10	16	20	25	35	50	63	80	100
Group 2	I_z in A	15	18	26	34	44	61	82	108	135	168
	I_n in A	10	10[2)]	20	25	35	50	63	80	100	125
Group 3	I_z in A	19	24	32	42	54	73	98	129	158	198
	I_n in A	20	20	25	35	50	63	80	100	125	160
Nominal Cu-wire diam. mm, approx.		1·1	1·4	1·8	2·3	2·8	3·6	multiwire			

Group 1: One or more single core cables laid in a conduit
2: Multi-core cables (including ribbon conductor)
3: Single-core cables in free space (spaced at least one wire-diameter apart).

1) The I_z value decreases (increases) by about 7% per 5 °C temperature increase (-decrease).
50 °C should not be exceeded!

2) For cables with only two current carrying conductors an over-load protection device. (L) or (gL) with I_n = 16 A should be used.

Switches

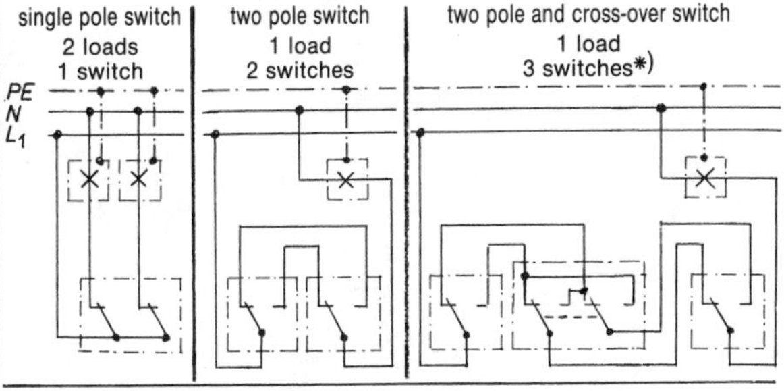

I_n : Nominal current rating of type gL fuses and type L automatic circuit-breakers.

I_z : Current rating of the cable. Also the maximum allowed value of overload protection devices of type B, C or K (for I_n I_z).

L_1 : Live *N*: Neutral *PE*: Earth

*) each additional switch requires an extra cross-over

Partly based IEC 364-5-523, edition 1983

RADIATION PHYSICS
Photometry and Optics

General

For every photometric quantity there is a corresponding radiation-physical quantity and the same relationships apply to both. They are differentiated by different suffixes, v for visual and e for energy.

	Photometry			Radiation physics		
	Quantity	Symbol	Units	Quantity	Symbol	Units
t 1	luminous intensity	I_v	candela cd	radiant intensity	I_e	$\frac{W}{sr}$
t 2	luminous flux	$\Phi_v = \Omega\, I_v$	lumen lm = cd sr	radiant power	$\Phi_e = \Omega\, I_e$	W = J/s
t 3	quantity of light, luminous energy	$Q_v = \Phi_v\, t$	lumen-second lm s, also lm h	radiant energy, quantity of radiation	$Q_e = \Phi_e\, t$	J = W s
t 4	luminance	$L_v = \frac{I_v}{A_1 \cos \varepsilon_1}$	$\frac{cd}{m^2}$	radiance	$L_e = \frac{I_e}{A_1 \cos \varepsilon_1}$	$\frac{W}{sr\, m^2}$
t 5	illuminance	$E_v = \frac{\Phi_v}{A_2}$	lux $lx = \frac{lm}{m^2}$	irradiance	$E_e = \frac{\Phi_e}{A_2}$	$\frac{W}{m^2}$
t 6	light exposure	$H_v = E_v\, t$	lx s	radiant exposure	$H_e = E_e\, t$	$\frac{W\, s}{m^2}$

Definition of the base unit "candela" (cd)

The luminous intensity of a surface of 1/600000 m^2 (= 1⅔ mm^2) of a black body at a temperature of 2042 K.

Photometric radiation equivalent

t 7 1 watt = 680 lm for wavelength 555 nm.

Luminous flux consumption for lighting (values see Z 25)

A surface A lit to an illumination E_v will require a luminous flux of

t 8 $$\Phi_v = \frac{A\, E_v}{\eta}$$

For symbols see T 2

RADIATION PHYSICS

Distance law, Refraction of light

T 2

Optical distance law

The illumination of a surface is inversely proportional to the square of its distance from the light source:

t 9 $$\frac{E_{v1}}{E_{v2}} = \frac{r_2^2}{r_1^2} = \frac{A_2}{A_1}$$

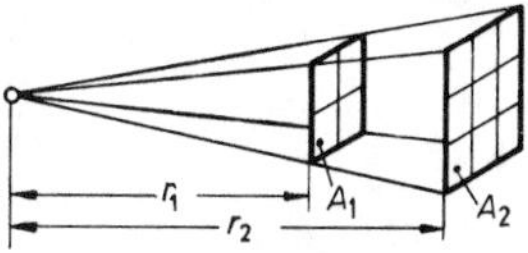

Where two light sources produce equal illumination of a surface, the ratio of the squares of their distances from the surface is equal to the ratio of their luminous intensities:

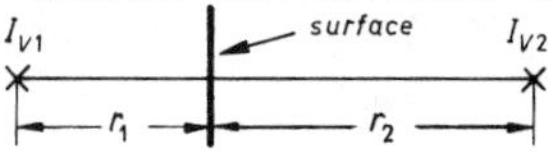

t10 $$\frac{I_{v1}}{I_{v2}} = \frac{r_1^2}{r_2^2}$$

Light refraction

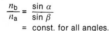

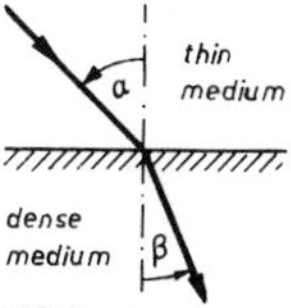

t11 $$\frac{n_b}{n_a} = \frac{\sin \alpha}{\sin \beta}$$

$= \text{const. for all angles.}$

t12 Where $\sin \beta \geqslant \frac{n_a}{n_b}$ total reflection occurs

t13 **Refractive index** for yellow sodium lightning $\lambda = 589{\cdot}3$ nm

solid matter in relation to atmosphere		fluid matter in relation to atmosphere		gasous matter in relation to vacuum	
plexiglas	1·49	water	1·33	hydrogen	1·000 292
quartz	1·54	alcohol	1·36	oxygen	1·000 271
crown glass	1·56	glycerine	1·47	atmosphere	1·000 292
diamond	2·41	benzol	1·50	nitrogen	1·000 297

A_1 : area of radiating surface

A_2 : area of illuminated or irradiated surface

$A_1 \cos \varepsilon_1$: projection of the radiating surface A_1 perpendicular to the direction of radiation

n_a, (n_b) : refractive index of thin (dense) medium

ε_1 : angle between emergent beam and normal to radiating surface A_1

Ω : solid angle Ω is the ratio of the area A_k intercepted on a sphere of radius r_k to the square of the radius: $\Omega = A_k/r_k^2$; unit sr = m^2/m^2.

t14 The solid angle of a point is $\Omega = 4\,\pi$ sr = 12·56 sr

t15 η : luminous efficacy (see table Z 25)

RADIATION PHYSICS

Wavelengths, Mirror

T 3

Wavelengths (in atmosphere)

t 16

Type of radiation		Wavelength $\lambda = c/f$
X-rays	hard	0·0057 nm ... 0·08 nm
	soft	0·08 nm ... 2·0 nm
	ultra-soft	2·0 nm ... 37·5 nm
optical radiation	UV-C ... IR-C	100 nm ... 1 mm
ultra-violet radiation	UV-C	100 nm ... 280 nm
	UV-B	280 nm ... 315 nm
	UV-A	315 nm ... 380 nm
visible radiation, light	violet	380 nm ... 420 nm
	blue	420 nm ... 490 nm
	green	490 nm ... 530 nm
	yellow	530 nm ... 650 nm
	red	650 nm ... 780 nm
infra-red radiation	IR-A	780 nm ... 1·4 µm
	IR-B	1·4 µm ... 3·0 µm
	IR-C	3·0 µm ... 1 mm

Mirrors

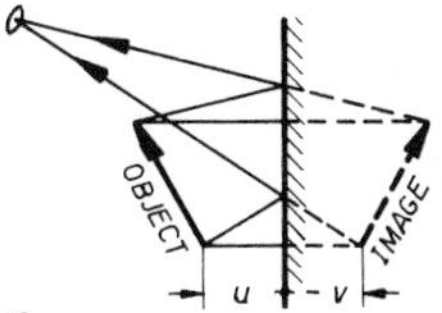

Plane mirrors

The image is at the same distance behind the mirror as the object is in front of it:

t 17 $u = -v$

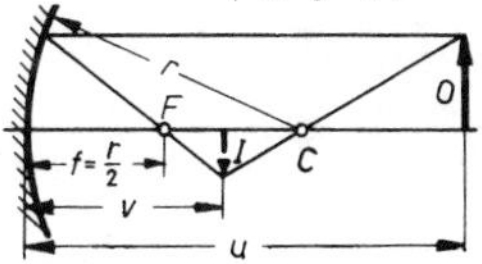

Concave mirrors

t 18 $$\frac{1}{f} = \frac{1}{u} + \frac{1}{v}$$

Depending upon the position of object, the image will be real or virtual:

u	v	image
∞	f	at focal point
$> 2f$	$f < v < 2f$	real, inverted, smaller
$2f$	$2f$	real, inverted, of equal size
$2f > u > f$	$> 2f$	real, inverted, larger
f	∞	no image
$< f$	negative	virtual, larger

Convex mirrors

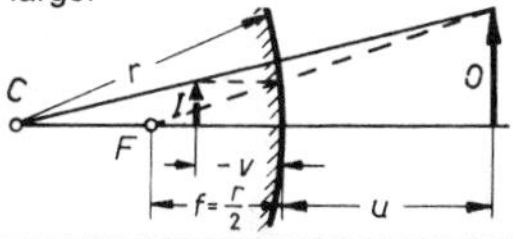

Produce only virtual and smaller images. Similar to concave mirror where:

$f = -r/2$

t 19 $c = 299\,792\,458$ m/s $\approx 0{\cdot}3 \times 10^9$ m/s (velocity of light)

Lenses

Refraction D of a lens

t 20 $D = \frac{1}{f}$; Unit: 1 dpt = 1 dioptrics = $\frac{1}{\text{m}}$

Lens equation (thin lenses only)

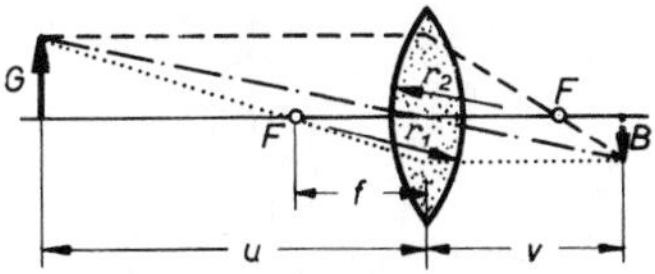

t 21 $\frac{1}{f} = \frac{1}{v} + \frac{1}{u}$

t 22 $= (n-1)\left(\frac{1}{r_1} + \frac{1}{r_2}\right)$

t 23 $m = \frac{B}{G} = \frac{v}{u}$

Where two lenses with focal depths f_1 and f_2 are placed immediately one behind the other, the equivalent focal length f, is given by

t 24 $\frac{1}{f} = \frac{1}{f_1} + \frac{1}{f_2}$

Magnifying lens

	general	where object is in focus
t 25	$m = \frac{s}{f} + 1$	$m = \frac{s}{f}$

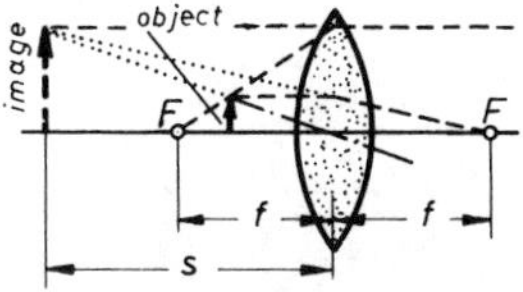

Microscope

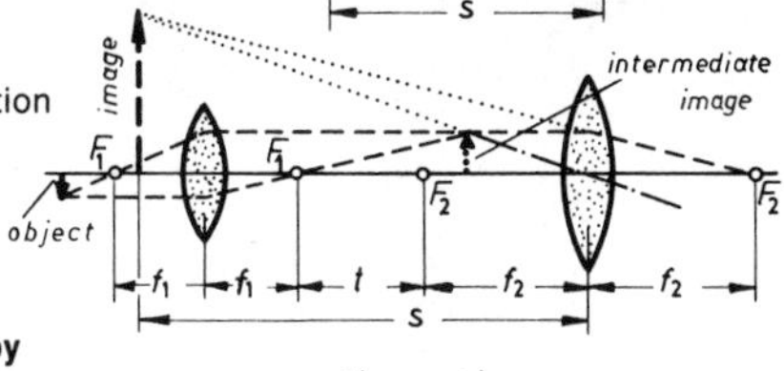

total magnification

t 26 $m = \frac{t\ s}{f_1\ f_2}$

t 27 $= m_1 \times m_2$

Macro photograpy

t 28 camera extension $a = f(m + 1)$

t 29 distance of object $c = \frac{a}{m} = f(1 + \frac{1}{m})$

B : size of image
F : focus
f : focal length
G : size of object
s : range of vision (= 25 cm for normal vision)

n : refractive index (see T2)
r : radius of curvature
t : optical length of tube
m : magnification factor

Ionizing radiation

Ionizing radiation is any radiation of charged particles which causes direct or indirect ionization or excitation of a permanent gas.

	Accumulated values	Units	time rate values	Units
t 30 t 31	amount of absorbed energy (measured value) $J = \dfrac{Q}{m}$	$1\,\dfrac{A\,s}{kg} = 1\,\dfrac{C}{kg}$ $\left(\begin{array}{l}1 \text{ röntgen} =\\ 1\,R = 258\,\dfrac{\mu C}{kg}\end{array}\right)$	rate of absorbing energy $j = \dfrac{J}{t} = \dfrac{I}{m}$	$1\,\dfrac{A}{kg}$ $\left(\begin{array}{l}1\,\dfrac{R}{s} = 258\,\dfrac{\mu A}{kg}\\[4pt]1\,\dfrac{R}{a} = 8{\cdot}2\,\dfrac{pA}{kg}\end{array}\right)$
t 32 t 33	absorbed dose $D = f\,J$ $= \dfrac{W}{m}$	$1 \text{ gray} = 1 \text{ gy}$ $1\,\dfrac{V\,A\,s}{kg} = 1\,\dfrac{W\,s}{kg}$ $= 1\,\dfrac{J}{kg}$ $\left(\begin{array}{l}1 \text{ rad} = 1 \text{ rd}\\[4pt]= \dfrac{cJ}{kg} = 0{\cdot}01 \text{ gy}\\[4pt]= 6{\cdot}242 \times 10^{16}\,\dfrac{eV}{kg}\end{array}\right)$	absorbed dose, -rate $\dot{D} = \dfrac{D}{t} = \dfrac{P}{m}$	$1\,\dfrac{gy}{s} = 1\,\dfrac{W}{kg}$ $= 31{\cdot}56 \times 10^{6}\,\dfrac{J}{kg\,a}$ $\left(\begin{array}{l}1\,\dfrac{rd}{s} = 10\,\dfrac{mW}{kg}\\[4pt]= 0{,}01\,\dfrac{gy}{s}\end{array}\right)$
t 34 t 35	dose equivalent (theoretical value) $H = D_q = q\,D$ $= q\,f\,J$	$1 \text{ sievert} = 1 \text{ sv}$ $= 1\,\dfrac{V\,A\,s}{kg} = 1\,\dfrac{W\,s}{kg}$ $= 1\,\dfrac{J}{kg}$ $[100 \text{ rem} = 1 \text{ sv}]$	dose equivalent, -rate $\dot{H} = \dot{D}_q = \dfrac{D_q}{t}$ $= q\,\dot{D}$	$1\,\dfrac{W}{kg} = 1\,\dfrac{gy}{s}$ $\left(\begin{array}{l}1\,\dfrac{rem}{s} = 10\,\dfrac{mW}{kg}\\[4pt]1\,\dfrac{rem}{a} = 317\,\dfrac{pW}{kg}\end{array}\right)$

Ionization current I: When air molecules are ionized by radiation and a voltage is applied, an ionization current I flows. (Instrument: the ionization chamber).

Charge Q: When an ionization current I flows for a time t it produces a charge

t 38
$$Q = I\,t$$

Units in ( ) are earlier units

continued on T 6

RADIATION PHYSICS
Ionizing radiation

T 6

t 37 **Dose** J: Dose J is a value related to mass m, e.g. $J = Q/m$.

Radiation energy W: W is the radiation energy necessary for ionization. Each pair of ions in the air molecule requires the energy

t 38 $$W_L = 33{\cdot}7 \text{ eV}$$

t 39 (Charge of one electron: $1 \text{ e} = 1{\cdot}602 \times 10^{-19}$ As)

t 40 (1 electron volt: $1 \text{ eV} = 1{\cdot}602 \times 10^{-19} \text{ As} \times 1 \text{ V} = 1{\cdot}602 \times 10^{-19}$ J)

Activity A: The activity A is the number of atoms of a radioactive substance that disintegrates per unit time.

t 41 $$A = -\mathrm{d}N/\mathrm{d}t = \lambda N.$$

Units: bq (becquerel) [1 curie = 1 ci = 37×10^9 bq]
1 bq is 1 disintegration of a radioactive atom per second.

t 42 **Decay** λ: $\lambda = \ln 2/T½$

The half life is $T½$ the time taken for one half of the radioactive mass to decay.
Units: s^{-1}, min^{-1}, h^{-1}, d^{-1}, a^{-1}.

Half lives of some natural and artificial isotopes

atomic [1] number Z	Element	relative atomic [2] mass A_r	half life $T½$	atomic [1] number Z	Element	relative atomic [2] mass A_r	half life $T½$
1	tritium	3	12 a	55	caesium	134	2·1 a
19	potassium	40	$1{\cdot}3 \times 10^9$ a	55	caesium	137	30 a
19	potassium	42	12·4 h	88	radium	226	1600 a
27	cobalt	60	5·3 a	90	thorium	232	14×10^9 a
38	strontium	90	29 a	92	uranium	238	$4{\cdot}5 \times 10^9$ a
53	iodine	131	8·0 d	94	plutonium	239	24000 a

Symbols used

m : mass (base unit)
$T½$: half life
N : number of radioactive atoms

t 43 q : quality factor for $\beta-$, $\gamma-$ und X-rays $q = 1$

t 44 for other radiation $q = 1 \ldots 20$

t 45 f : ionization constant for tissue $f = f_L$

t 46 for bone $f = (1 \ldots 4) f_L$

t 47 f_L : ionization constant for air $f_L = W_L/e = 33{\cdot}7$ V)

Notation of units used

t 48 A: ampere | C: coulomb | J: joule | a: annum (1 annum = 1 a = $31{\cdot}56 \times 10^6$ s)
d: dies (1 dies = 1 d = 86 400 s)

Exposure to radiation (dose equivalent): In 1982 the average person in the Federal Republic of Germany would have been exposed to the follwing radiation:

Type	H in m sv	H in [m rem]
from natural sources	1·1	110
for medical reasons	0·5	50
other artificial radiation*	< 0·1	< 10
*permitted by law	≤ 0·3	≤ 30

[1] number of protons | [2] number of protons and neutrons

CHEMISTRY
Elements

element	symbol	atomic mass in u	element	symbol	atomic mass in u
aluminum	Al	26·9815	neodymium	Nd	144·240
antimony	Sb	121·75	neon	Ne	20·183
argon	Ar	39·948	nickel	Ni	58·71
arsenic	As	74·9216	niobium	Nb	92·906
barium	Ba	137·34	nitrogen	N	14·0067
beryllium	Be	9·0122	osmium	Os	190·2
bismuth	Bi	208·980	oxygen	O	15·9994
boron	B	10·811	palladium	Pd	106·4
bromine	Br	79·909	phosphorus	P	30·9738
cadmium	Cd	112·40	platinum	Pt	195·09
caesium	Cs	132·905	potassium	K	39·102
calcium	Ca	40·08	praseodymium	Pr	140·907
carbon	C	12·0112	radium	Ra	226·04
cerium	Ce	140·12	rhodium	Rh	102·905
chlorine	Cl	35·453	rubidium	Rb	85·47
chromium	Cr	51·996	ruthenium	Ru	101·07
cobalt	Co	58·9332	samarium	Sm	150·35
copper	Cu	63·54	scandium	Sc	44·956
erbium	Er	167·26	selenium	Se	78·96
fluorine	F	18·9984	silicon	Si	28·086
gadolinium	Gd	157·25	silver	Ag	107·870
gallium	Ga	69·72	sodium	Na	22·9898
germanium	Ge	72·59	strontium	Sr	87·62
gold	Au	196·967	sulfur	S	32·064
helium	He	4·0026	tantalum	Ta	180·948
hydrogen	H	1·008	tellurium	Te	127·6
indium	In	114·82	thallium	Tl	204·37
iodine	I	126·9044	thorium	Th	232·038
iridium	Ir	192·2	thulium	Tm	168·934
iron	Fe	55·847	tin	Sn	118·69
krypton	Kr	83·80	titanium	Ti	47·90
lanthanum	La	138·91	tungsten	W	183·85
lead	Pb	207·19	uranium	U	238·03
lithium	Li	6·939	vanadium	V	50·942
magnesium	Mg	24·312	xenon	Xe	131·30
manganese	Mn	54·9381	yttrium	Y	88·905
mercury	Hg	200·59	zinc	Zn	65·37
molybdenum	Mo	95·94	zirconium	Zr	91·22

u : atomic mass unit ($1 \text{ u} = 1{\cdot}66 \times 10^{-27}$ kg)

CHEMISTRY

Chemicals

Chemical terms

trade name	chemical name	chemical formula
acetone	acetone	$(CH_3)_2 \cdot CO$
acetylene	acetylene	C_2H_2
ammonia	ammonia	NH_3
ammonium (hydroxide of)	ammonium hydroxide	NH_4OH
aniline	aniline	$C_6H_5 \cdot NH_2$
bauxite	hydrated aluminium oxides	$Al_2O_3 \cdot 2\ H_2O$
bleaching powder	calcium hypochlorite	$CaCl\ (OCl)$
blue vitriol	copper sulfate	$CuSO_4 \cdot 5\ H_2O$
borax	sodium tetraborate	$Na_2B_4O_7 \cdot 10\ H_2O$
butter of zinc	zinc chloride	$ZnCl_2 \cdot 3\ H_2O$
cadmium sulfate	cadmium sulfate	$CdSO_4$
calcium chloride	calcium chloride	$CaCl_2$
carbide	calcium carbide	CaC_2
carbolic acid	phenol	C_6H_5OH
carbon dioxide	carbon dioxide	CO_2
carborundum	silicon carbide	SiC
caustic potash	potassium hydroxide	KOH
caustic soda	sodium hydroxide	$NaOH$
chalk	calcium carbonate	$CaCO_3$
cinnabar	mercuric sulfide	HgS
ether	di-ethyl ether	$(C_2H_5)_2O$
fixing salt or hypo	sodium thiosulfate	$Na_2S_2O_3 \cdot 5\ H_2O$
glauber's salt	sodium sulfate	$Na_2SO_4 \cdot 10\ H_2O$
glycerine or glycerol	glycerine	$C_3H_5\ (OH)_3$
graphite	crystaline carbon	C
green vitriol	ferrous sulfate	$FeSO_4 \cdot 7\ H_2O$
gypsum	calcium sulfate	$CaSO_4 \cdot 2\ H_2O$
heating gas	propane	C_3H_8
hydrochloric acid	hydrochlorid acid	HCl
hydrofluoric acid	hydrofluoric acid	HF
hydrogen sulfide	hydrogen sulfide	H_2S
iron chloride	ferrous chloride	$FeCl_2 \cdot 4\ H_2O$
iron sulfide	ferrous sulfide	FeS
laughing gas	nitrous oxide	N_2O
lead sulfide	lead sulfide	PbS

continued on U 3

CHEMISTRY
Chemicals.

continued from U 2

trade name	chemical name	chemical formula
limestone	calcium carbonate	$CaCO_3$
magnesia	magnesium oxide	MgO
marsh gas	methane	CH_4
minimum or red lead	plumbate	$2\ PbO \cdot PbO_2$
nitric acid	nitric acid	HNO_3
phosphoric acid	ortho phosphoric acid	H_3PO_4
potash	potassium carbonate	K_2CO_3
potassium bromide	potassium bromide	KBr
potassium chlorate	potassium chlorate	$KClO_3$
potassium chloride	potassium chloride	KCl
potassium chromate	potassium chromate	K_2CrO_4
potassium cyanide	potassium cyanide	KCN
potassium dichromate	potassium dichromate	$K_2Cr_2O_7$
potassium iodide	potassium iodide	KI
prussic acid	hydrogen cyanide	HCN
pyrolusite	manganese dioxide	MnO_2
quicklime	calcium monoxide	CaO
red prussiate of potassium	potassium ferrocyan.	$K_3Fe(CN)_6$
salammoniac	ammonium chloride	NH_4Cl
silver bromide	silver bromide	$AgBr$
silver nitrate	silver nitrate	$AgNO_3$
slaked lime	calcium hydroxide	$Ca(OH)_2$
soda ash	hydrated sodium carb.	$Na_2CO_3 \cdot 10\ H_2O$
sodium monoxide	sodium oxide	Na_2O
soot	amorphous carbon	C
stannous chloride	stannous chloride	$SnCl_2 \cdot 2\ H_2O$
sulphuric acid	sulphuric acid	H_2SO_4
table salt	sodium chloride	$NaCl$
tinstone, tin putty	stannic oxide	SnO_2
trilene	trichlorethylene	C_2HCl_3
urea	urea	$CO(NH_2)_2$
white lead	basic lead carbonate	$2\ PbCO_3 \cdot Pb\ (OH)_2$
white vitriol	zinc sulphate	$ZnSO_4 \cdot 7\ H_2O$
yellow prussiate of potass.	potass. ferrocyanide	$K_4Fe(CN)_6 \cdot 3\ H_2O$
zinc blende	zinc sulphide	ZnS
zinc or chinese white	zink oxide	ZnO

pH values

The negative log of the hydrogen-ion-concentration c_{H+} indicates its *pH* value:

$$pH = -\log c_{H+}$$ u 1

c_{H+}	1	10^{-1}	10^{-2}	...	10^{-7}	...	10^{-12}	10^{-13}	10^{-14}
pH–value	0	1	2		7		12	13	14
	← acid →				neutral	← alkaline →			

Establishing *pH* values by using suitable indicators.

Acid-base-indicators

Indicator	*pH*-Range	Colour change from	to
thymol blue [benz.	1·2...2·8	red	yellow
p-dimethylamino-azo-	2·9...4·0	red	orange-yellow
bromophenolblue	3·0...4·6	yellow	red-violet
congo red	3·0...4·2	blue-violet	red-orange
methyl orange	3·1...4·4	red	yellow-(orange)
brom cresol green	3·8...5·4	yellow	blue
methyl red	4·4...6·2	red	(orange)-yellow
litmus	5·0...8·0	red	blue
bromocresol purple	5·2...6·8	yellow	purple
brom phenol red	5·2...6·8	orange yell.	purple
bromothymol blue	6·0...7·6	yellow	blue
phenol red	6·4...8·2	yellow	red
neutral red	6·4...8·0	(blue)-red	orange-yellow
cresol red	7·0...8·8	yellow	purple
meta cresol purple	7·4...9·0	yellow	purple
thymol blue	8·0...9·6	yellow	blue
phenolphtalein	8·2...9·8	colourless	red-violet
alizarin yellow 66	10·0...12·1	light-yellow	light brown-yellow

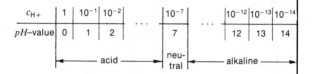

CHEMISTRY

Reagents, Equations, Freezing mixtures

U 5

Reagents

	reagent	indicator	colouration
u 2	acids	blue litmus paper	red
u 3	acids	red phenolphthalein	colourless
u 4	acids	yellow methylorange	red
u 5	bases	red litmus paper	blue
u 6	bases	colourless phenolphthalein	red
u 7	bases	red methylorange	yellow
u 8	ozone	potassium-iodide starch paper	blue-black
u 9	H_2S	lead iodide paper	brown-black
u 10 u 11	ammonia solution	hydrochloric acid	white fumes
	carbonic acid	calcium hydroxide	sediment

Preparation of chemicals

	to prepare	use reaction
u 12	ammonia	$CO(NH_2)_2 + H_2O \rightarrow 2\,NH_3 + CO_2$
u 13	ammonium chloride	$NH_4OH + HCl \rightarrow NH_4Cl + H_2O$
u 14	ammonium hydroxide	$NH_3 + H_2O \rightarrow NH_4OH$
u 15	cadmium sulfide	$CdSO_4 + H_2S \rightarrow Cds + H_2SO_4$
u 16	carbon dioxide	$CaCO_3 + 2\,HCl \rightarrow CO_2 + CaCl_2 + H_2O$
u 17	chlorine	$CaOCl_2 + 2\,HCl \rightarrow Cl_2 + CaCl_2 + H_2O$
u 18	hydrogen	$H_2SO_4 + Zn \rightarrow H_2 + ZnSO_4$
u 19	hydrogen sulfide	$FeS + 2\,HCl \rightarrow H_2S + FeCl_2$
u 20	lead sulfide	$Pb(NO_3)_2 + H_2S \rightarrow PbS + 2\,HNO_3$
u 21	oxygen	$2\,KClO_3 \rightarrow 3\,O_2 + 2\,KCl$
u 22	sodium hydroxide	$Na_2O + H_2O \rightarrow 2\,NaOH$
u 23	zinc sulfide	$ZnSO_4 + H_2S \rightarrow ZnS + H_2SO_4$

Freezing mixtures

	Drope in temperature from °C	to °C	Mixture (The figures stand for proportions by mass)
u 24	+ 10	− 12	$4\,H_2O + 1\,KCl$
u 25	+ 10	− 15	$1\,H_2O + 1\,NH_4NO_3$
u 26	+ 8	− 24	$1\,H_2O + 1\,NaNO_3 + 1\,NH_4Cl$
u 27	0	− 21	3·0 ice (crushed) + 1 NACl
u 28	0	− 39	1·2 ice (crushed) + $2\,CaCl_2 \cdot 6\,H_2O$
u 29	0	− 55	1·4 ice (crushed) + $2\,CaCl_2 \cdot 6\,H_2O$
u 30	+ 15	− 78	1 methyl alcohol + $1\,CO_2$ solid

CHEMISTRY

Moisture, Drying agents, Water-hardness

Atmospheric relative humidity in closed containers

Relative humidity above the solution (%) 20°C = 65°F	Supersaturated aqueous solution	
92	$Na_2CO_3 \cdot 10\,H_2O$	u 31
86	KCl	u 32
80	$(NH_4)_2SO_4$	u 33
76	$NaCl$	u 34
63	NH_4NO_3	u 35
55	$Ca(NO_3)_2 \cdot 4\,H_2O$	u 36
45	$K_2CO_3 \cdot 2\,H_2O$	u 37
35	$CaCl_2 \cdot 6\,H_2O$	u 38

Drying agents (desiccants) for desiccators

Water remaining after drying at 25°C (77°F), g/m³ air	desiccant name	desiccant formula	
1·4	copper sulfate, dehydrated	$CuSO_4$	u 39
0·8	zinc chloride	$ZnCl_2$	u 40
0·14 … 0·25	calcium chloride	$CaCl_2$	u 41
0·16	sodium hydroxide	$NaOH$	u 42
0·008	magnesium oxide	MgO	u 43
0·005	calcium sulfate, dehydrated	$CaSO_4$	u 44
0·003	hydrated aluminum	Al_2O_3	u 45
0·002	potassium hydroxide	KOH	u 46
0·001	silica gel	$(SiO_2)_x$	u 47
0·000025	phosphorus pentoxide	P_2O_5	u 48

Hardness of a water

$$1^\circ \text{German hardness} \triangleq 1^\circ\text{d} \triangleq \frac{10\text{ mg CaO}}{1\text{ l water}} \triangleq \frac{7{\cdot}19\text{ mg MgO}}{1\text{ l water}}$$ u 49

1°d = $1{\cdot}25^\circ$ English hardness = $1{\cdot}78^\circ$ French hardness u 50

= $17{\cdot}8^\circ$ American hardness (1·00 ppm $CaCO_3$) u 51

Classification of hardness

0 … 4°d	very soft	12 … 18°d	rather hard	u 52
4 … 8°d	soft	18 … 30°d	hard	u 53
8 … 12°d	slightly hard	above 30°d	very hard	u 54

Mixture rule for fluids (mixture cross)

a	capa-	starting	fluid	in weight-%
c	city	mixed		
b	of the	admixture		

for water is b = 0.

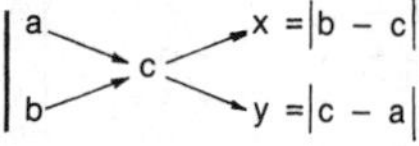

$x = |b - c|$ u 55

$y = |c - a|$ u 56

Example: a = 54%; b = 92%; c shall become 62%.
One should mix thus 30 weight-sharings of a with 8 of b.

TABLES
Properties of solids

Z 1

Reference conditions

Density ϱ at $t = 20°C$

Boiling point t: The values in brackets refer to sublimation, i.e. direct transition from the solid to the gaseous state.

Thermal conductivity λ at $t = 20°C$

Specific heat c for the temperature range $0 < t < 100°C$

Substance	density ϱ	melting point t	boiling point t	thermal conductivity λ	specific heat c
	kg/dm³	°C	°C	W/(m k)[1]	kJ/(kg K)[2]
agate	2·6	1600	2600	10·89	0·80
aluminum bronze	7·7	1040	2300	127·9	0·436
aluminum cast	2·6	658	2200	209·4	0·904
aluminum, rolled	2·7	658	2200	209·4	0·904
amber	1·0	300	·	·	·
antimony	6·67	630	1440	22·53	0·209
arsenic	5·72	815	·	·	0·348
artificial wool	1·5	·	·	·	1·357
asbestos	2·5	1300	·	0·17	0·816
barium	3·59	704	1700	·	0·29
barytes	4·5	1580	·	·	0·46
beryllium	1·85	1280	2970	165	1·02
bismuth	9·8	271	1560	8·1	0·13
boiler scale	2·5	1200	2800	1·2...3·5	0·80
borax	1·72	740	·	·	0·996
brass, cast	8·4	900	1100	113	0·385
brass, rolled	8·5	900	1100	113	0·385
brick	1·8	·	·	1·0	0·92
bromine	3·14	−7·3	63	·	·
bronce (Cu Sn 6)	8·83	910	2300	64	0·37
brown iron ore	5·1	1570	·	0·58	0·67
cadmium	8·64	321	765	92·1	0·234
calcium	1·55	850	1439	·	0·63
carbon	3·51	3600	·	8·9	0·854
cast iron	7·25	1200	2500	58	0·532
cerium	6·77	630	·	·	·
chalk	1·8	·	·	0·92	0·84

[1] 1 W/(m K) = 0·8589 kcal/(h m K) | [2] 1 kJ/(kg K) = 0·2388 kcal/(kg K)

TABLES

Properties of solids

Substance	density ϱ	melting point t	boiling point t	thermal conductivity λ	specific heat c
	kg/dm³	°C	°C	W/(m K)[1]	kJ/(kg K)[2]
charcoal	0·4	·	·	0·084	0·84
chromium	7·1	1800	2700	69	0·452
clay	1·8…2·1	1600	2980	1	0·88
cobalt	8·8	1490	3100	69·4	0·435
coke	1·4	·	·	0·184	0·84
concrete reinforce	2·4	·	·	0·8…1·7	0·88
constantan	8·89	1600	2400	23·3	0·41
copper, cast	8·8	1083	2500	384	0·394
copper, rolled	8·9	1083	2500	384	0·394
cork	0·2…0·3	·	·	0·05	2·0
diamond	3·5	·	[3540]	·	0·52
dripping, beef	0·9…1·0	40…50	350	·	0·88
duralium	2·8	650	2000	129·1	0·92
ebonite	1·2…1·8	·	·	0·17	·
electron	1·8	650	1500	162·8	1·00
emery	4·0	2200	3000	11·6	0·96
fire brick	1·8…2·2	2000	2900	0·47	0·88
glass, window	2·5	700	·	0·81	0·84
glass-wool	0·15	·	·	0·04	0·84
gold	19·29	1063	2700	310	0·130
graphite	2·24	3800	4200	168	0·71
ice	0·92	0	100	2·3	2·09
ingot iron	7·9	1460	2500	47…58	0·49
iodine	4·95	113·5	184	0·44	0·218
iridium	22·5	2450	4800	59·3	0·134
iron, cast	7·25	1200	2500	58	0·532
iron, forged	7·8	1200	·	46…58	0·461
iron-oxide	5·1	1570	·	0·58	0·67
lead	11·3	327·4	1740	34·7	0·130
leather	0·9…1·0	·	·	0·15	1·5
limestone	2·6	·	·	2·2	0·909
lithium	0·53	179	1372	301·2	0·36
magnesia	3·2…3·6	·	·	·	·
magnesium	1·74	657	1110	157	1·05
magnesium, alloy	1·8	650	1500	70…145	1·01

[1] 1 W/(m K) = 0·8598 kcal/(h m K)

[2] 1 kJ/(kg K) = 0·2388 kcal/(kg K)

TABLES
Properties of solids

Substance	density ϱ	melting point t	boiling point t	thermal conductivity λ	specific heat c
	kg/dm³	°C	°C	W/(m K)[1]	kJ/(kg K)[2]
manganese	7·43	1221	2150	·	0·46
marble	2·0…2·8	·	·	2·8	0·84
mica	2·8	·	·	0·35	0·87
molybdenum	10·2	2600	5500	145	0·272
nickel	8·9	1452	2730	59	0·461
osmium	22·48	2500	5300	·	0·130
oxide of chrom	5·21	2300	·	0·42	0·75
palladium	12·0	1552	2930	70·9	0·24
paper	0·7…1·1	·	·	0·14	1·336
paraffin	0·9	52	300	0·26	3·26
peat	0·2	·	·	0·08	1·9
phosphorbronce	8·8	900	·	110	0·36
phosphorus	1·82	44	280	·	0·80
pig iron, white	7·0…7·8	1560	2500	52·3	0·54
pinchbeck	8·65	1000	1300	159	0·38
pitch	1·25	·	·	0·13	·
pit coal	1·35	·	·	0·24	1·02
platinum	21·5	1770	4400	70	0·13
porcelain	2·2…2·5	1650	·	0·8…1·0	0·92
potassium	0·86	63	762·2	1	1
quartz	2·5	1470	2230	9·9	0·80
radium	5	960	1140	·	·
red lead	8·6…9·1	·	·	0·7	0·092
red metal	8·8	950	2300	127·9	0·381
rhenium	21·4	3175	5500	71	0·14
rhodium	12·3	1960	2500	88	0·24
rosin	1·07	100...300	·	0·32	1·30
rubber, raw	0·95	125	·	0·2..0·35	·
rubidium	1·52	39	700	58	0·33
sand, dry	1·4…1·6	1550	2230	0·58	0·80
sandstone	2·1…2·5	1500	·	2·3	0·71
selenium	4·4	220	688	0·20	0·33
silicon	2·33	1420	2600	83	0·75
silicon, carbide	3·12	·	·	15·2	0·67
silver	10·5	960	2170	407	0·234

[1] 1 W/(m K) = 0·8598 kcal/(h m K)
[2] 1 kJ/(kg K) = 0·2388 kcal/(kg K)

TABLES

Properties of solids

Substance	density ϱ	melting point t	boiling point t	thermal conductivity λ	specific heat c
	kg/dm³	°C	°C	W/(m K)[1]	kJ/(kg K)[2]
slate	2·6...2·7	2000	·	0·5	0·76
snow	0·1	0	100	·	4·187
sodium	0·98	97·5	880	126	1·26
soot	1·6...1·7	·	·	0·07	0·84
steatite	2·6...2·7	1600	·	2	0·83
steel	7·85	1460	2500	47...58	0·49
sulfur, cryst.	2·0	115	445	0·20	0·70
tantalum	16·6	2990	4100	54	0·138
tar	1·2	− 15	300	0·19	·
tellurium	6·25	455	1300	4·9	0·201
thorium	11·7	1800	4000	38	0·14
timber, alder	0·55	·	·	0·17	1·4
" , ash	0·75	·	·	0·16	1·6
" , birch	0·65	·	·	0·14	1·9
" , larch	0·75	·	·	0·12	1·4
" , maple	0·75	·	·	0·16	1·6
" , oak	0·85	·	·	0·17	2·4
" , pitchpine	0·75	·	·	0·14	1·3
" , pockwood	1·28	·	·	0·19	2·5
" , red beech	0·8	·	·	0·14	1·3
" , red pine	0·65	·	·	0·15	1·5
" , white pine	0·75	·	·	0·15	1·5
" , walnut	0·65	·	·	0·15	1·4
tin, cast	7·2	232	2500	64	0·24
tin, rolled	7·28	232	2500	64	0·24
titanium	4·5	1670	3200	15·5	0·47
tungsten	19·2	3410	5900	130	0·13
uranium	19·1	1133	3800	28	0·117
vanadium	6·1	1890	3300	31·4	0·50
wax	0·96	60	·	0·084	3·43
welding iron	7·8	1600	2500	54·7	0·515
white metal	7·5...10	300...400	2100	35...70	0·147
zinc, cast	6·86	419	906	110	0·38
zinc, die-cast	6·8	393	1000	140	0·38
zinc, rolled	7·15	419	906	110	0·38

1) 1 W/(m K) = 0·8598 kcal/(h m K)

2) 1 kJ/(kg K) = 0·2388 kcal/(kg K)

Reference conditions

Density ϱ at $t = 20°C$ and $p = 1{\cdot}0132$ bar.

Melting point and boiling point t at $p = 1{\cdot}0132$ bar.

Thermal conductivity λ at $t = 20°C$. For other temperatures see Z 15.

Specific heat c for the temperature range $0 < t < 100°C$.

Substance	density ϱ	melting point t	boiling point t	thermal conductivity λ	specific heat c
	kg/dm³	°C	°C	W/(m K)[3]	kJ/(kg K)[2]
acetic acid	1·08	16·8	118	·	·
acetone	0·79	−95	56·1	·	·
alcohol	0·79	−130	78·4	0·17...0·2	2·43
benzene	0·89	5·4	80	0·137	1·80
benzine	0·7	−150	50...200	0·16	2·1
chloroform	1·53	−70	61	·	·
diesel oil	0·88	−5	175	0·13	·
ether	0·73	117	35	0·14	2·26
gas oil	0·86	−30	200...300	0·15	·
glycerine	1·27[3]	−20	290	0·29	2·43
heating oil	0·92	−5	175...350	0·12	·
hydrochlor acid 10%	1·05	−14	102	0·50	3·14
hydrochlor acid 40%	1·20	·	·	·	·
hydrofluoric acid	0·99	−92·5	19·5	2·33	2·09
linseed oil	0·96	−20	316	0·15	·
machine oil	0·91	−5	380...400	0·126	1·68
mercury	13·6	−38·9	357	8·4	0·138
methyl alcohol	0·8	−98	66	·	2·51
nitric acid	1·56[3]	−41	86	0·26	1·72
oil of resin	0·94	−20	150...300	0·15	·
oil of turpentine	0·87	−10	160	0·10	1·80
perchlor ethylene	1·62	−20	119	·	0·905
petroleum	0·80	−70	150...300	0·159	2·09
petroleum ether	0·67	−160	40...70	0·14	1·76
sulfuric acid conc.	1·84	−10	338	0·5	1·38
sulfuric acid 50%	1·40	·	·	·	·
sulfurus acid	1·49[4]	−73	−10	·	1·34
toluene	0·88	−94·5	110	0·14	1·59
trichlor ethylene	1·47	−86	87	0·16	1·30
water	1 at 4°	0	100	0·58	4·183

[1] 1 W/(m K) = 0·8598 kcal/(h m K)
[2] 1 kJ/(kg K) = 0·2388 kcal/(kg K)

[3] at t = 0°C
[4] at t = −20°C

TABLES

Properties of gases

Reference conditions

Density ϱ at $t = 0°C$ and $p = 1{\cdot}0132$ bar. For perfect gases ϱ can be calculated for other pressures and/or temperatures from: $\varrho = p/(R \times T)$.

Melting point and boiling point t at $p = 1{\cdot}0132$ bar.

Thermal conductivity λ at $t = 0°C$ and $p = 1{\cdot}0132$ bar. For other temperatures see Z 15.

Specific heat c_p and c_v at $t = 0°C$ and $p = 1{\cdot}0132$ bar. c_p at other temperatures see Z 13.

Substances	density ϱ	melting point t	boiling point t	thermal conductivity λ	specific heat c_p	specific heat c_v
	kg/m³	°C	°C	W/(m K)[1]	kJ/(kg K)[2]	
acetylene	1·17	– 83	– 81	0·018	1·616	1·300
air, atmosphere	1·293	– 213	– 192·3	0·02454	1·005	0·718
ammonia	0·77	– 77·9	– 33·4	0·022	2·056	1·568
argon	1·78	– 189·3	– 185·9	0·016	0·52	0·312
blast furnace gas	1·28	– 210	– 170	0·02	1·05	0·75
butane, iso-	2·67	– 145	– 10	·	·	·
butane, n-	2·70	– 135	1	·	·	·
carbon di-oxide	1·97	– 78·2	– 56·6	0·015	0·816	0·627
carbon disulfide	3·40	– 111·5	46·3	0·0069	0·582	0·473
carbon monoxide	1·25	– 205·0	– 191·6	0·023	1·038	0·741
chlorine	3·17	– 100·5	– 34·0	0·0081	0·473	0·36
coal gas	0·58	– 230	– 210	·	2·14	1·59
ethylene	1·26	– 169·3	– 103·7	0·017	1·47	1·173
helium	0·18	– 270·7	– 268·9	0·143	5·20	3·121
hydrochlor acid	1·63	– 111·2	– 84·8	0·013	0·795	0·567
hydrogen	0·09	– 259·2	– 252·8	0·171	14·05	9·934
hydrogen sulfide	1·54	– 85·6	– 60·4	0·013	0·992	0·748
krypton	3·74	– 157·2	– 153·2	0·0088	0·25	0·151
methane	0·72	– 182·5	– 161·5	0·030	2·19	1·672
neon	0·90	– 248·6	– 246·1	0·046	1·03	0·618
nitrogen	1·25	– 210·5	– 195·7	0·024	1·038	0·741
oxigen	1·43	– 218·8	– 182·9	0·024	0·909	0·649
ozone	2·14	– 251	– 112	·	·	·
propane	2·01	– 187·7	– 42·1	0·015	1·549	1·360
sulfur dioxide	2·92	– 75·5	– 10·0	0·0086	0·586	0·456
water vapor[3]	0·77	0·00	100·00	0·016	1·842	1·381
xenon	5·86	– 111·9	– 108·0	0·0051	0·16	0·097

[1] 1 W/(m K) = 0·8598 kcal/(h m K)

[2] 1 kJ/(kg K) = 0·2388 kcal/(kg K)

[3] at $t = 100°C$

TABLES
Friction numbers

Coefficients of sliding and static friction

material	on material	sliding friction μ: dry	sliding friction μ: on water	sliding friction μ: with lubrication	static friction μ_0: dry	static friction μ_0: on water	static friction μ_0: with lubrication
bronze	bronze	0·20	0·10	0·06			0·11
	cast iron	0·18		0·08			
	steel	0·18		0·07	0·19		0·10
oak	oak ∥	0·20…0·40	0·10	0·05…0·15	0·40…0·60		0·18
	oak ⧺	0·15…0·35	0·08	0·04…0·12	0·50		
cast iron	cast iron		0·31	0·10			0·16
	steel	0·17…0·24		0·02…0·05	0·18…0·24		0·10
rubber	asphalt	0·50	0·30	0·20			
	concrete	0·60	0·50	0·30			
hemp rope	timber				0·50		
leather belt	oak	0·40			0·50		
	cast iron		0·40	0·2…0·7	0·40	0·50	0·12
steel	wood	0·20…0·50	0·26	0·02…0·10	0·50…0·60		0·11
	ice	0·014			0·027		
	steel	0·10…0·30		0·02…0·08	0·15…0·30		0·10
	PE-W [1]	0·40…0·50					
	PTFE [2]	0·03…0·05					
	PA 66 [3]	0·30…0·50		0·10			
	POM [4]	0·35…0·45					
PE–W [1]	PE-W [1]	0·50…0·70					
PTFE [2]	PTFE [2]	0·035…0·055					
POM [4]	POM [4]	0·40…0·50					

Rolling friction

(for section K 12 and L 9)

material on material	lever arm *f* of frictional force in mm
rubber on asphalt	0·10
rubber on concrete	0·15
lignum vitae on lignum vitae	0·50
steel on steel (hard: ball bearing)	0·005…0·01
steel on steel (soft)	0·05
elm on lignum vitae	0·8

∥ : movement with grain of both materials

⧺ : movement perpendicular to grain of sliding body

[1] polyethylene with plasticizer (e. g. Lupolen from BASF)
[2] polytetrafluorethylene (e. g. Teflon C 126 from Dupont)
[3] polyamide (e. g. Ultramit CA from BASF)
[4] polyoxymethylene (e. g. Hostaflon C 2520 from Hoechst)

TABLES

Friction factors for flow in pipes $\zeta = f(Re, \frac{k}{d})$

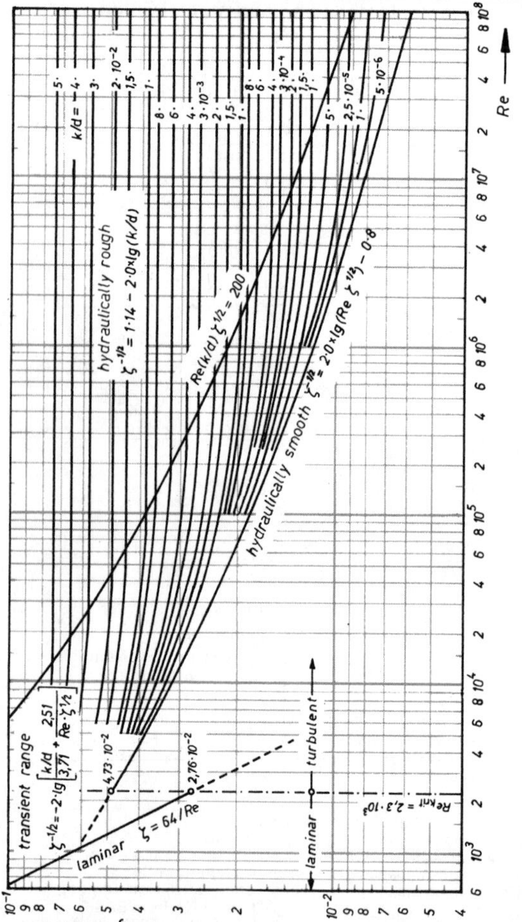

Note: For non-circular pipes k/d is to be replaced by k/d_h.

Galvanized Steel Tubes, suitable

for Screwing to B. S. 21 Pipe Threads
(Approximate values for Medium pipes, colour code – blue, to B. S. 1389)

Nominal bore (inches)	1/8	1/4	3/8	1/2	3/4	1	1 1/4	1 1/2	2
Threads per inch	28	19	19	14	14	11	11	11	11
Outside diam. of pipe (mm)	10·2	13·5	17·2	21·3	26·9	33·7	42·4	48·3	60·3
Inside diam. of pipe (mm)	6·2	8·8	12·3	16	21·6	27·2	35·9	41·8	53
Flow area (mm^2)	30	61	119	201	366	581	1012	1371	2206
Ration of flow area (mm^2) to nominal bore (inches)	242	243	317	402	488	581	810	914	1103

Roughness k

(according to Richter, Hydraulics of Pipes)

Material and kind of pipe	Condition of pipe	k in mm
new seamless rolled or drawn (commercial) steel pipes	typical rolled finish	0·02...0·06
	pickled	0·03...0·04
	cleanly galvanized (dipping process)	0·07...0·10
	commercial galvanized	0·10...0·16
used steel pipes	uniform corrosion pits	about 0·15
	medium corrosion, light incrustation	0·15...0·4
	medium incrustation	about 1·5
	heavy incrustation	2...4
	cleaned after long use	0·15...0·20
cast iron pipes	new, typical cast finish	0·2 ...0·6
	new, bituminized	0·1 ...0·13
	used, corroded	1 ...1·5
	incrusted	1·5 ...4
	cleaned after several years of use	0·3 ...1·5
	mean value in urban sewerage installations	1·2
	heavily corroded	4·5
pipes folded and riveted of sheet steel	new, folded	about 0·15
	new, depending on kind and quality of riveting	
	light riveting	about 1
	heavy riveting	to 9
	25 years old, heavily incrusted, riveted pipe	12·5

Latent heat of fusion per unit mass l_f

material	$\frac{kJ}{kg}$	material	$\frac{kJ}{kg}$	material	$\frac{kJ}{kg}$
aluminum	377	glycerine	176	paraffin	147
antimony	164	gold	67	phenol	109
brass	168	ice	335	platinium	113
cadmium	46	iron	205	potassium	59
cast iron	126	lead	23	silver	109
chromium	134	manganese	155	sulfur	38
cobalt	243	mercury	11·7	tin	59
copper	172	naphthaline	151	Wood's alloy	33·5
ethyl ether	113	nickel	234	zinc	117

Latent heat of evaporation per unit mass l_d
at 101·32 kN/m² (= 760 torr)

material	$\frac{kJ}{kg}$	material	$\frac{kJ}{kg}$	material	$\frac{kJ}{kg}$
alcohol	880	hydrogen	503	oxygen	214
ammonia	1410	mercury	281	sulfur dioxide	402
carbon dioxide	595	methyl chloride	406	toluene	365
chlorine	293	nitrogen	201	water	2250

Calorific value *Hu*
(average values)

Solids	*Hu* MJ/kg	Liquids	*Hu* MJ/kg	Gases	*Hu* MJ/kg
anthracite	33·4	alcohol	26·9	acethylen	48·2
bituminous coal	31·0	benzene	40·2	butane	45·3
brown coal	9·6	Diesel fuel oil	42·1	coal gas	4·1
furnace coke	30·1	gasoline	42·5	hydrogen	119·9
gas coke	29·2	heating oil	41·8	methane	50·0
non coking coal	31·0	methyl alcohol	19·5	municipial gas	18·3
peat, dry	14·6	methylspirit (95%)	25·0	natural gas, dry	43·9
wood, dry	13·3	petroleum	40·8	propane	46·3

1 kWh = 3·6 MJ (cf. A 3)

TABLES
Heat values

Linear coefficient of expansion α in 1/K
at $t = 0 \ldots 100°C$

material	$\alpha/10^{-6}$	material	$\alpha/10^{-6}$	material	$\alpha/10^{-6}$
aluminium	23·8	German silver	18·0	porcelain	4·0
bismuth	13·5	gold	14·2	quartz glass	0·5
brass	18·5	lead	29·0	silver	19·7
bronze	17·5	molybdenum	5·2	steatite	8·5
cadmium	30·0	nickel	13·0	steel, mild	12·0
cast iron	10·5	nickel steel		tin	23·0
constantan	15·2	= Invar 36% Ni	1·5	tungsten	4·5
copper	16·5	platinum	9·0	zinc	30·0

Cubic coefficient of expansion γ in 1/K
at $t = 15°C$

material	$\gamma/10^{-3}$	material	$\gamma/10^{-3}$	material	$\gamma/10^{-3}$
alcohol	1·1	glycerine	0·5	petroleum	1·0
benzene	1·0	mercury	0·18	toluene	1·08
ether	1·6	oil of turpentine	1·0	water	0·18

Coefficient of heat transfer k in $W/(m^2K)$
(Approx. values, natural convection on both sides)

material	thickness of insulating layer in mm								
	3	10	20	50	100	120	250	380	510
reinforced concr.				4·3	3·7	3·5	2·4		
insulat. cement blocks (e.g. thermalite)									
σ_c = 2·45 N/mm²						1·2	0·7	0·5	
σ_c = 4·90 N/mm²						1·6	0·9	0·7	
σ_c = 7·35 N/mm²						1·7	1·0	0·7	
glass	5·8	5·3							
glass-, mineral-wool, hard foam	4·1	2·4	1·5	0·7	0·4				
timber wall			3·8	2·4	1·8	1·7			
chalky sandstone						3·1	2·2	1·7	1·4
gravel concrete				4·1	3·6	3·4	2·3		
slag concrete						2·7	1·7	1·4	1·0
brick						2·9	2·0	1·5	1·3

double or treble glazing	2·6 or 1·9
single window, puttied	5·8
double window, 20 mm spacing, puttied*)	2·9
double window, 120 mm spacing, puttied*)	2·3
tiled roof without/with joint packing	11·6/5·8

*) also for windows with sealed air gaps

Gas constant R and molecular mass M

material	R $\frac{\text{J}}{\text{kg K}}$	M $\frac{\text{kg}}{\text{kmol}}$	material	R $\frac{\text{J}}{\text{kg K}}$	M $\frac{\text{kg}}{\text{kmol}}$
acetylene	319	26	hydrogen	4124	2
air	287	29	nitrogen	297	28
ammonia	488	17	oxygen	260	32
carbonic acid	189	44	sulfuric acid	130	64
carbon monoxide	297	28	water vapor	462	18

Radiation constant C at 20°C

material	C $\text{W/(m}^2\text{K}^4)$	material	C $\text{W/(m}^2\text{K}^4)$
silver, polished	$0{\cdot}17 \times 10^{-8}$	copper, oxidated	$3{\cdot}60 \times 10^{-8}$
aluminum, polished	$0{\cdot}23 \times 10^{-8}$	water	$3{\cdot}70 \times 10^{-8}$
copper, polished	$0{\cdot}28 \times 10^{-8}$	timber, planed	$4{\cdot}40 \times 10^{-8}$
brass, polished	$0{\cdot}28 \times 10^{-8}$	porcelain, glaz	$5{\cdot}22 \times 10^{-8}$
zinc, polished	$0{\cdot}28 \times 10^{-8}$	glass	$5{\cdot}30 \times 10^{-8}$
iron, polished	$0{\cdot}34 \times 10^{-8}$	brickwork	$5{\cdot}30 \times 10^{-8}$
tin, polished	$0{\cdot}34 \times 10^{-8}$	soot, smooth	$5{\cdot}30 \times 10^{-8}$
aluminum,unpolished	$0{\cdot}40 \times 10^{-8}$	zinc, unpolished	$5{\cdot}30 \times 10^{-8}$
nickel, polished	$0{\cdot}40 \times 10^{-8}$	iron, unpolished	$5{\cdot}40 \times 10^{-8}$
brass, unpolished	$1{\cdot}25 \times 10^{-8}$	absolutely	
ice	$3{\cdot}60 \times 10^{-8}$	black surface	$5{\cdot}67 \times 10^{-8}$

Dynamic viscosity η of motor oils in N s/m^2 $\times 10^{-5}$ *)

t in °C		0	20	50	100
SAE	10	0·31	0·079	0·020	0·005
	20	0·72	0·170	0·033	0·007
	30	1·53	0·310	0·061	0·010
	40	2·61	0·430	0·072	0·012
	50	3·82	0·630	0·097	0·015

*) 1 N s/m^2 = 1 kg/(m s) = 1 Pa s = 1000 cP

Mean specific heat $c_{pm}\Big|_0^t$ of various gases in kJ/(kg K) as a function of temperature

t °C	CO	CO_2	H_2	H_2O[1]	N_2 pure	N_2[2]	O_2	SO_2	air
0	1·039	0·8205	14·38	1·858	1·039	1·026	0·9084	0·607	1·004
100	1·041	0·8689	14·40	1·874	1·041	1·031	0·9218	0·637	1·007
200	1·046	0·9122	14·42	1·894	1·044	1·035	0·9355	0·663	1·013
300	1·054	0·9510	14·45	1·918	1·049	1·041	0·9500	0·687	1·020
400	1·064	0·9852	14·48	1·946	1·057	1·048	0·9646	0·707	1·029
500	1·075	1·016	14·51	1·976	1·066	1·057	0·9791	0·721	1·039
600	1·087	1·043	14·55	2·008	1·076	1·067	0·9926	0·740	1·050
700	1·099	1·067	14·59	2·041	1·087	1·078	1·005	0·754	1·061
800	1·110	1·089	14·64	2·074	1·098	1·088	1·016	0·765	1·072
900	1·121	1·109	14·71	2·108	1·108	1·099	1·026	0·776	1·082
1000	1·131	1·126	14·78	2·142	1·118	1·108	1·035	0·784	1·092
1100	1·141	1·143	14·85	2·175	1·128	1·117	1·043	0·791	1·100
1200	1·150	1·157	14·94	2·208	1·137	1·126	1·051	0·798	1·109
1300	1·158	1·170	15·03	2·240	1·145	1·134	1·058	0·804	1·117
1400	1·166	1·183	15·12	2·271	1·153	1·142	1·065	0·810	1·124
1500	1·173	1·195	15·21	2·302	1·160	1·150	1·071	0·815	1·132
1600	1·180	1·206	15·30	2·331	1·168	1·157	1·077	0·820	1·138
1700	1·186	1·216	15·39	2·359	1·174	1·163	1·083	0·824	1·145
1800	1·193	1·225	15·48	2·386	1·181	1·169	1·089	0·829	1·151
1900	1·198	1·233	15·56	2·412	1·186	1·175	1·094	0·834	1·156
2000	1·204	1·241	15·65	2·437	1·192	1·180	1·099	0·837	1·162
2100	1·209	1·249	15·74	2·461	1·197	1·186	1·104		1·167
2200	1·214	1·256	15·82	2·485	1·202	1·191	1·109		1·172
2300	1·218	1·263	15·91	2·508	1·207	1·195	1·114		1·176
2400	1·222	1·269	15·99	2·530	1·211	1·200	1·118		1,181
2500	1·226	1·275	16·07	2·552	1·215	1·204	1·123		1·185
2600	1·230	1·281	16·14	2·573	1·219	1·207	1·127		1·189
2700	1·234	1·286	16·22	2·594	1·223	1·211	1·131		1·193
2800	1·237	1·292	16·28	2·614	1·227	1·215	1·135		1·196
2900	1·240	1·296	16·35	2·633	1·230	1·218	1·139		1·200
3000	1·243	1·301	16·42	2·652	1·233	1·221	1·143		1·203

[1] at low pressures [2] derived from air

Calculated from figures given in E. Schmidt:
Einführung in die Technische Thermodynamik, 11. Auflage, Berlin/Göttingen/Heidelberg: Springer 1975.

Z 14

TABLES

Heat values

Liquids*)

Substance	t	ϱ	c_p	λ	$10^6\eta$	Pr
	°C	$\frac{kg}{m^3}$	$\frac{kJ}{kg\ K}$	$\frac{W}{m\ K}$	Pa s	–
water	0	999·8	4·217	0·5620	1791,8	13·44
	20	998·3	4·182	0·5996	1002,6	6·99
	50	988·1	4·181	0·6405	547,1	3·57
	100	958·1	4·215	0·6803	281,7	1·75
	200	864·7	4·494	0·6685	134,6	0·90
octane C_8H_{18}	–25	738	2·064	0·144	1020	14·62
	0	719	2·131	0·137	714	11·11
ethane C_2H_5OH	–25	–	2·093	0·183	3241	37·07
	0	806	2·232	0·177	1786	22·52
	20	789	2·395	0·173	1201	16·63
	50	763	2·801	0·165	701	11·90
	100	716	3·454	0·152	326	7·41
benzene C_6H_6	20	879	1·729	0·144	649	7·79
	50	847	1·821	0·134	436	5·93
	100	793	1·968	0·127	261	4·04
	200	661	–	0·108	113	–
toluene C_7H_8	0	885	1·612	[illegible]	773	8·65
	20	867	[illegible]	[illegible]	586	7·14
	50	839	1·800	0·136	419	5·55
	100	793	1·968	0·128	269	4·14
	200	672	2·617	0·108	133	3·22
sulfur dioxide SO_2	0	1435	1·33	0·212	368	2·31
	20	1383	1·37	0·199	304	2·09
	50	1296	1·48	0·177	234	1·96
ammonia NH_3	–50	695	4·45	0·547	317	2·58
	0	636	4·61	0·540	169	1·44
	20	609	4·74	0·521	138	1·26
	50	561	5·08	0·477	103	1·10
spindle-oil	20	871	1·85	0·144	13060	168
	50	852	2·06	0·143	5490	79
	100	820	2·19	0·139	2000	32
insulating oil	20	866	–	0·124	31609	482
	60	842	2·29	0·122	7325	125
	100	818	2·29	0·119	3108	60
mercury Hg	0	13546	0·139	9·304	1558	0·02
glycerine $C_3H_8O_3$	20	1260	2·366	0·286	$15\cdot10^6$	$1\cdot24\times10^{11}$

*) Explanation of the symbols cf. O 11

Gases (at 1000 mbar)*)

Substance	t	ϱ	c_p	λ	$10^6\eta$	Pr
	°C	$\frac{kg}{m^3}$	$\frac{kJ}{kg\ K}$	$\frac{W}{m\ K}$	Pa s	–
air, dry	–20	1·377	1·006	0·023	16·15	0·71
	0	1·275	1·006	0·025	17·10	0·70
	20	1·188	1·007	0·026	17·98	0·70
	100	0·933	1·012	0·032	21·60	0·69
	200	0·736	1·026	0·039	25·70	0·68
	400	0·517	1·069	0·053	32·55	0·66
carbon dioxide CO_2	–30	2·199	0·800	0·013	12·28	0·78
	0	1·951	0·827	0·015	13·75	0·78
	25	1·784	0·850	0·016	14·98	0·78
	100	1·422	0·919	0·022	18·59	0·77
	200	1·120	0·997	0·030	26·02	0·76
chlorine Cl	0	3·13	0·473	0·0081	12·3	0·72
	25	2·87	0·477	0·0093	13·4	0·69
	100	2·29	0·494	0·012	16·8	0·69
ammonia NH_3	0	0·76	2·056	0·022	9·30	0·87
	25	0·70	2·093	0·024	10·0	0·87
	100	0·56	2·219	0·033	12·8	0·86
	200	0·44	2·366	0·047	16·5	0·83
oxygen O_2	–50	1·73	0·903	–	16·3	–
	0	1·41	0·909	0·024	19·2	0·73
	25	1·29	0·913	0·026	20·3	0·71
	100	1·03	0·934	0·032	24·3	0·71
	200	0·81	0·963	0·039	28·8	0·71
sulfur dioxide SO_2	0	2·88	0·586	0·0086	11·7	0·80
	25	2·64	0·607	0·0099	12·8	0·78
	100	2·11	0·662	0·014	16·3	0·77
nitrogen N_2	0	1·23	1·038	0·024	16·6	0·72
	25	1·13	1·038	0·026	17·8	0·71
	100	0·90	1·038	0·031	20·9	0·70
	200	0·71	1·047	0·037	24·7	0·70
hydrogen H_2	–50	0·11	13·50	0·141	7·34	0·70
	0	0·09	14·05	0·171	8·41	0·69
	25	0·08	14·34	0·181	8·92	0·71
	100	0·07	14·41	0·211	10·4	0·71
	200	0·05	14·41	0·249	12·2	0·71
water vapour (at saturation)	0	0·0049	1·864	0·0165	9·22	1·041
	50	0·0830	1·907	0·0203	10·62	0·999
	100	0·5974	2·034	0·0248	12·28	1·007
	200	7·865	2·883	0·0391	15·78	1·163
	300	46·255	6·144	0·0718	19·74	1·688

*) Explanation of the symbols cf. O 11

TABLES

Strength values in N/mm^2 *)

Modulus of elasticity (Young's modulus) of steel

$E = 210000$ N/mm^2

BS-Standard specifica-tions from	BS-Standard Material	USA-Standard specifica-tions from	USA-Standard Material	Tensile strength R_m	Yield point 0·2 proof stress R_e; $R_{p0 \cdot 2}$	Fatigue strength torsion com-pression A p_{tA}	Fatigue strength torsion com-pression U p_{tU}	Fatigue strength bending A p_{btA}	Fatigue strength bending U p_{btU}	Fatigue strength torsion A p_{qA}	Fatigue strength torsion U p_{qU}	Notes sizes in mm
BS 970	050 A 20	ASTM	A 283 Grade C	340	225	153	225	170	283	99	131	$16 < d \leq 40$
	060 A 22		A 284 Grade D	410	265	185	265	205	342	119	154	
	080 M 30		A 572 Grade 55	470	285	212	285	235	392	136	165	
	080 M 40		A 572 Grade 65	570	325	257	325	285	455	165	189	
	080 M 50	SAE	1044	670	355	302	355	335	497	194	206	
	080 A 47	SAE	1045	697	481	314	481	349	582	202	279	$d \leq 16$
				657	412	296	412	329	548	191	239	$16 < d \leq 40$
				618	373	278	373	309	515	179	216	$40 < d \leq 100$
	708 A 42	SAE	4140	1079	883	486	810	540	900	313	512	$d \leq 16$
				981	765	441	735	491	818	285	444	$16 < d \leq 40$
				883	638	397	638	442	737	256	370	$40 < d \leq 100$
BS 970/2-70	823 M 30	UNS	K 31820	1226	1030	552	920	613	1022	356	593	$16 < d \leq 40$
BS 2789	SGC I 420/12	ASTM A 536	60–40–18	420	250	180	250	200	333	116	145	
	SGC I 600/3		80–55–06	600	350	245	350	270	450	155	200	
	SGC I 700/2		100–70–03	700	400	280	400	315	525	180	230	

A: Alternating
U: Undulating } see P 2

*) Allow a safety factor for the permissible stresses (see P 2 and P 18)

Note: The strength depends on the diameter, especially with heat-treated steel

Allowable bending and torsional stresses; E and G moduli for elastic materials in N/mm²

Material	Modulus of elasticity E	Type of loading[1]	p_{bt} A	p_{bt} B	p_{bt} C	Modulus of rigidity G	p_{qt}
Spring Steel SAE1078; hard. + temp.	210000	I	1000	500	150	80000	650
		II	750	350	120		500
		III	500	250	80		350
Yellow Brass ASTM–B134(274) HV150	110000	I	200	100	40	42000	120
		II	150	80	30		100
		III	100	50	20		80
Nickel Silver 65–18 HV160 ASTM–B122(752)	142000	I	300	150	50	55000	200
		II	250	120	40		180
		III	200	100	30		150
Tin Bronze CDA–419 HV190	110000	I	200	100	40	42000	120
		II	150	80	30		100
		III	100	50	20		80
Phosphor Bronze CDA–529 HV190	117000	I	300	150	50	45000	200
		II	220	110	40		180
		III	150	80	30		150

A: for simple springs (safety factor ≏ 1·5)
B: for bent and shaped springs (" " ≏ 3)
C: for springs with no hysteresis effect (" " ≏ 10)

[1] For explanation refer to P 1.
[2] For cylindrical helical springs use diagram on page Q 9.

Characteristic quantities for machining
(for turning outside longitudinally)

Material	Strength in N/mm² or hardness	m_c	$1 - m_c$	$k_{c\,1.1}$ N/mm²
ASTM – A572 (Grade	520	0·26	0·74	1990
UNS – K04600	720	0·30	0·70	2260
SAE – 1045	670	0·14	0·86	2220
SAE – 1060	770	0·18	0·82	2130
SAE – 5120	770	0·26	0·74	2100
SAE – 3140	630	0·30	0·70	2260
SAE – 4135	600	0·21	0·79	2240
SAE – 4140	730	0·26	0·74	2500
SAE – 6150	600	0·26	0·74	2220
SAE – L6 annealed	940	0·24	0·76	1740
SAE – L6 tempered	ASTM E18–74–HRD54	0·24	0·76	1920
Mehanite A	360	0·26	0·74	1270
Chilled cast iron	ASTM E18–74–HRD60	0·19	0·81	2060
ASTM – A48–40B	ASTM E18–74–HRD33	0·26	0·74	1160

Specified values apply directly for turning with carbide tip
Cutting speed $v = 90 \ldots 125$ m/min
Chip thickness $h = 0{\cdot}05\ \text{mm} \leq h \leq 2{\cdot}5\ \text{mm}$ | Ratio of slenderness $\varepsilon_s = 4$
Normal side-rake angle $\gamma = 6^\circ$ for steel, 2° for cast iron

Permissible contact pressure p_b in N/mm²

Bearing pressure of joint bolts (Building construction DIN 1050)

Load characteristic	material	p_b	material	p_b
main load	ASTM–	206	ASTM–	304
main and additional load	A 283 Grade C	235	A 440	343

Journals and bearings, bearing plates (see q 13)
Hydrodynamic lubrication see q 47.
Mixed lubrication, shaft hardened and ground:[1), 2)]

Material	v m/s	p_b
gray cast iron		5
ASTM-B30 Cast- (836) Leaded Red Brass		8…12
	1	20[3)]
ASTM-B30 Cast- (938) Leaded Tin Bronze	0·3…1	15[3)]
sintered iron	<1	6
	3	1
sintered iron with copper	<1	8
	3	3
sintered bronze	<1	12
	3	6
	5	4
tin-bronze graphite (DEVA metal)	<1	20 … 90[4)]

Material	v m/s	p_b
Cast Tin Bronze CDA 902		
grease lubrication	<0·03	4…12
quality bearings	<1	60
PA 66 (polyamide)	→0	15
dry[5)]	1	0·09
grease lubrication[5)]	1	0·35
HDPE (high-density polyethylene)	→0	2…4
	1	0·02
PTFE (polytetrafluorethylene enclosed)	→0	30
	1	0·06
PTFE + lead + bronze (GLACIER-DU)	<0·005	80…140[4)]
	0·5…5	1

General, non-sliding surfaces: Max. values are possible up to the compressive yield point at the material ($\sigma_{dF} \approx R_e$). But normal values for good p_b are lower.

Material	Normal values of p_b under dead load	undulating load	shock load
bronze	30 … 40	20 … 30	10 … 15
cast iron	70 … 80	45 … 55	20 … 30
gunmetal	25 … 35	15 … 25	8 … 12
malleable iron	50 … 80	30 … 55	20 … 30
steel	80 … 150	60 … 100	30 … 50

1) $(p \times v)_{perm}$ are closely related to heat dissipation, load, bearing pressure, type of lubrication.
2) Sometimes a much higher load capacity with hydrodynamic lubrication is possible.
3) Limited life (wearing parts).
4) Specially developed metals
5) For shell thickness 1 mm

Properties of friction materials
(Q 15 ... Q 17)

	Material pairing	Sliding friction coefficient μ_{slide}[4)] —	Max. temperature continuous °C	Max. temperature transient °C	Contact pressure p N/mm²	Thermal capacity per unit area q_b kW/m²
dry	organic friction lining/steel or cast iron					
dry	general[5)]	0·2 .. 0·65	150 ... 300 (K ... H)	300 ... 600 (K ... H)	0·1 ... 10	2·2 ... 23
dry	single plate friction-clutch	0·35 ... 0·4	150 ... 300	400	1	12 ... 23
dry	automobile drum brake	0·2 ... 0·3	250 ... 300	350 ... 450	0·5 ... 1·5 2·0	
dry	automobile disc brake	0·3 ... 0·4	400	600	10 (emergency braking)	
dry	cast iron/steel	0·15 ... 0·2	300		0·8 ... 1·4	
dry	sintered bronze/steel	0·05 ... 0·3	400 ... 450	500 ... 600	1	5·5
wet[6)]	sintered bronze/steel	0·05 ... 0·1[1)]	180	500 ... 600	3	12 ... 23
wet[6)]	steel/steel	0·06 ... 0·1[2)]	200 ... 250		1	3·5 ... 5·5[3)]

1) $\mu_{static} = (1·3 \dots 1·5)\ \mu_{slide}$
2) $\mu_{static} = (1·8 \dots 2·0)\ \mu_{slide}$
3) splash lubrication lower, internal lubrication higher
4) often: $\mu_{static} \approx 1·25\ \mu_{slide}$
5) K = rubber bond; H = synthetic resin bond
6) running in oil

TABLES

Work w and yield strength k_f

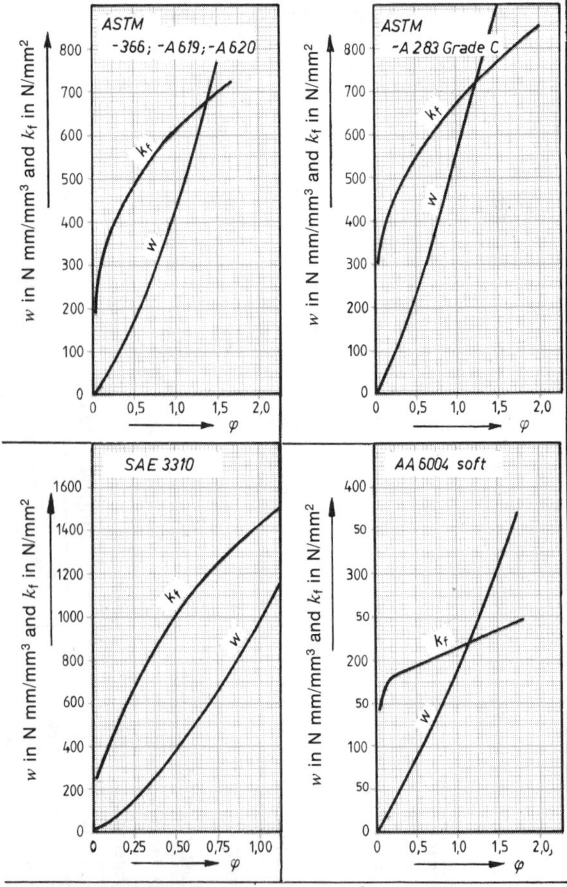

φ : logarithmic deformation ratio

w : strain energy per unit volume

For other materials see VDI 3200

k_f : yield strength

Electrical specific resistance ϱ and specific conductance γ of conductors at $t = 20^\circ C$

material	ϱ $\frac{\Omega\, mm^2}{m}$	γ $\frac{m}{\Omega\, mm^2}$	material	ϱ $\frac{\Omega\, mm^2}{m}$	γ $\frac{m}{\Omega\, mm^2}$
aluminum	0·0278	36	iron (pure)	0·10	10
antimony	0·417	2·4	lead	0·208	4·8
brass – 58% Cu	0·059	17	magnesium	0·0435	23
brass – 63% Cu	0·071	14	manganese	0·423	2·37
cadmium	0·076	13·1	mercury	0·941	1·063
carbon	40	0·025	mild steel	0·13	7·7
cast iron	1	1	nickel	0·087	11·5
chromium-Ni-Fe	0·10	10	nickeline	0·5	2·0
constantan	0·48	2·08	platinum	0·111	9
copper	0·0172	58	silver	0·016	62·5
German silver	0·369	2·71	tin	0·12	8·3
gold	0·0222	45	tungsten	0·059	17
graphite	8·00	0·125	zinc	0·061	16·5

Electrical resistance ϱ of insulators

material	ϱ Ω cm	material	ϱ Ω cm
bakelite	10^{14}	plexiglass	10^{15}
glass	10^{15}	polystyrene	10^{18}
marble	10^{10}	porcelain	10^{14}
mica	10^{17}	pressed amber	10^{18}
paraffin oil	10^{18}	vulcanite	10^{16}
paraffin wax (pure)	10^{18}	water, distilled	10^{7}

Electric temperature coefficient α_{20} at $t = 20^\circ C$

material	α_{20} 1/K or 1/°C	material	α_{20} 1/K or 1/°C
aluminum	+0·00390	mercury	+0·00090
brass	+0·00150	mild steel	+0·00660
carbon	−0·00030	nickel	+0·00400
constantan	−0·00003	nickeline	+0·00023
copper	+0·00380	platinum	+0·00390
German silver	+0·00070	silver	+0·00377
graphite	−0·00020	tin	+0·00420
manganese	±0·00001	zinc	+0·00370

TABLES

Electrical properties

Dielectric constant ε_r

insulant	ε_r	insulant	ε_r	insulant	ε_r
araldite	3·6	mica	6	quartz	4·5
atmosphere	1	micanite	5	shellac	3·5
bakelite	3·6	nylon	5	slate	4
casting compound	2·5	oil paper	4	soft rubber	2·5
caster oil	4·7	olive oil	3	steatite	6
ebonite	2·5	paper	2·3	sulfur	3·5
glass	5	paper, impregnated	5	teflon	2
guttapercha	4	paraffin oil	2·2	transformer oil mineral	2·2
hard paper (laminated)	4·5	paraffin wax	2·2	transformer oil vegetable	2·5
insulation of high voltage cables	4·2	petroleum	2·2	turpentine	2·2
insulation of telephone cables	1·5	phenolic resin	8	vulcanised fibres	2·5
marble	8	plexiglass	3·2	vulcanite	80
		polystyrene	3	water	
		porcelain	4·4		
		pressed board	4		

Electro-motive series

(potential difference with respect to hydrogen electrode)

material	V volt	material	V volt	material	V volt
potassium	−2·93	chromium	−0·74	hydrogen	0·00
calcium	−2·87	tungsten	−0·58	antimony	+0·10
sodium	−2·71	iron	−0·41	cooper	+0·34
magnesium	−2·37	cadmium	−0·40	silver	+0·80
beryllium	−1·85	cobalt	−0·28	mercury	+0·85
aluminum	−1·66	nickel	−0·23	platinum	+1·20
manganese	−1·19	tin	−0·14	gold	+1·50
zinc	−0·76	lead	−0·13	fluorine	+12·87

Standardized numbers using progression ratio according to E-series

(Shown for E 6 ... E 24)

E 6 series ($\approx \sqrt[6]{10}$)			E 12 series ($\approx \sqrt[12]{10}$)			E 24 series ($\approx \sqrt[24]{10}$)		
1·0	2·2	4·7	1·0	2·2	4·7	1·0	2·2	4·7
						1·1	2·4	5·1
			1·2	2·7	5·6	1·2	2·7	5·6
						1·3	3·0	6·2
1·5	3·3	6·8	1·5	3·3	6·8	1·5	3·3	6·8
						1·6	3·6	7·5
			1·8	3·9	8·2	1·8	3·9	8·2
						2·0	4·3	9·1
10	22	47	10	22	47	10	22	47
	etc.			etc.			etc.	

Magnetic field strength H and relative permeability μ_r as a function of induction B

induction B		cast iron		steel casting and dynamo allow sheet and strips steel $P\,1{\cdot}0 = 3{\cdot}6\frac{W}{kg}$		alloyed dynamo sheet and strips steel $P\,1{\cdot}0 = 1{\cdot}3\frac{W}{kg}$	
$T = \frac{V\,s}{m^2}$ tesla	[G gauss]	H A/m	μ_r –	H A/m	μ_r –	H A/m	μ_r –
0·1	1000	440	181	30	2650	8,5	9390
0·2	2000	740	215	60	2650	25	6350
0·3	3000	980	243	80	2980	40	5970
0·4	4000	1250	254	100	4180	65	4900
0·5	5000	1650	241	120	3310	90	4420
0·6	6000	2100	227	140	3410	125	3810
0·7	7000	3600	154	170	3280	170	3280
0·8	8000	5300	120	190	3350	220	2900
0·9	9000	7400	97	230	3110	280	2550
1·0	10000	10300	77	295	2690	355	2240
1·3	11000	14000	63	370	2360	460	1900
1·2	12000	19500	49	520	1830	660	1445
1·3	13000	29000	36	750	1380	820	1260
1·4	14000	42000	29	1250	890	2250	495
1·5	15000	65000	18	2000	600	4500	265
1·6	16000			3500	363	8500	150
1·7	17000			7900	171	13100	103
1·8	18000			12000	119	21500	67
1·9	19000			19100	79	39000	39
2·0	20000			30500	52	115000	14
2·1	21000			50700	33		
2·2	22000			130000	13		
2·3	23000			218000	4		

—— practical limit

Note: $P\,1{\cdot}0$ see Z 24

Dynamo sheet properties

type			mild sheet and strips steel	allow sheet and strips steel			
				low	medium	high	
stray power at 1·0 T in W/kg			3·6	3·0	2·3	1·5	1·3
thickness mm			0·5				0·35
density kg/dm³			7·8	7·75	7·65	7·6	
core losses per unit mass at $f = 50\,\text{Hz}$ W/kg (max.)	P 1·0		3·6	3·0	2·3	1·5	1·3
	P 1·5		8·6	7·2	5·6	3·7	3·3
induction (min.)	B_{25}	V s/m² [gauss]	1·53 [15300]	1·50 [15000]	1·47 [14700]	1·43 [14300]	
	B_{50}	V s/m² [gauss]	1·63 [16300]	1·60 [16000]	1·57 [15700]	1·55 [15500]	
	B_{100}	V s/m² [gauss]	1·73 [17300]	1·71 [17100]	1·69 [16900]	1·65 [16500]	
	B_{300}	V s/m² [gauss]	1·98 [19800]	1·95 [19500]	1·93 [19300]	1·85 [18500]	

Explanations

$B_{25} = 1{\cdot}53\ \text{V s/m}^2$ indicates that a minimum induction of 1·53 V s/m² [or 15300 gauss] is reached with a field strength of 25 A/cm. Thus a flux length of e.g. 5 cm requires a circulation of $5 \times 25\ \text{A} = 125\ \text{A}$.

P 1·0	describes the core losses per unit mass at $f = 50$ Hz and an induction of	1·0 V s/m² = [10000 G]
P 1·5		1·5 V s/m² = [15000 G]

Guide values for illumination E_V in lx = lm/m²

Type of establishment or location		General lighting only	General and spec. lighting: bench	General and spec. lighting: general
workshops according to work done	rough	100	50	200
	medium	200	100	500
	precise	300	200	1000
	very precise	500	300	1500
offices	normal	500		
	open	750		
living rooms, lighting	medium	200		
	bright	500		
streets and squares with traffic	light	20		
	medium	50		
	heavy	100		
factory yards with traffic	light	20		
	heavy	50		

Luminous efficacy η

Type of lighting	Colour of illuminated surface: light	medium	dark
direct	0·60	0·45	0·30
indirect	0·35	0·25	0·15

Type of lighting	deep bowl reflector	widespread reflector
street and square lighting	0·45	0·40

Luminous flux Φ_V of lamps

Standard lamps with single coiled filament (at operating voltage)		P_{el}	W	15	25	40	60	75	100
		Φ_V	klm	0·12	0·23	0·43	0·73	0·96	1·39
		P_{el}	W	150	200	300	500	1000	2000
		Φ_V	klm	2·22	3·15	5·0	8·4	18·8	40·0
Fluores-cent lamps values for 'Warmwhite' 'Daylight'	tubular diameter								
	26 mm	P_{el}	W		18	36	58		
		Φ_V	klm		1·45	3·47	5·4		
	38 mm	P_{el}	W	15	20	40	65		
		Φ_V	klm	0·59	1·20	3·1	5·0		
High-pressure lamps filled with mercury vapour		P_{el}	W	125	250	400	700	1000	2000
		Φ_V	klm	6·5	14	24	42	60	125

TABLES

Statistics

$$\varphi(x) = \frac{1}{\sqrt{2\pi}} e^{-\frac{x^2}{2}}; \quad \Phi_0(x) = \frac{2}{\sqrt{2\pi}} \int_0^x e^{-\frac{t^2}{2}} \cdot dt; \quad \operatorname{erf}(x) = \frac{2}{\sqrt{\pi}} \int_0^x e^{-t^2} \cdot dt$$

x	$\varphi(x)$	$\Phi_0(x)$	erf (x)	x	$\varphi(x)$	$\Phi_0(x)$	erf (x)
0,00	0,398 942	0,000 000	0,000 000	0,50	0,352 065	0,382 925	0,520 500
0,01	0,398 922	0,007 979	0,011 283	0,51	0,350 292	0,389 949	0,529 244
0,02	0,398 862	0,015 957	0,022 565	0,52	0,348 493	0,396 936	0,537 899
0,03	0,398 763	0,023 933	0,033 841	0,53	0,346 668	0,403 888	0,546 464
0,04	0,398 623	0,031 907	0,045 111	0,54	0,344 818	0,410 803	0,554 939
0,05	0,398 444	0,039 878	0,056 372	0,55	0,342 944	0,417 681	0,563 323
0,06	0,398 225	0,047 845	0,067 622	0,56	0,341 046	0,424 521	0,571 616
0,07	0,397 966	0,055 806	0,078 858	0,57	0,339 124	0,431 322	0,579 816
0,08	0,397 668	0,063 763	0,090 078	0,58	0,337 180	0,438 085	0,587 923
0,09	0,397 330	0,071 713	0,101 281	0,59	0,335 213	0,444 809	0,595 937
0,10	0,396 953	0,079 656	0,112 463	0,60	0,333 225	0,451 494	0,603 856
0,11	0,396 536	0,087 591	0,123 623	0,61	0,331 215	0,458 138	0,611 681
0,12	0,396 080	0,095 517	0,134 758	0,62	0,329 184	0,464 742	0,619 412
0,13	0,395 585	0,103 434	0,145 867	0,63	0,327 133	0,471 306	0,627 047
0,14	0,395 052	0,111 340	0,156 947	0,64	0,325 062	0,477 828	0,634 586
0,15	0,394 479	0,119 235	0,167 996	0,65	0,322 972	0,484 308	0,642 029
0,16	0,393 868	0,127 119	0,179 012	0,66	0,320 864	0,490 746	0,649 377
0,17	0,393 219	0,134 990	0,189 992	0,67	0,318 737	0,497 142	0,656 628
0,18	0,392 531	0,142 847	0,200 936	0,68	0,316 593	0,503 496	0,663 782
0,19	0,391 806	0,150 691	0,211 840	0,69	0,314 432	0,509 806	0,670 840
0,20	0,391 043	0,158 519	0,222 702	0,70	0,312 254	0,516 073	0,677 801
0,21	0,390 242	0,166 332	0,233 522	0,71	0,310 060	0,522 296	0,684 666
0,22	0,389 404	0,174 129	0,244 296	0,72	0,307 851	0,528 475	0,691 433
0,23	0,388 529	0,181 908	0,255 022	0,73	0,305 627	0,534 610	0,698 104
0,24	0,387 617	0,189 670	0,265 700	0,74	0,303 389	0,540 700	0,704 678
0,25	0,386 668	0,197 413	0,276 326	0,75	0,301 137	0,546 745	0,711 156
0,26	0,385 683	0,205 136	0,286 900	0,76	0,298 872	0,552 746	0,717 537
0,27	0,384 663	0,212 840	0,297 418	0,77	0,296 595	0,558 700	0,723 822
0,28	0,383 606	0,220 522	0,307 880	0,78	0,294 305	0,564 609	0,730 010
0,29	0,382 515	0,228 184	0,318 283	0,79	0,292 004	0,570 472	0,736 103
0,30	0,381 388	0,235 823	0,328 627	0,80	0,289 692	0,576 289	0,742 101
0,31	0,380 226	0,243 439	0,338 908	0,81	0,287 369	0,582 060	0,748 003
0,32	0,379 031	0,251 032	0,349 126	0,82	0,285 036	0,587 784	0,753 811
0,33	0,377 801	0,258 600	0,359 279	0,83	0,282 694	0,593 461	0,759 524
0,34	0,376 537	0,266 143	0,369 365	0,84	0,280 344	0,599 092	0,765 143
0,35	0,375 240	0,273 661	0,379 382	0,85	0,277 985	0,604 675	0,770 668
0,36	0,373 911	0,281 153	0,389 330	0,86	0,275 618	0,610 211	0,776 100
0,37	0,372 548	0,288 617	0,399 206	0,87	0,273 244	0,615 700	0,781 440
0,38	0,371 154	0,296 054	0,409 009	0,88	0,270 864	0,621 141	0,786 687
0,39	0,369 728	0,303 463	0,418 739	0,89	0,268 477	0,626 534	0,791 843
0,40	0,368 270	0,310 843	0,428 392	0,90	0,266 085	0,631 880	0,796 908
0,41	0,366 782	0,318 194	0,437 969	0,91	0,263 688	0,637 178	0,801 883
0,42	0,365 263	0,325 514	0,447 468	0,92	0,261 286	0,642 427	0,806 768
0,43	0,363 714	0,332 804	0,456 887	0,93	0,258 881	0,647 629	0,811 563
0,44	0,362 135	0,340 063	0,466 225	0,94	0,256 471	0,652 782	0,816 271
0,45	0,360 527	0,347 290	0,475 482	0,95	0,254 059	0,657 888	0,820 891
0,46	0,358 890	0,354 484	0,484 656	0,96	0,251 644	0,662 945	0,825 424
0,47	0,357 225	0,361 645	0,493 745	0,97	0,249 228	0,667 954	0,829 870
0,48	0,355 533	0,368 773	0,502 750	0,98	0,246 809	0,672 914	0,834 231
0,49	0,353 812	0,375 866	0,511 668	0,99	0,244 390	0,677 826	0,838 508

$$\varphi(x) = \frac{1}{\sqrt{2\pi}} e^{-\frac{x^2}{2}}; \quad \Phi_0(x) = \frac{2}{\sqrt{2\pi}} \int_0^x e^{-\frac{t^2}{2}} \, dt; \quad \operatorname{erf}(x) = \frac{2}{\sqrt{\pi}} \int_0^x e^{-t^2} \, dt$$

x	$\varphi(x)$	$\Phi_0(x)$	erf (x)	x	$\varphi(x)$	$\Phi_0(x)$	erf (x)
1·00	0·241 971	0·682 689	0·842 701	1·50	0·129 518	0·866 336	0·966 105
1·01	0·239 551	0·687 505	0·846 810	1·51	0·127 583	0·868 957	0·967 277
1·02	0·237 132	0·692 272	0·850 838	1·52	0·125 665	0·871 489	0·968 414
1·03	0·234 714	0·696 990	0·854 784	1·53	0·123 763	0·873 983	0·969 516
1·04	0·232 297	0·701 660	0·858 650	1·54	0·121 878	0·876 440	0·970 586
1·05	0·229 882	0·706 282	0·862 436	1·55	0·120 009	0·878 858	0·971 623
1·06	0·227 470	0·710 855	0·866 144	1·56	0·118 157	0·881 240	0·972 628
1·07	0·225 060	0·715 381	0·869 773	1·57	0·116 323	0·883 585	0·973 603
1·08	0·222 653	0·719 858	0·873 326	1·58	0·114 505	0·885 893	0·974 547
1·09	0·220 251	0·724 287	0·876 803	1·59	0·112 704	0·888 165	0·975 462
1·10	0·217 852	0·728 668	0·880 205	1·60	0·110 921	0·890 401	0·976 348
1·11	0·215 458	0·733 001	0·883 533	1·61	0·109 155	0·892 602	0·977 207
1·12	0·213 069	0·737 286	0·886 788	1·62	0·107 406	0·894 768	0·978 038
1·13	0·210 686	0·741 524	0·889 971	1·63	0·105 675	0·896 899	0·978 843
1·14	0·208 308	0·745 714	0·893 082	1·64	0·103 961	0·898 995	0·979 622
1·15	0·205 936	0·749 856	0·896 124	1·65	0·102 265	0·901 057	0·980 376
1·16	0·203 571	0·753 951	0·899 096	1·66	0·100 586	0·903 086	0·981 105
1·17	0·201 214	0·757 999	0·902 000	1·67	0·098 925	0·905 081	0·981 810
1·18	0·198 863	0·762 000	0·904 837	1·68	0·097 282	0·907 043	0·982 493
1·19	0·196 520	0·765 953	0·907 608	1·69	0·095 657	0·908 972	0·983 153
1·20	0·194 186	0·769 861	0·910 314	1·70	0·094 049	0·910 869	0·983 790
1·21	0·191 860	0·773 721	0·912 956	1·71	0·092 459	0·912 734	0·984 407
1·22	0·189 543	0·777 535	0·915 534	1·72	0·090 887	0·914 568	0·985 003
1·23	0·187 235	0·781 303	0·918 050	1·73	0·089 333	0·916 370	0·985 578
1·24	0·184 937	0·785 024	0·920 505	1·74	0·087 796	0·918 141	0·986 135
1·25	0·182 649	0·788 700	0·922 900	1·75	0·086 277	0·919 882	0·986 672
1·26	0·180 371	0·792 331	0·925 236	1·76	0·084 776	0·921 592	0·987 190
1·27	0·178 104	0·795 915	0·927 514	1·77	0·083 293	0·923 273	0·987 691
1·28	0·175 847	0·799 455	0·929 734	1·78	0·081 828	0·924 924	0·988 174
1·29	0·173 602	0·802 949	0·931 899	1·79	0·080 380	0·926 546	0·988 641
1·30	0·171 369	0·806 399	0·934 008	1·80	0·078 950	0·928 139	0·989 090
1·31	0·169 147	0·809 804	0·936 063	1·81	0·077 538	0·929 704	0·989 524
1·32	0·166 937	0·813 165	0·938 065	1·82	0·076 143	0·931 241	0·989 943
1·33	0·164 740	0·816 482	0·940 015	1·83	0·074 766	0·932 750	0·990 347
1·34	0·162 555	0·819 755	0·941 914	1·84	0·073 407	0·934 232	0·990 736
1·35	0·160 383	0·822 984	0·943 762	1·85	0·072 065	0·935 687	0·991 111
1·36	0·158 225	0·826 170	0·945 562	1·86	0·070 740	0·937 115	0·991 472
1·37	0·156 080	0·829 313	0·947 313	1·87	0·069 433	0·938 516	0·991 821
1·38	0·153 948	0·832 413	0·949 016	1·88	0·068 144	0·939 892	0·992 156
1·39	0·151 831	0·835 471	0·950 673	1·89	0·066 871	0·941 242	0·992 479
1·40	0·149 727	0·838 487	0·952 285	1·90	0·065 616	0·942 567	0·992 790
1·41	0·147 639	0·841 460	0·953 853	1·91	0·064 378	0·943 867	0·993 090
1·42	0·145 564	0·844 392	0·955 376	1·92	0·063 157	0·945 142	0·993 378
1·43	0·143 505	0·847 283	0·956 857	1·93	0·061 952	0·946 393	0·993 656
1·44	0·141 460	0·850 133	0·958 297	1·94	0·060 765	0·947 620	0·993 922
1·45	0·139 431	0·852 941	0·959 695	1·95	0·059 595	0·948 824	0·994 179
1·46	0·137 417	0·855 710	0·961 054	1·96	0·058 441	0·950 004	0·994 426
1·47	0·135 418	0·858 438	0·962 373	1·97	0·057 304	0·951 162	0·994 664
1·48	0·133 435	0·861 127	0·963 654	1·98	0·056 183	0·952 297	0·994 892
1·49	0·131 468	0·863 776	0·964 898	1·99	0·055 079	0·953 409	0·995 111

INDEX